CALVERT MATH

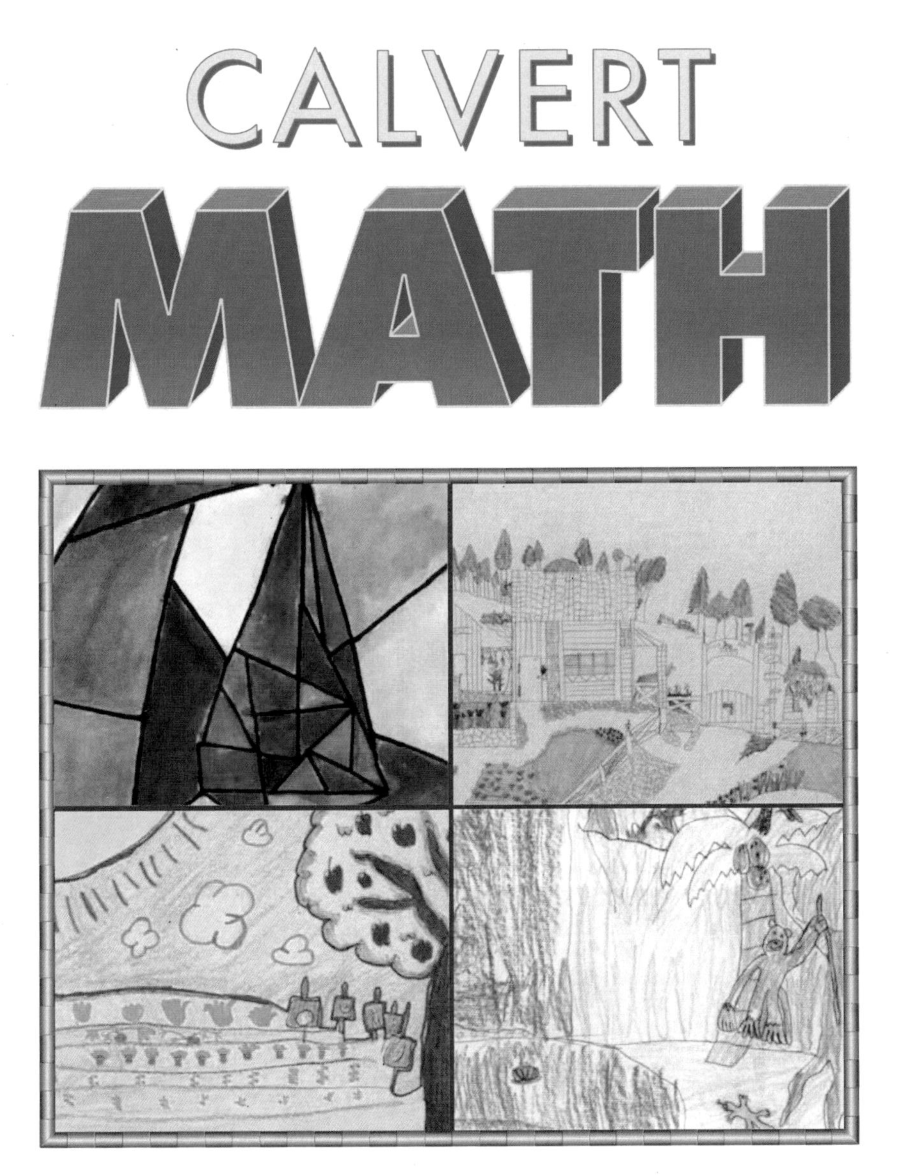

Calvert Math is based upon a previously published textbook series. Calvert School has customized the textbooks using the mathematical principles developed by the original authors. Calvert School wishes to thank the authors for their cooperation. They are:

Audrey V. Buffington
Mathematics Teacher
Wayland Public Schools
Wayland, Massachusetts

Alice R. Garr
Mathematics Department Chairperson
Herricks Middle School
Albertson, New York

Jay Graening
Professor of Mathematics
and Secondary Education
University of Arkansas
Fayetteville, Arkansas

Philip P. Halloran
Professor, Mathematical Sciences
Central Connecticut State University
New Britain, Connecticut

Michael Mahaffey
Associate Professor,
Mathematics Education
University of Georgia
Athens, Georgia

Mary A. O'Neal
Mathematics Laboratory Teacher
Brentwood Unified Science
Magnet School
Los Angeles, California

John H. Stoeckinger
Mathematics Department Chairperson
Carmel High School
Carmel, Indiana

Glen Vannatta
Former Mathematics Supervisor
Special Mathematics Consultant
Indianapolis Public Schools
Indianapolis, Indiana

ISBN-13: 978-1-888287-76-9

Printed in the U.S.A.
1 2 3 4 5 6 7 8 9 10 12 11

Chief Learning Officer
Gloria D. Julius, Ed.D

Senior Consultant/Project Coordinator
Jessie C. Sweeley, *Associate Director of Research and Design*

Curriculum Consultant
Robin Fiastro, *Manager of Curriculum*

Project Facilitator/Lead Researcher
Nicole M. Henry, *Math Curriculum Specialist*
Hannah Miriam Delventhal, *Math Curriculum Specialist*

Field Test Teachers
The following Calvert teachers used the Calvert Math prototype in their classrooms and added materials and teaching techniques based on Calvert philosophy and methodology.

Andrew S. Bowers
James O. Coady Jr.
Roman A. Doss
Shannon C. Frederick
Julia T. Holt
Willie T. Little III
John M. McLaughlin
Brian J. Mascuch
Mary Ellen Nessler
Margaret B. Nicolls
Michael E. Paul
Diane E. Proctor
Patricia G. Scott
E. Michael Shawen
Ada M. Stankard
Virginia P.B. White
Jennifer A. Yapsuga
Andrea L. Zavitz

Calvert Education Services (CES) Consultants
The following CES staff worked with the Day School faculty in developing the mathematical principles for Calvert Math.

Eileen M. Byrnes
Nicole M. Henry
Linda D. Hummel
Christina L. Mohs
Kelly W. Painter
Mary-Louise Stenchly
Ruth W. Williams

Revision Authors
The following Day School faculty and CES staff contributed to the correlation of the math materials to national standards and to the authoring of revisions based on the correlations.

Nicole M. Henry
Julia T. Holt
Barbara B. Kirby
Willie T. Little III
Mary Ellen Nessler
Mary Ethel Vinje
Jennifer A. Yapsuga

Editors
Anika Trahan, *Managing Editor*
Bernadette Burger, *Senior Editor*
Eileen Baylus
Holly Bohart
Maria R. Casoli
Danica K. Crittenden
Pam Eisenberg
Sarah E. Hedges
Mary E. Pfeiffer
Megan L. Snyder

Graphic Designers
Lauren Loran, *Manager of Graphic Design*
Vickie M. Johnson
Steven M. Burke
Bonnie R. Kitko
Vanessa Ann Panzarino
Caitlin Rose Brown
Deborah A. Sharpe
Joshua Steves
Teresa M. Tirabassi

To the Student

Being successful in math depends on having the right tools, skills, and mental attitude. Your math book is a tool to help you discover and master the skills you will need to become powerful in math. These thinking skills will prepare you to solve everyday problems.

About the Art in this Book

Calvert homeschooling students from all over the world and Calvert Day School students have contributed their original art for this book. We hope you enjoy looking at their drawings as you study mathematics.

Contents

1 Number Patterns and Algebra

2 Adding and Subtracting Decimals

3 Multiplying and Dividing Decimals

4 Fractions and Decimals

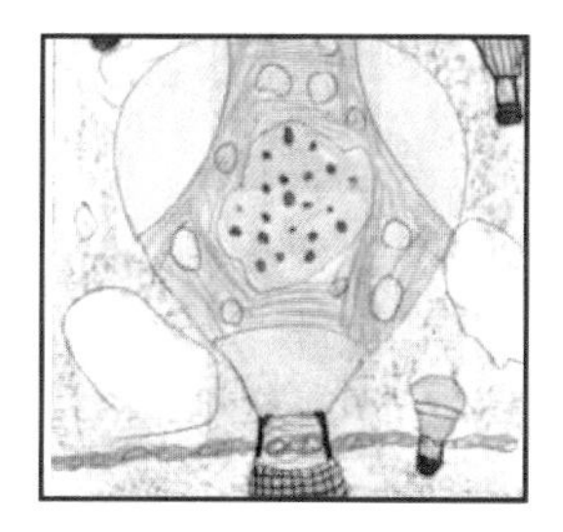

5 Adding and Subtracting Fractions

6 Multiplying and Dividing Fractions

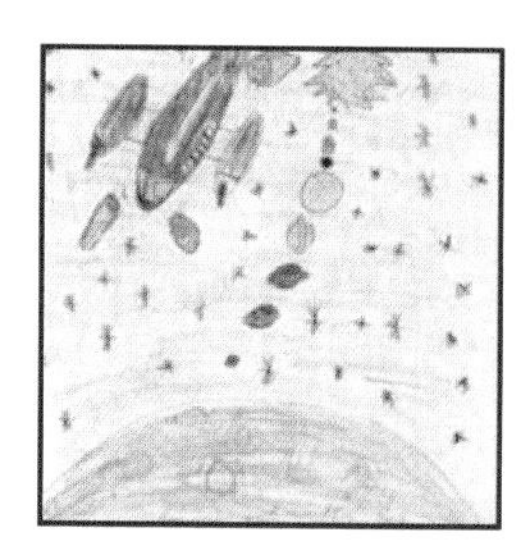

7 Ratios, Proportions, and Percents

8 Probability

9 Data Analysis and Graphs

10 Geometric Figures

11 Geometric Measurement

12 Solving Equations

13 Integers

14 Coordinate Graphing

Diagnosing Readiness

In this preliminary chapter you will take a diagnostic skills pretest, practice skills that you need to review, and take a diagnostic skills posttest. This preliminary chapter ensures that you have the skills necessary to begin Chapter 1.

Diagnostic Skills Pretest

Complete the diagnostic skills pretest. f you get one or more problems in a section incorrect, please refer to the section number in parentheses for review. After reviewing this chapter, take the diagnostic skills posttest at the end of the chapter.

What is the value of each of the following digits in 7,482,609? (Section 1)

1. 8   2. 6   3. 7   4. 2

Write in expanded form. (Section 1)

5. 905   6. 8,721   7. 27,555   8. 780,438

Compare using <, >, or = . (Section 2)

9. 68 ■ 608   10. 3,924 ■ 3,924   11. 48,000 ■ 4,800

Order from least to greatest. (Section 2)

12. 189, 891, 819, 198, 888   13. 12,342; 11,432; 14,222; 11,434

Round to the nearest thousandth. (Section 3)

14. 4,759   15. 23,409   16. 893,939   17. 555,555

Round to the underlined place-value position. (Section 3)

18. $\underline{5}6$ 19. $\underline{3}48$ 20. $6,\underline{7}40$ 21. $\underline{3},958$

Write in Roman numerals. (Section 4)

22. 30   23. 40   24. 465   25. 4,789

Write in standard form. (Section 4)

26. MMI   27. CXL   28. XXXVII   29. CCLVII

Add. (Section 5)

30. 68 + 7   31. \$954 + \$281   32. 6,974 + 586

33. 84 + 16 + 27 + 53   34. \$76,360 + \$35,640   35. 3,660 + 4,283 + 599

Subtract. (Section 6)

36. 789 – 394

37. 32,323 – 7,654

38. \$67,541 – \$28,684

39. \$10,000 – \$974

40. 523,217 – 184,985

41. 400,000 – 320,796

Multiply. (Section 7)

42. 42×3

43. 648×5

44. \$208 × 7

45. \$9,387 × 6

46. $5{,}663 \times 5$

47. $37{,}489 \times 7$

Multiply. (Section 8)

48. \$70 × 40

49. 24×32

50. \$948 × 81

51. $3{,}745 \times 38$

52. $4{,}073 \times 54$

53. $27{,}514 \times 23$

Divide. (Section 9)

54. $3\overline{)88}$

55. $6\overline{)328}$

56. $5\overline{)754}$

57. $7\overline{)8{,}436}$

58. \$848 ÷ 4

59. \$2,346 ÷ 6

Divide. (Section 10)

60. $16\overline{)868}$

61. $49\overline{)2{,}989}$

62. $25\overline{)75}$

63. $23\overline{)483}$

64. \$126 ÷ 18

65. \$1,742 ÷ 26

1 Whole Numbers and Place Value

The area of the United States is 3,623,461 square miles.

The digits 0, 1, 2, 3, 4, 5, 6, 7, 8, and 9 are used to name the whole numbers zero through nine.

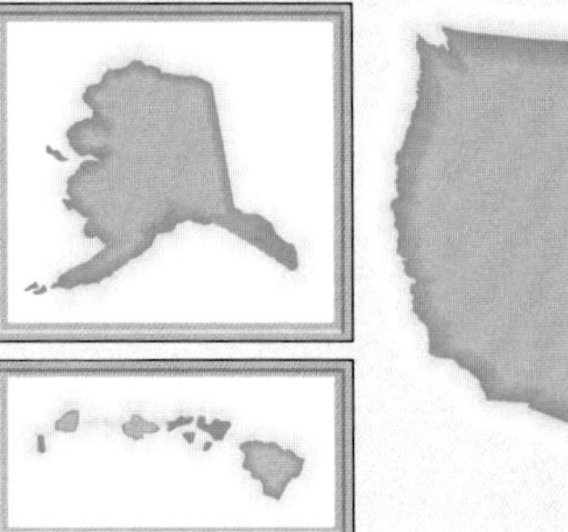

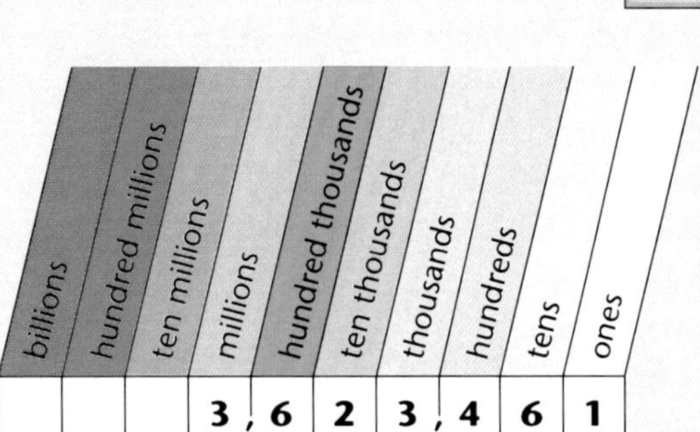

For any whole number, the place-value position determines the number named by each digit. In 3,623,461, the digit 4 and its place-value position name the number four hundred or 400. The number 3,623,461 is written in *standard form.*

Examples

A. Numbers written in *expanded form* show the number named by each digit. **2,495 → 2,000 + 400 + 90 + 5**

B. Place value can be used to rename numbers such as 600 and 45,000.

600 → 6 hundreds → 60 tens → 600 ones

45,000 → 45 thousands → 450 hundreds → _?_ tens → _?_ ones

What are two more ways to rename 45,000?

Try THESE

Name the digit in each place-value position in 6,149,308,275.

1. thousands
2. ten millions
3. billions
4. ten thousands
5. ones
6. hundred millions
7. hundreds
8. tens
9. millions

What is the value of each of the following digits in 472,365?

10. 2
11. 6
12. 5
13. 4

Exercises

Write the place-value position for each of the following digits in 8,371,426,059.

1. 1 **2.** 0 **3.** 8

4. 5 **5.** 4 **6.** 7

What is the value of each of the following digits in 5,842,173?

7. 1 **8.** 2 **9.** 3

10. 5 **11.** 7 **12.** 4

Write in expanded form.

13. 307 **14.** 4,123 **15.** 29,546

16. 40,098 **17.** 190,057 **18.** 703,508

Replace each ■ with a number to make a true equation.

19. 40 = ■ tens **20.** 720 = ■ tens **21.** 8,400 = ■ hundreds

22. 3,600 = ■ tens **23.** 2,000 = ■ tens

24. 27,000 = ■ hundreds **25.** 510,000 = ■ tens

26. 300,000 = ■ thousands **27.** 70,000,000 = ■ ten thousands

28. 6,000,000 = ■ hundreds

Rename each of the following in three different ways.

29. 500 **30.** 12,000,000

31. The area of Rhode Island is 1,049 square miles. What is the place-value position of the digit 4?

32. The area of Maryland is 9,775 square miles. What is the place-value position of the digit 9?

2 Comparing and Ordering

The Colorado River is 1,450 miles long. The Yukon River is 1,770 miles long. A number line can be used to compare the lengths.

1,200 1,300 1,400 1,500 1,600 1,700 1,800 1,900 2,000

1,450 1,770

1,450 is to the left of 1,770.

1,450 is less than 1,770.

1,450 < 1,770

1,770 is to the right of 1,450.

1,770 is greater than 1,450.

1,770 > 1,450

The symbol points toward the lesser number.

The Yukon River is longer. (Or the Colorado River is shorter.)

More Examples

A. 2,306 < 2,341

In the tens place, 0 < 4.

B. 824 = 824

All the digits are the same.

C. 1,406 > 988

Note that 1,406 has four digits.

What steps can you use to compare two whole numbers?

Start with the greatest place-value position first. Then compare each place-value position until the numbers differ.

Try THESE

Compare using <, >, or = .

1. 56 ● 65
2. 172 ● 171
3. 905 ● 509
4. 94 ● 940
5. 7,140 ● 7,104
6. 48,905 ● 48,903
7. 54,179 ● 54,179
8. 501,999 ● 502,000
9. 183,572 ● 138,572
10. 1,996,889 ● 1,996,789
11. 948,204 ● 947,204
12. 739,005 ● 739,005
13. 139,000 ● 1,390,000

Exercises

Order from least to greatest.

1. 34, 53, 28, 46, 87, 43
2. 749; 386; 1,005; 683; 479; 501
3. 10,111; 10,101; 11,000; 10,000; 1,000; 11,111; 101; 20,101
4. 143,341; 341,143; 14,000; 58,000; 43,043; 431,000

Order from greatest to least.

5. 46, 64, 406, 460, 56, 65
6. 10,900; 9,900; 19,000; 91,000; 9,999; 19,101
7. 501,007; 105,000; 700,000; 751,001; 51,000; 700,105

★ 8. Does $89 > 67 < 25$ or $25 < 67 < 89$ show the correct order relation for 25, 67, and 89?

Problem SOLVING

9. Which state had the most people?
10. Which state had the least people?
11. Which state had a population of one million, two hundred forty-two thousand?
12. Rank the states in order from 1 to 10. The state with the most people gets a rank of 1.

State	Population Estimate 1996
Georgia	7,486,000
Florida	14,654,000
Iowa	2,852,000
Maine	1,242,000
Maryland	5,094,000
New York	18,137,000
Nevada	1,677,000
Oregon	3,243,000
North Dakota	641,000
California	32,268,000

3 Rounding Whole Numbers

New York has 127 miles of Atlantic coastline. One source reports that New York has 130 miles of coastline. In this case, the number of miles of coastline is rounded to the nearest ten.

A number line helps explain what *rounding* means.

To the nearest *ten,* 127 rounds to 130 because 127 is closer to 130 than 120.

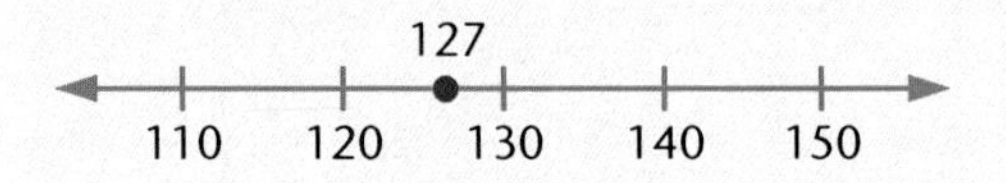

To the nearest *hundred,* 127 rounds to 100 because 127 is closer to 100 than 200.

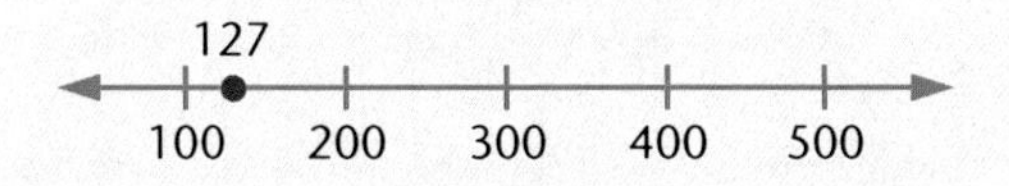

You can round without using a number line. Round 83,451 to the nearest thousand. Look at the digit to the right of the place being rounded. Then round as follows.

Round up if the digit to the right is 5, 6, 7, 8, or 9.

To the nearest thousand, 83,451 rounds to 83,000.

The underlined digit remains the same if the digit to the right is 0, 1, 2, 3, or 4.

Try THESE

Round to the nearest ten. Use a number line to help explain each answer.

1. 43 **2.** 68 **3.** 739 **4.** 987

5. 6,054 **6.** 9,738 **7.** 42,757 **8.** 56,798

Round to the nearest hundred.

9. 780 **10.** 430 **11.** 2,630 **12.** 7,881

13. 3,660 **14.** 9,340 **15.** 87,694 **16.** 43,286

Exercises

Choose the correct answer for rounding to the underlined place-value position.

1. 46 a. 40 b. 50
2. 92 a. 90 b. 100
3. 138 a. 100 b. 200
4. 785 a. 700 b. 800
5. 4,940 a. 4,940 b. 4,950
6. 63,273 a. 63,200 b. 63,300

Round to the nearest thousand.

7. 1,400
8. 3,946
9. 48,596
10. 4,007,506
11. 30,605
12. 448,234
13. 29,902
14. 17,009,504

Round to the underlined place-value position.

15. 18,904
16. 123,949
17. 650,208
18. 54,872
19. 6,499
20. 906,582
21. 707,295
22. 70,453
23. 29,368
24. 12,399,878
25. 6,999,888

Problem SOLVING

★ 26. If 115,642 is rounded to 120,000, to what place-value position is this number rounded?

★ 27. Round the diameter of Earth, 856,000 meters, to the nearest hundred thousand.

28. When the height of a building has been rounded to the nearest ten feet, the result is 160 feet. What are four possible heights of the building?

29. Mount McKinley in Alaska is 20,321 feet in elevation. What is this number rounded to the nearest thousand? Based on your rounded number, what are the least and greatest heights possible?

4 Roman Numerals

The ancient Romans developed their own numeration system. They used letters to name numbers, but they had no symbol for zero.

Roman Numeral	I	V	X	L	C	D	M
Standard Form	1	5	10	50	100	500	1,000

The Romans did not use place value. Their system used addition, subtraction, and multiplication principles.

Addition Principle

VIII → 5 + 3 or 8

XVI → 10 + 5 + 1 or 16

LXXX → 50 + 10 + 10 + 10 or 80

CCC → 100 + 100 + 100 or 300

MDC → 1,000 + 500 + 100 or 1,600

Notice that no symbol is used more than three times in a row.

Subtraction Principle

IV → 5 − 1 or 4

IX → 10 − 1 or 9

XL → 50 − 10 or 40

XC → 100 − 10 or 90

CD → 500 − 100 or 400

CM → 1,000 − 100 or 900

These are the only pairings allowed for the subtraction principle.

To write larger numbers, the ancient Romans used a line above the letters. A bar above a letter means multiply by 1,000.

$\overline{\text{V}}$ = 5,000 $\overline{\text{L}}$ = 50,000

$\overline{\text{X}}$ = 10,000 $\overline{\text{C}}$ = 100,000

More Examples

Write in standard form.

A. XLIII
40 + 3 = 43

B. CXCIV
100 + 90 + 4 = 194

C. CCLXIV
200 + 60 + 4 = 264

D. **$\overline{\text{VI}}$ = 6,000**

Try THESE

Write in standard form.

1. XXX
2. LXV
3. XIII
4. DCCC
5. XC
6. XLV
7. MCMLXIII
8. MCMXCIX

Exercises

Write in standard form.

1. LX
2. CLXXXIV
3. LXXVII
4. XXV
5. DCXLIII
6. MDLV
7. MMCDV
8. $\overline{\text{IX}}$

Express using Roman numerals.

9. 15
10. 27
11. 95
12. 230
13. 267
14. 838
15. 3,007
16. 32,000
17. Add XLIX and LXXXVII. Write your answer in Roman numerals.

★ 18. Multiply XIV by IX. Write your answer in Roman numerals.

Problem SOLVING

19. Write the Roman numeral that comes before C.
20. Write the Roman numeral for one less than the number M.
21. Why do you think the subtraction principle may have been developed by clock makers?
22. Mount Sunflower, the highest point in Kansas, is MXXXIV feet in elevation. Write this in standard form.

5 Adding Whole Numbers

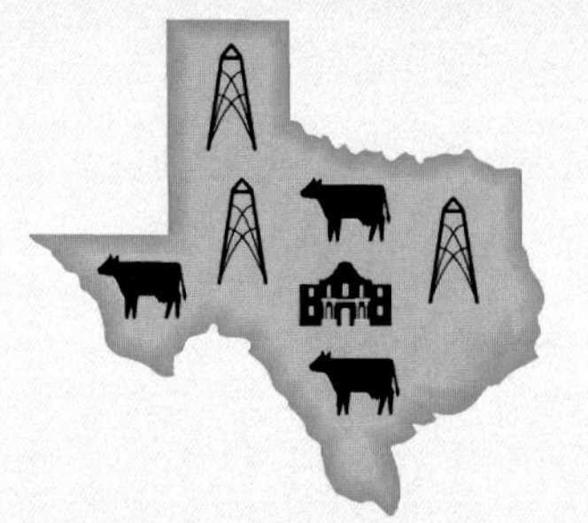

Traveling through Texas, the Jamisons spent $356 for hotels, $227 for food, and $164 for entertainment. To find how much they spent in all, add $356, $227, and $164.

Add in each place-value position from least to greatest.

Estimate 356 + 227 + 164 as 400 + 200 + 200 or 800.

Step 1	Step 2	Step 3
Add the ones. $\begin{array}{r} {}^{1} \\ \$356 \\ 227 \\ +\ 164 \\ \hline 7 \end{array}$ Look for sums of ten. 6 + 4 = 10 10 + 7 = 17 6 + 7 + 4 = 17 Rename the 10 ones as 1 ten.	Add the tens. $\begin{array}{r} {}^{1\,1} \\ \$356 \\ 227 \\ +\ 164 \\ \hline 47 \end{array}$ 1 + 5 + 2 + 6 = 14 Rename the 10 tens as 1 hundred.	Add the hundreds. $\begin{array}{r} {}^{1\,1} \\ \$356 \\ 227 \\ +\ 164 \\ \hline \$747 \end{array}$ 1 + 3 + 2 + 1 = 7

The Jamisons spent $747 on their trip.

Compared to the estimate, 747 is reasonable.

More Examples

Check: Add in the reverse order.

A. $\begin{array}{r} {}^{1} \\ 78 \\ +\ 42 \\ \hline 120 \end{array}$

B. $\begin{array}{r} {}^{1\,1\,1} \\ 8{,}067 \\ +\ 946 \\ \hline 9{,}013 \end{array}$

C. $\begin{array}{r} {}^{1\,2\,1\,1} \\ 40{,}208 \\ 2{,}950 \\ 682 \\ +\ 27{,}800 \\ \hline 71{,}640 \end{array}$ ⟶ $\begin{array}{r} {}^{1\,2\,1\,1} \\ 27{,}800 \\ 682 \\ 2{,}950 \\ +\ 40{,}208 \\ \hline 71{,}640 \end{array}$ ✔

Try THESE

Add.

1. $\begin{array}{r} \$48 \\ +\ 25 \\ \hline \end{array}$

2. $\begin{array}{r} 37 \\ +\ 54 \\ \hline \end{array}$

3. $\begin{array}{r} 216 \\ +\ 46 \\ \hline \end{array}$

4. $\begin{array}{r} 473 \\ +\ 367 \\ \hline \end{array}$

5. $\begin{array}{r} 2{,}415 \\ +\ 908 \\ \hline \end{array}$

6. $\begin{array}{r} 156 \\ 57 \\ +\ 94 \\ \hline \end{array}$

7. $\begin{array}{r} \$\ 91 \\ 307 \\ +\ 711 \\ \hline \end{array}$

8. $\begin{array}{r} 16 \\ 37 \\ 24 \\ +\ 13 \\ \hline \end{array}$

9. $\begin{array}{r} 45 \\ 164 \\ 93 \\ +\ 235 \\ \hline \end{array}$

10. $\begin{array}{r} 897 \\ 280 \\ 62 \\ +\ 544 \\ \hline \end{array}$

Exercises

Add.

1. $\begin{array}{r} 20 \\ +\ 52 \\ \hline \end{array}$
2. $\begin{array}{r} \$53 \\ +\ 37 \\ \hline \end{array}$
3. $\begin{array}{r} 709 \\ +\ 87 \\ \hline \end{array}$
4. $\begin{array}{r} 79 \\ +\ 530 \\ \hline \end{array}$
5. $\begin{array}{r} \$207 \\ +\ 506 \\ \hline \end{array}$
6. $\begin{array}{r} 346 \\ +\ 229 \\ \hline \end{array}$
7. $\begin{array}{r} \$365 \\ +\ 1{,}259 \\ \hline \end{array}$
8. $\begin{array}{r} 6{,}859 \\ +\ 8{,}541 \\ \hline \end{array}$
9. $\begin{array}{r} 64{,}395 \\ +\ 43{,}857 \\ \hline \end{array}$
10. $\begin{array}{r} 387{,}297 \\ +\ 814{,}548 \\ \hline \end{array}$
11. $\begin{array}{r} 467 \\ 295 \\ +\ 535 \\ \hline \end{array}$
12. $\begin{array}{r} \$3{,}479 \\ 6{,}114 \\ +\ 7{,}809 \\ \hline \end{array}$
13. $\begin{array}{r} 673 \\ 3{,}348 \\ 937 \\ +\ 765 \\ \hline \end{array}$
14. $\begin{array}{r} 65{,}873 \\ 2{,}316 \\ 27{,}407 \\ +\ 46{,}549 \\ \hline \end{array}$
15. $\begin{array}{r} 786 \\ 11{,}904 \\ 28{,}565 \\ +\ 645{,}020 \\ \hline \end{array}$

16. 6,772 + 349
17. 11,240 + 667
18. 327,394 + 54,129
19. 636,354 + 228,799
20. 59 + 201 + 760 + 1,887
21. 245 + 20,950 + 67,700 + 186,500
22. Find the sum of $46,519 and $88,695.

Problem SOLVING

23. A radio station had 15,000 bumper stickers to hand out on Saturday and Sunday at the County Fair. Were there enough stickers for each person on those days? Explain your answer.
24. Tug-of-war team members weigh 161, 138, 131, 143, and 158 pounds. The total team weight must be less than 750 pounds. Does the team qualify? Explain your answer.

County Fair Attendance	
Monday	5,433
Tuesday	4,951
Wednesday	4,015
Thursday	3,370
Friday	2,499
Saturday	7,204
Sunday	7,976

6 Subtracting Whole Numbers

Ahmad Mosa's hometown has a population of 3,005 people. His cousin's hometown has 1,438 people. To find out how many more people live in Ahmad's hometown, subtract 1,438 from 3,005.

Estimate: 3,000 − 1,000 = 2,000

Subtract in each place-value position as follows.

Step 1	Step 2	Step 3	Step 4
Subtract the ones.	Subtract the tens.	Subtract the hundreds.	Subtract the thousands.
2 9 9 15 3,005 − 1,438 7	2 9 9 15 3,005 − 1,438 67	2 9 9 15 3,005 − 1,438 1,567	2 9 9 15 3,005 − 1,438 1,567

Ahmad Mosa's hometown has 1,567 more people.

Compared to the estimate, 1,567 is reasonable.

Check: 1 11
1,438 + 1,567 = 3,005 ✔

More Examples

A. (15; 3 5 11) $461 − 82 = $379

B. (3 13) 96,437 − 20,260 = 76,177

C. (14; 2 9 4 9 10) 30,500 − 14,699 = 15,801

D. (3 9 9 9 9 10) 400,000 − 131,206 = 268,794

Explain the renaming done for each example.

Try THESE

Subtract.

1. 67 − 36 (no renaming)
2. 92 − 17 (8 12: 92 − 17)
3. 50 − 5 (4 10: 50 − 5)
4. 807 − 298 (7 9 17: 807 − 298)
5. $9,747 − 6,369
6. 794 − 57
7. 671 − 563
8. 8,007 − 348
9. 3,000 − 294
10. 40,300 − 7,629

Exercises

Subtract.

1. 85 − 54
2. 90 − 78
3. $55 − 49
4. 700 − 673
5. 972 − 95

6. 901 − 563
7. $851 − 708
8. 5,834 − 3,458
9. 656 − 584
10. 7,006 − 5,479

11. $85,040 − 7,652
12. 73,200 − 67,400
13. $36,681 − 10,794
14. 163,227 − 24,759
15. 250,000 − 129,709

16. 7,754 − 4,531
17. 28,000 − 6,340
18. 37,554 − 26,876
19. 60,200 − 8,379
20. 500,000 − 329,654

Problem SOLVING

21. How much farther is it from Croton to River Glen than from Croton to Peach Grove?

22. Which distance is farther, Galena to Peach Grove or Galena to Willowview? How much farther is it?

23. How much farther is it from Galena to Lake City than from Willowview to Croton?

Mileage Chart

	Willowview	Lake City	River Glen	Galena	Croton	Peach Grove
Willowview		129	112	73	195	17
Lake City	129		33	282	40	80
River Glen	112	33		12	100	7
Galena	73	282	12		68	64
Croton	195	40	100	68		44
Peach Grove	17	80	7	64	44	

7 Multiplying by 1-Digit Numbers

Joe Chang places price tags on new books at the bookstore. He can price about 236 books in 1 hour. How many can he price in 4 hours?

To find out, multiply 4 by 236.
Think of this as combining 4 groups of 236.

$$\begin{array}{r} 200 \\ \times \quad 4 \\ \hline 800 \end{array}$$

An estimate is 800.

Multiply as shown below.

Step 1	Step 2	Step 3
Multiply the ones.	Multiply the tens.	Multiply the hundreds.
$\begin{array}{r} {\scriptstyle 2} \\ 236 \\ \times \quad 4 \\ \hline 4 \end{array}$	$\begin{array}{r} {\scriptstyle 1\ 2} \\ 236 \\ \times \quad 4 \\ \hline 44 \end{array}$	$\begin{array}{r} {\scriptstyle 1\ 2} \\ 236 \\ \times \quad 4 \\ \hline 944 \end{array}$
24 is 2 tens and 4 ones. The 2 is recorded in the tens place.	12 tens + 2 tens is 14 tens. Where is the 1 recorded?	How is 9 hundreds computed?

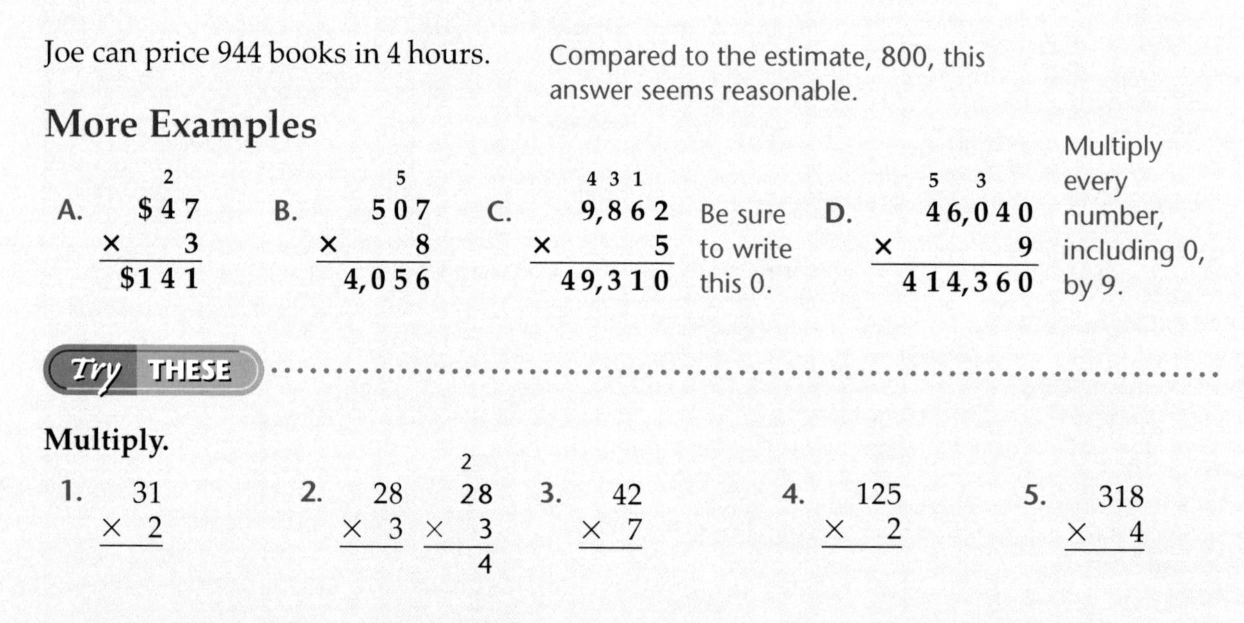

Joe can price 944 books in 4 hours.

Compared to the estimate, 800, this answer seems reasonable.

More Examples

A. $\begin{array}{r} {\scriptstyle 2} \\ \$47 \\ \times \quad 3 \\ \hline \$141 \end{array}$

B. $\begin{array}{r} {\scriptstyle 5} \\ 507 \\ \times \quad 8 \\ \hline 4{,}056 \end{array}$

C. $\begin{array}{r} {\scriptstyle 4\ 3\ 1} \\ 9{,}862 \\ \times \quad 5 \\ \hline 49{,}310 \end{array}$ Be sure to write this 0.

D. $\begin{array}{r} {\scriptstyle 5\ \ 3} \\ 46{,}040 \\ \times \quad 9 \\ \hline 414{,}360 \end{array}$ Multiply every number, including 0, by 9.

Try THESE

Multiply.

1. $\begin{array}{r} 31 \\ \times\ 2 \\ \hline \end{array}$

2. $\begin{array}{r} 28 \\ \times\ 3 \\ \hline \end{array}$ $\begin{array}{r} {\scriptstyle 2} \\ 28 \\ \times\ 3 \\ \hline 4 \end{array}$

3. $\begin{array}{r} 42 \\ \times\ 7 \\ \hline \end{array}$

4. $\begin{array}{r} 125 \\ \times\ 2 \\ \hline \end{array}$

5. $\begin{array}{r} 318 \\ \times\ 4 \\ \hline \end{array}$

Exercises

Multiply.

1. $\begin{array}{r} 23 \\ \times\ 4 \\ \hline \end{array}$
2. $\begin{array}{r} 47 \\ \times\ 3 \\ \hline \end{array}$
3. $\begin{array}{r} 98 \\ \times\ 2 \\ \hline \end{array}$
4. $\begin{array}{r} 64 \\ \times\ 7 \\ \hline \end{array}$
5. $\begin{array}{r} 86 \\ \times\ 3 \\ \hline \end{array}$
6. $\begin{array}{r} 274 \\ \times\ 6 \\ \hline \end{array}$
7. $\begin{array}{r} \$809 \\ \times\ 7 \\ \hline \end{array}$
8. $\begin{array}{r} 483 \\ \times\ 8 \\ \hline \end{array}$
9. $\begin{array}{r} 563 \\ \times\ 5 \\ \hline \end{array}$
10. $\begin{array}{r} \$945 \\ \times\ 9 \\ \hline \end{array}$
11. $\begin{array}{r} 11{,}811 \\ \times\ 6 \\ \hline \end{array}$
12. $\begin{array}{r} 25{,}090 \\ \times\ 8 \\ \hline \end{array}$
13. $\begin{array}{r} \$45{,}024 \\ \times\ 5 \\ \hline \end{array}$
14. $\begin{array}{r} 52{,}986 \\ \times\ 7 \\ \hline \end{array}$
15. $\begin{array}{r} 78{,}912 \\ \times\ 6 \\ \hline \end{array}$
16. $\begin{array}{r} 7{,}238 \\ \times\ 9 \\ \hline \end{array}$
17. $\begin{array}{r} 45{,}811 \\ \times\ 3 \\ \hline \end{array}$
18. $\begin{array}{r} 18{,}002 \\ \times\ 4 \\ \hline \end{array}$
19. $\begin{array}{r} 90{,}765 \\ \times\ 2 \\ \hline \end{array}$
20. $\begin{array}{r} 82{,}957 \\ \times\ 4 \\ \hline \end{array}$
21. $\begin{array}{r} 48{,}583 \\ \times\ 7 \\ \hline \end{array}$
22. $\begin{array}{r} 5{,}965 \\ \times\ 3 \\ \hline \end{array}$
23. $\begin{array}{r} 9{,}931 \\ \times\ 6 \\ \hline \end{array}$
24. $\begin{array}{r} 81{,}245 \\ \times\ 9 \\ \hline \end{array}$
25. $\begin{array}{r} 52{,}255 \\ \times\ 5 \\ \hline \end{array}$
26. $6 \times 4{,}015$
27. $\$8{,}473 \times 8$
28. $27{,}850 \times 9$
29. Find the product of 6 and 297.
30. Mr. Javit typed 7 three-page reports in the morning and 9 four-page reports in the afternoon. How many pages in all did he type that day?

8 Multiplying by 2-Digit Numbers

A pet shop uses 45 pounds of dog food each day. How many pounds are used in one year? (There are 365 days in a year).

Estimate:

$$\begin{array}{r} 400 \\ \times \ \ 50 \\ \hline 20{,}000 \end{array}$$

To find the pounds of dog food used in one year, multiply as follows.

Step 1	Step 2	Step 3
Multiply by the ones.	Multiply by the tens.	Add.
$\begin{array}{r} ^{3\ 2} \\ 365 \\ \times \ \ 45 \\ \hline 1825 \end{array}$	$\begin{array}{r} ^{2\ 2} \\ ^{\not 3\ \not 2} \\ 365 \\ \times \ \ 45 \\ \hline 1825 \\ 14600 \\ \hline \end{array}$	$\begin{array}{r} ^{2\ 2} \\ ^{\not 3\ \not 2} \\ 365 \\ \times \ \ 45 \\ \hline 1{,}825 \\ +14{,}600 \\ \hline 16{,}425 \end{array}$
5 × 365	4 tens × 365	

In one year, the pet shop uses 16,425 pounds of dog food.

Compared to the estimate, this answer makes sense.

More Examples

A. $\begin{array}{r} ^{4} \\ ^{\not 1} \\ 76 \\ \times \ \ 83 \\ \hline 228 \\ 6{,}080 \\ \hline 6{,}308 \end{array}$

B. $\begin{array}{r} 32 \\ \times \ \ 24 \\ \hline 128 \\ 640 \\ \hline 768 \end{array}$

C. $\begin{array}{r} ^{3\ 2} \\ ^{\not 1\ \not 1} \\ 364 \\ \times \ \ 53 \\ \hline 1{,}092 \\ 18{,}200 \\ \hline 19{,}292 \end{array}$

D. $\begin{array}{r} ^{1\ \ \ 5} \\ ^{\not 1\ \ \ \not 7} \\ 1{,}209 \\ \times \ \ 68 \\ \hline 9{,}672 \\ 72{,}540 \\ \hline 82{,}212 \end{array}$

Try THESE

Multiply.

1. $\begin{array}{r} 12 \\ \times\ 24 \\ \hline \end{array}$

2. $\begin{array}{r} 74 \\ \times\ 14 \\ \hline \end{array}$

3. $\begin{array}{r} 107 \\ \times\ \ 68 \\ \hline \end{array}$

4. $\begin{array}{r} 256 \\ \times\ \ 70 \\ \hline \end{array}$

5. $\begin{array}{r} 743 \\ \times\ \ 59 \\ \hline \end{array}$

Exercises

Multiply.

1. 43 × 32	2. 28 × 17	3. 34 × 29	4. 75 × 31	5. 83 × 52
6. 37 × 29	7. 65 × 52	8. 82 × 76	9. 59 × 42	10. 36 × 55
11. 132 × 23	12. 217 × 15	13. 323 × 27	14. $436 × 41	15. 623 × 72
16. 258 × 97	17. 374 × 68	18. 568 × 34	19. 207 × 56	20. $490 × 83
21. 865 × 34	22. $374 × 58	23. 2,476 × 73	24. 4,083 × 62	25. 26,207 × 54

26. 37 × 63

27. 82 × $294

28. 45 × 3,274

29. 5,938 × 19

30. 41,724 × 73

Problem SOLVING

31. Puppies sell for $299 each. In one year, the pet store sells 85 puppies. What are the total sales for puppies for one year?

9 Dividing by 1-Digit Numbers

A bus ticket to the Center of Science and Industry costs \$9. To earn money for tickets, the sixth grade students at Luna School collected cans for recycling. They earned \$488. How many tickets can they buy? Divide \$488 by \$9.

Estimate: $\$450 \div \$9 = 50$

Step 1	**Step 2**	**Step 3**
Divide the hundreds. $\$9\overline{)\$488}$ $9 \times ■ = 4$ There are not enough hundreds to divide without renaming. Why is 0 not written in the hundreds place?	Divide the tens. $\begin{array}{r} 5 \\ \$9\overline{)\$488} \\ -\ 45 \\ \hline 3 \end{array}$ THINK $9 \times 5 = 45$ $9 \times 6 = 54$ Why is 5 written in the tens place?	Divide the ones. $\begin{array}{r} 54\ R2 \\ \$9\overline{)\$488} \\ -\ 45 \\ \hline 38 \\ -\ 36 \\ \hline \text{remainder}\ \ 2 \end{array}$ THINK $9 \times 4 = 36$ $9 \times 5 = 45$ Why is 4 written in the ones place?

The students can buy 54 tickets. The remainder means \$2 will be left.

Compare 54 R2 to the estimate.

Use multiplication to check division.

$$\begin{array}{r} \overset{3}{5}4 \\ \times\ \ \$9 \\ \hline \$486 \end{array} \qquad \begin{array}{r} \$486 \\ +\ \ \ 2 \\ \hline \$488 \end{array} ✔$$

Division follows these steps.

1. Divide. $\begin{array}{r} 4 \\ 4\overline{)194} \end{array}$
2. Multiply. $4 \times 4 = 16$
3. Subtract. $\begin{array}{r} 4 \\ 4\overline{)194} \\ -\ 16 \\ \hline 3 \end{array}$
4. Compare. Is $3 < 4$?
5. Bring down the next digit and continue dividing. $\begin{array}{r} 48R2 \\ 4\overline{)194} \\ -\ 16\downarrow \\ \hline 34 \\ -\ 32 \\ \hline 2 \end{array}$

More Examples

A. $\begin{array}{r} 487 \\ 4\overline{)1,948} \\ -\ 16 \\ \hline 34 \\ -\ 32 \\ \hline 28 \\ -\ 28 \\ \hline 0 \end{array}$ ← This remainder is 0.

B. $\begin{array}{r} 207 \\ 3\overline{)621} \\ -\ 6 \\ \hline 2 \\ -\ 0 \\ \hline 21 \\ -\ 21 \\ \hline 0 \end{array}$ Why is 0 written in the tens place?

C. $\begin{array}{r} \$390 \\ 8\overline{)\$3,120} \\ -\ 24 \\ \hline 72 \\ -\ 72 \\ \hline 0 \\ -\ 0 \\ \hline 0 \end{array}$ Why is 0 written in the ones place?

Try THESE

Divide.

1. $4\overline{)84}$ $\begin{array}{r}2\\4\overline{)84}\end{array}$
2. $5\overline{)80}$
3. $6\overline{)624}$
4. $8\overline{)984}$
5. $9\overline{)5{,}013}$

Exercises

Divide.

1. $3\overline{)31}$
2. $2\overline{)86}$
3. $4\overline{)79}$
4. $7\overline{)\$91}$
5. $5\overline{)54}$
6. $5\overline{)525}$
7. $3\overline{)935}$
8. $6\overline{)222}$
9. $7\overline{)354}$
10. $4\overline{)960}$
11. $6\overline{)9{,}195}$
12. $5\overline{)2{,}095}$
13. $3\overline{)\$6{,}291}$
14. $4\overline{)2{,}220}$
15. $9\overline{)2{,}837}$
16. $6\overline{)810}$
17. $5\overline{)670}$
18. $5\overline{)615}$
19. $6\overline{)570}$
20. $3\overline{)1{,}758}$
21. $5\overline{)7{,}720}$
22. $6\overline{)28{,}506}$
23. $3\overline{)64{,}482}$
24. $9\overline{)8{,}613}$
25. $8\overline{)38{,}800}$
26. $74 \div 7$
27. $247 \div 6$
28. $\$1{,}053 \div 3$
29. $6{,}273 \div 5$

Problem SOLVING

30. How many trips must an eight-passenger helicopter make to transport 232 passengers?

10 Dividing by 2-Digit Numbers

A cable car can carry 28 passengers each trip. One week, there were 5,950 passengers. To find the least number of trips made, divide 5,950 by 28.

Estimate $\begin{array}{r} 200 \\ 30\overline{)6{,}000} \end{array}$ Use estimation to determine the number of digits in the quotient.

Step 1	Step 2	Step 3
Divide the hundreds. $\begin{array}{r} 2 \\ 28\overline{)5{,}950} \\ -\,5\,6 \\ \hline 3 \end{array}$ THINK 28 × 2 = 56 28 × 3 = 84 Why is 2 written in the hundreds place?	Divide the tens. $\begin{array}{r} 21 \\ 28\overline{)5{,}950} \\ -\,5\,6 \\ \hline 35 \\ -\,28 \\ \hline 7 \end{array}$ THINK 28 × 1 = 28 28 × 2 = 56 Why is 1 written in the tens place?	Divide the ones. $\begin{array}{r} 212 \text{ R14} \\ 28\overline{)5{,}950} \\ -\,5\,6 \\ \hline 35 \\ -\,28 \\ \hline 70 \\ -\,56 \\ \hline 14 \end{array}$ THINK 28 × 2 = 56 28 × 3 = 84

The answer, 212 R14, means 212 full trips with 14 people left. One more trip is needed to carry all the people, for a total of at least 213 trips. 213 is close to the estimate.

More Examples

A. Sometimes the quotient is too high.

$\begin{array}{r} 2 \\ 33\overline{)638} \\ -\,66 \end{array}$ $\quad$ $\begin{array}{r} 19 \text{ R11} \\ 33\overline{)638} \\ -\,33 \\ \hline 308 \\ -\,297 \\ \hline 11 \end{array}$

Estimate:
600 ÷ 30 = 20

Since 66 > 63,
change 2 to 1.

B. Sometimes the quotient is too low.

$\begin{array}{r} 7 \\ 28\overline{)2{,}296} \\ -\,1\,96 \\ \hline 33 \end{array}$ $\quad$ $\begin{array}{r} 82 \\ 28\overline{)2{,}296} \\ -\,2\,24 \\ \hline 56 \\ -\,56 \\ \hline 0 \end{array}$

Estimate:
2,100 ÷ 30 = 70

Since 33 > 28,
change 7 to 8.

Try THESE

Tell if the first digit of the quotient is too high, too low, or correct.

1. $\begin{array}{r} 2 \\ 21\overline{)438} \end{array}$
2. $\begin{array}{r} 3 \\ 23\overline{)499} \end{array}$
3. $\begin{array}{r} \$2 \\ 25\overline{)\$7{,}600} \end{array}$
4. $\begin{array}{r} 3 \\ 39\overline{)9{,}879} \end{array}$
5. $\begin{array}{r} 5 \\ 44\overline{)2{,}154} \end{array}$

Exercises

Divide.

1. $91\overline{)929}$
2. $59\overline{)828}$
3. $88\overline{)742}$
4. $62\overline{)378}$
5. $29\overline{)464}$
6. $45\overline{)827}$
7. $27\overline{)991}$
8. $56\overline{)2{,}083}$
9. $42\overline{)2{,}573}$
10. $81\overline{)7{,}559}$
11. $51\overline{)1{,}070}$
12. $99\overline{)4{,}487}$
13. $12\overline{)6{,}113}$
14. $11\overline{)\$5{,}621}$
15. $19\overline{)4{,}617}$
16. $22\overline{)9{,}020}$
17. $18\overline{)2{,}178}$
18. $27\overline{)4{,}833}$
19. $83\overline{)1{,}328}$
20. $55\overline{)2{,}860}$
21. $4{,}231 \div 41$
22. $9{,}994 \div 38$
23. $8{,}971 \div 29$
24. $7{,}114 \div 32$
25. $156 \div 26$
26. $940 \div 47$
27. $2{,}135 \div 61$
28. $2{,}623 \div 43$

Problem SOLVING

29. If you can move 30 pounds of gravel at a time, how many trips do you make to move 250 pounds? How much does the last load weigh?
30. An automobile company produced 174,276 cars annually. On average, how many cars were produced each month?
31. A newspaper prints 6,380 papers each day for home delivery. There are 49 carriers. What is the average number of papers delivered by each carrier?

Diagnostic Skills Posttest

Write in expanded form. (Section 1)

1. 850 2. 3,415 3. 90,001 4. 605,869

What number is named by each of the following digits in 1,401,759? (Section 1)

5. 9 6. 0 7. 4 8. 7

Compare using <, >, or = . (Section 2)

9. 9,004 ■ 9,000 10. 565 ■ 556 11. 13,481 ■ 13,481

Order from least to greatest. (Section 2)

12. 531, 499, 153, 431, 134 13. 1,856; 4,456; 5,646; 999

Round to the nearest thousand. (Section 3)

14. 4,580 15. 18,090 16. 59,498 17. 39,603

Round to the underlined place-value position. (Section 3)

18. $5{,}8\underline{6}3$ 19. $43{,}0\underline{0}9$ 20. $9\underline{3}1{,}434$ 21. $\underline{9}{,}554{,}732$

Write in standard form. (Section 4)

22. XX 23. XVII 24. LXIV 25. MCM

Write in Roman numerals. (Section 4)

26. 178 27. 505 28. 747 29. 8,404

Add. (Section 5)

30. 46 + 29 31. $829 + $274 32. $54,789 + $37,670

33. 65,937 + 3,230 + 489 34. 739,087 + 384,852 35. 56 + 1,987 + 35,006

Subtract. (Section 6)

36. 72 – 31 37. 484 – 399 38. $5,634 – 2,785

39. 345,000 – 127,637 40. $78,005 – $19,438 41. 420,000 – 352,965

Multiply. (Section 7)

42. 964×7
43. $\$2,499 \times 8$
44. $\$989 \times 7$
45. $8,450 \times 5$
46. $81,744 \times 6$
47. $22,599 \times 4$

Multiply. (Section 8)

48. 15×61
49. $\$25 \times 74$
50. 501×92
51. $\$6,027 \times 84$
52. $23,697 \times 57$
53. $89,237 \times 65$

Divide. (Section 9)

54. $3\overline{)162}$
55. $7\overline{)651}$
56. $6\overline{)4,980}$
57. $\$570 \div 6$
58. $\$774 \div 9$
59. $8,613 \div 9$

Divide. (Section 10)

60. $15\overline{)948}$
61. $57\overline{)4,831}$
62. $45\overline{)9,315}$
63. $\$940 \div 47$
64. $2,135 \div 61$
65. $\$2,860 \div 55$

CHAPTER 1

Number Patterns and Algebra

Katherine MacMartin
New Hampshire

1.1 Problem-Solving Strategy: Four-Step Plan

Cora is building a bird feeder. From a board she will cut pieces into the lengths shown. Should Cora buy a board that is 72 in., 96 in., or 120 in. long?

Bird Feeder Piece	Length in Inches
side	10
side	10
bottom	18
back	16
roof	20

You can use the four-step plan to help you solve problems.

Read the problem carefully and slowly. Ask yourself questions such as "What facts do I know?" and "What do I need to find?" You may wish to write down important facts.

See how the facts relate to each other. Decide how to solve the problem. Estimate the answer.

Then write an equation using a letter for the unknown.

Compute or work through your plan. Answer the problem.

Reread the problem. Ask yourself "Does the answer make sense for the problem?" and "Compared to the estimate, is the answer reasonable?" If not, try another way to solve the problem.

Now apply this four-step plan to the problem given above.

You need to find how long a board Cora needs. You know the length of each piece.

To find the total length, add.
Estimate: $\mathbf{10 + 10 + 20 + 20 + 20 = 80}$

Write an equation: $\boldsymbol{n} = \mathbf{10 + 10 + 18 + 16 + 20}$

$10 + 10 + 18 + 16 + 20 = 74$

Cora can buy a board that is 96 inches long.

A 96-inch board is reasonable because a 72-inch board is too short and a 120-inch board produces more waste.

Try THESE

Use the four-step plan to solve.

1. Jamie is 56 inches tall. Kathy is 60 inches tall. How many inches taller is Kathy?

2. Gino collects 18 pounds of cans. Paula collects 24 pounds of cans. Together, how many pounds did they collect?

3. Jim wants 5 bird feeders. A bird feeder costs $14. How much money does Jim need for 5 bird feeders?

4. Mrs. Snell had 4 gallons of paint. She used 2 gallons. How much paint does she have now?

Solve

1. Danny needs a board that is 74 inches. He cuts it 6 inches too short. How long is the board that Danny cut?

2. Camden Yards baseball stadium in Baltimore has 48,876 seats. Fenway Park in Boston seats 36,298. How many more seats are in Camden Yards?

3. During a 7-day week, a waitress served 135 people. On average, how many people did she serve each day?

4. The same waitress received $2 in tips from each person she served. How much money did she earn from tips during the week?

5. Four bowlers are on a team. In one game, their scores are 159, 96, 143, and 171. What is the total for the game?

6. If Jim's heart beats 60 times per minute, how many times will his heart beat in 30 minutes?

7. Sally is reading a 526-page book. She has already read 321 pages. How many pages are left?

8. What three consecutive odd numbers add up to 999?

Constructed RESPONSE

9. Olivia will visit her cousin and her grandmother. Her visits will total 3 weeks and 2 days. She will spend 5 more days with her grandmother than with her cousin. How much time will she spend with each relative? Explain how you found your answer.

1.2 Powers and Exponents

Objective: to represent numbers by using exponents

Each aisle of Earl's Market has 5 sections. In each section there are 5 shelves. To find the total number of shelves in an aisle, multiply 5×5. Another way to express 5×5 is 5^2. In 5^2 the *exponent* is 2 and the *base* is 5. The exponent tells how many times the base is used as a factor of the product.

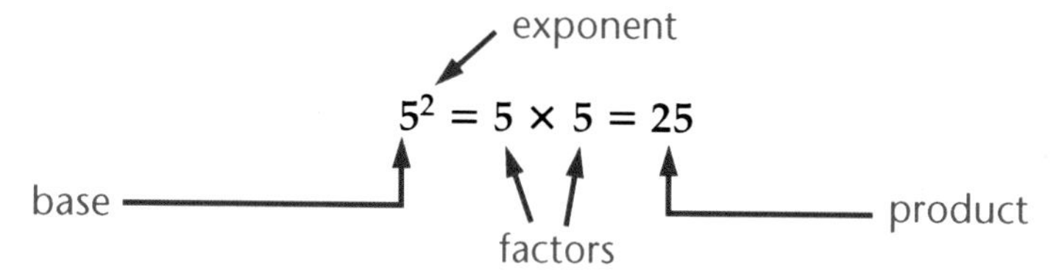

Numbers that are written using exponents are called **powers.**

Write: 3^2 **Say:** 3 to the second power or 3 squared

7^3 7 to the third power or 7 cubed

6^4 6 to the fourth power

Look at the powers of two and ten below. Describe the patterns you see.

In powers of ten, what do you notice about the exponent and the number of zeros in standard form?

$2^0 \rightarrow 1$

$2^1 \rightarrow 2$

$2^2 \rightarrow 2 \times 2 = 4$

$2^3 \rightarrow 2 \times 2 \times 2 = 8$

$2^4 \rightarrow 2 \times 2 \times 2 \times 2 = 16$

$2^5 \rightarrow 2 \times 2 \times 2 \times 2 \times 2 = 32$

$10^0 \rightarrow 1$

$10^1 \rightarrow 10$

$10^2 \rightarrow 10 \times 10 = 100$

$10^3 \rightarrow 10 \times 10 \times 10 = 1{,}000$

$10^4 \rightarrow 10 \times 10 \times 10 \times 10 = 10{,}000$

$10^5 \rightarrow 10 \times 10 \times 10 \times 10 \times 10 = 100{,}000$

Examples

A. $8^3 = 8 \times 8 \times 8 = 512$

B. $3^4 = 3 \times 3 \times 3 \times 3 = 81$

C. $7^2 = 7 \times 7 = 49$

D. $4^2 = 4 \times 4 = 16$

Replace each ■ with an exponent to make a true equation.

1. $4 \times 4 \times 4 = 4^{■}$
2. $6 \times 6 = 6^{■}$
3. $3 \times 3 \times 3 \times 3 = 3^{■}$
4. $8 \times 8 \times 8 = 8^{■}$
5. $9 \times 9 \times 9 \times 9 = 9^{■}$
6. $7 = 7^{■}$

Exercises

Write each expression using exponents.

1. $3 \times 3 \times 3 \times 3$
2. $6 \times 6 \times 6$
3. 8 to the ninth power

Write using factors. For example, $2^4 = 2 \times 2 \times 2 \times 2$.

4. 6^2
5. 5 cubed
6. 4^5
7. 2^6

Write in standard form.

8. 2 squared
9. 4 cubed
10. 8 to the first power
11. 5^1
12. 9^3
13. 6^2
14. 1^8
15. 72^1
16. 3^4
17. 10^6
18. 50^3
19. 7^2
20. 2^8
21. Write the sum in standard form for $3^4 + 4^3$.
22. If $A = 15$, then $A^2 = ?$

★23. Which is greater, 2^3 or 3^2? Use $<$, $>$, or $=$ in your answer.

Problem SOLVING

24. Rosa and Raquel wrote 5^4 using factors. Who was correct?

Rosa
$5^4 = 5 \bullet 5 \bullet 5 \bullet 5$

Raquel
$5^4 = 4 \bullet 4 \bullet 4 \bullet 4 \bullet 4$

Constructed RESPONSE

★25. Describe the pattern of the ones digit in successive powers of 3. What is the ones digit in 3^{100}?

$3^1 = 3$ $\quad$ $3^4 = 81$
$3^2 = 9$ $\quad$ $3^5 = 243$
$3^3 = 27$ $\quad$ $3^6 = 729$

1.3 Order of Operations

Objective: to use the order of operations

In life, there are many things that must be completed in the correct order. This is especially true in baking. When you prepare the batter for chocolate chip cookies, you begin with butter, then add brown sugar and granulated sugar, and then add vanilla. Next, you add eggs, then add flour, and finally mix in the chocolate chips. Imagine if the recipe for chocolate chip cookies asked you to begin with chocolate chips, and then add eggs. Beat well. Would the cookies turn out the same?

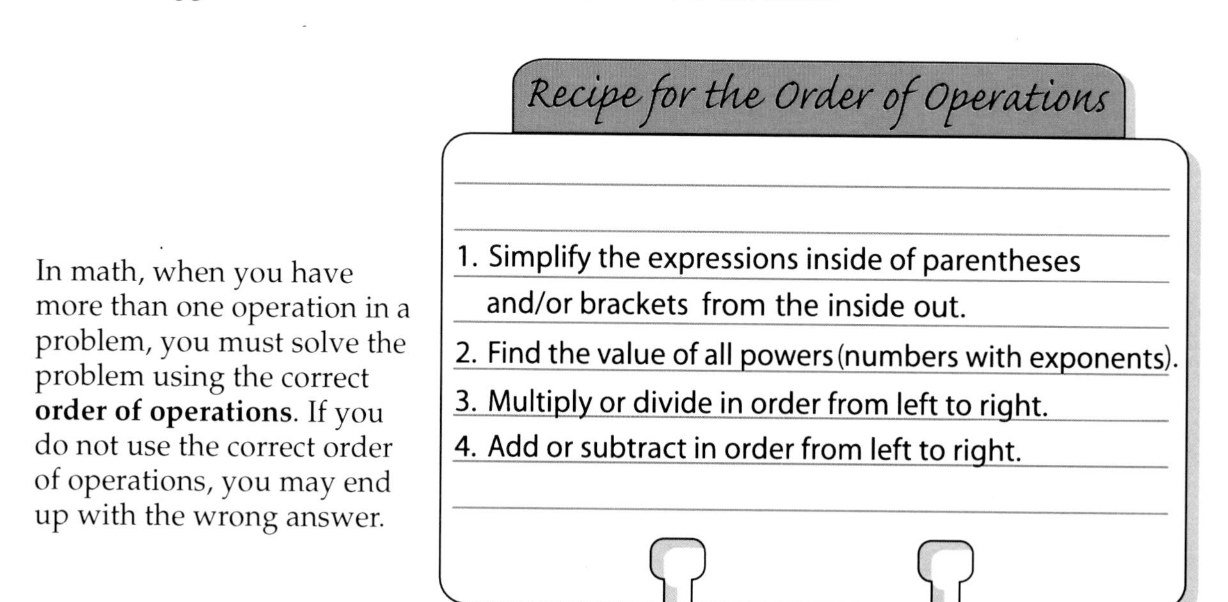

In math, when you have more than one operation in a problem, you must solve the problem using the correct **order of operations**. If you do not use the correct order of operations, you may end up with the wrong answer.

Bill and Miranda both write this expression, $2 \times 185 + 3 \times 20$, but each simplifies it differently.

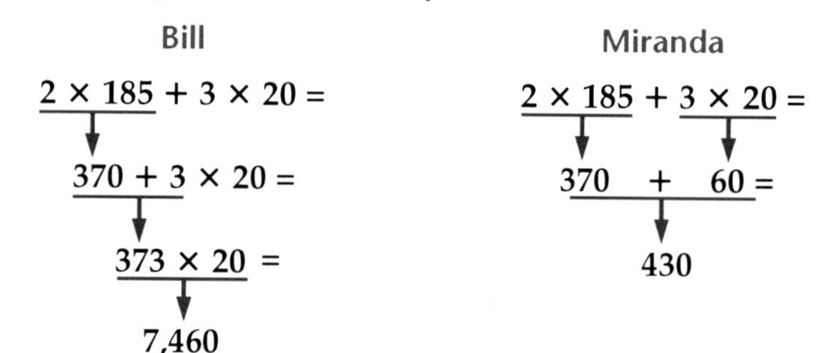

Bill found his answer by working from left to right. Miranda multiplied first and then added. Who is correct? Since Miranda followed the order of operations, her answer is correct.

Examples

A. $(1 + 2) + 3 \times 4 =$

$3 + 3 \times 4 =$

$3 + 12 =$

15

B. $2^2 + 5 \times 4 \div 2 =$

$4 + 5 \times 4 \div 2 =$

$4 + 20 \div 2 =$

$4 + 10 =$

14

C. $3^3 - 15 \div 3 \times 2 =$

$27 - 15 \div 3 \times 2 =$

$27 - 5 \times 2 =$

$27 - 10 =$

17

Try THESE

Choose the correct answer.

1. $8 + 9 \times 1 \times 2$ a. 34 b. 26 c. 72
2. $6 + 10 \div 2 \times 4$ a. 32 b. 44 c. 26

Exercises

Solve using the order of operations.

1. $8 \times 7 + 3$
2. $8 \times (7 + 3)$
3. $3 + 2 \times 7$
4. $7 \times 8 + 4 + 1$
5. $[8 + (9 \times 1)] \times 2$
6. $3 \times 2 + 7 \times 0$
7. $2^4 - 8 \times 2 \div 4$
8. $5 + 2^3 \times 2$
9. $5 + (2 \times 8 + 4)$
10. $5^2 + 4 \times (2 + 3)$

★11. $(5 - 2 \times 2) \div (3^3 - 5^2)$

★12. $(4^2 - 3 \times 4)^2$

Problem SOLVING

13. Regular airfare to Boston is $125 one way, but as a special discount the fare was lowered to $105 one way. How much money will 4 people save altogether using the lower rate?
14. A store sells DVDs for $15 and CDs for $12. What is the total cost of 5 DVDs and 6 CDs?

1.4 Variables and Expressions

Objective: to evaluate expressions with variables; to translate word phrases into variables and operations

Kris Davis has a poster collection. Her brother gave her 4 more posters for her birthday. Each poster Kris has can be represented by the shape ■.

If Kris had 8 posters already, you could represent ■ as follows.

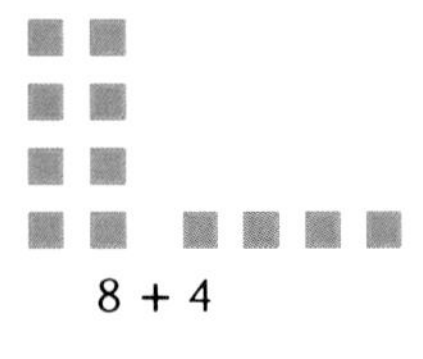

Instead of having to use symbols, such as ■, the number of posters Kris has can be represented by the **algebraic expression** $p + 4$.

p	$+$	4
number of posters	plus	

In the expression $p + 4$, p is the **variable**. A variable is used to represent some quantity. In this case, p stands for the total number of posters Kris had at first.

A variable can represent more than one quantity.

If p is 32, then $p + 4 = 32 + 4$.
$p + 4 = 36$

If p is 36, then $p + 4 = 36 + 4$.
$p + 4 = 40$

Examples

A. Evaluate $p + 5$ if p is 9.

$9 + 5$	Replace p with 9.
$9 + 5 = 14$	Add 9 and 5.

B. Evaluate $a - 4$ if a is 7.

$7 - 4$	Replace a with 7.
$7 - 4 = 3$	Subtract 4 from 7.

Translate each into an algebraic expression.

C. 4 is multiplied by a

$4 \times a$ or $4a$ — Both expressions represent multiplication.

D. the quotient of m and 6

$\frac{m}{6}$ — *Quotient* indicates division.

Try THESE

Evaluate each expression if $a = 2$, $b = 5$, and $c = 10$.

1. $a + 3$
2. $b - 1$
3. $c - 8$
4. $b \times 3$
5. $\frac{c}{2}$
6. $a + b$
7. $b - a$
8. $\frac{c}{b}$
9. $a \times b$
10. $a + b + c$
11. $a^2 + b^2$
12. $2a + c^3$

Exercises

Evaluate each expression if $w = 13$, $x = 9$, $y = 1.9$, and $z = 0.5$.

1. $w - 6$
2. $w + 5$
3. $x \times 4$
4. $\frac{x}{3}$
5. $w + 6 - 15$
6. $x + 20 - 1$
7. $y + 10$
8. $y - 1$
9. $w + x$
10. $x \times y$
11. $x^2 + w^2$
12. $w^2 - (x^2 + y^2)$
13. $x^2 + 4$
14. $2x + y^4$
15. $w + x + y$
16. $w + x + z$

Choose the correct answer.

17. the sum of 9.1 and v — a. $9.1 + v$ b. $9.1 \times v$ c. $9.1v$
18. 14 less d — a. $d - 14$ b. $\frac{d}{14}$ c. $14 - d$
19. 35 divided by h — a. $35 \times h$ b. $\frac{35}{h}$ c. $\frac{h}{35}$

Translate each into an expression.

20. the sum of k and 8
21. the difference of t and 19
22. 4 less s
23. 7 added to r
24. the product of m and 3
25. the quotient of b and 5
26. Write an expression for 4 more than y.
27. Name the variable in the expression $a + 6$.

Problem SOLVING

Solve. Write an expression for problems 28–29.

28. Barry has s pairs of shoes. His mother bought him two more pairs of shoes. How many pairs of shoes does Barry have now?

29. The math club is selling brownies to raise money for their annual trip to Math Land. The club earns $2 for each brownie ($b$) sold. Write the expression that represents the amount of money the math club will earn. Complete the table to the right for the given number of brownies sold.

Number of Brownies Sold	Amount of Money Earned
b	
20	
35	
75	

Constructed RESPONSE

30. State two methods to answer the following problem. Susie wants to send a package. It costs $2 to send the package plus 50¢ for each pound. How much will it cost to send a 6-pound package? Explain how you know that your answer is correct.

Test PREP

31. What is the value of $x + y$ when $x = 3$ and $y = 7$?

a. 21 b. 10 c. 4 d. 6

32. Which expression represents x decreased by 8?

a. $\frac{x}{8}$ b. $x + 8$ c. $8x$ d. $x - 8$

Mid-Chapter REVIEW

Write in standard form.

1. 5 squared **2.** 3 cubed **3.** 6^2 **4.** 20^2

Write using exponents.

5. $2 \times 2 \times 2 \times 2$ **6.** $6 \times 6 \times 6 \times 6 \times 6$ **7.** $8 \times 8 \times 8 \times 8 \times 8$

Solve using the order of operations.

8. $6 + 8 \times 2$ **9.** $3 + 4^2 - 5$ **10.** $5 \times 7 + (6^2 + 2)$

Cumulative Review

Name the digit in each place-value position in 826,749.

1. hundreds
2. ten thousands
3. ones
4. hundred thousands
5. tens
6. thousands

Write in standard form.

7. five
8. sixteen
9. thirty
10. eighty
11. seventy-nine
12. forty-three
13. five hundred
14. nine hundred
15. three thousand
16. ninety thousand
17. eighty thousand, eight
18. five hundred five thousand
19. twenty-nine million
20. sixty-seven billion
21. sixteen billion, six thousand, six hundred sixty-six

Write in expanded form.

22. 832
23. 2,764
24. 724,005
25. 1,204,070

Write in words.

26. 705
27. 328,014
28. 6,007,993

Compare using <, >, or = .

29. 498 ● 500
30. 706 ● 760
31. 14,958 ● 14,758
32. 700,000 ● 699,995
33. 2,007,541 ● 2,007,541
34. 53,000,000 ● 530,000

Round to the underlined place-value position.

35. $6\underline{0}3$
36. $1{,}2\underline{3}9$
37. $17{,}\underline{8}08$
38. $5\underline{3}{,}600{,}879$
39. $8{,}\underline{9}95{,}548$
40. $1\underline{7}{,}846{,}320{,}115$
41. $\underline{3}94{,}728{,}041{,}652$

Write in standard form.

42. CCCLXXXV
43. DXLIV
44. MCMXCII
45. $\overline{\text{LXII}}$

1.5 Distributive Property

Objective: to use the Distributive Property

The **Distributive Property** states that the product of a number and a sum is equal to the sum of the products.

Look at the following example:

$3 \times (10 + 6) =$	The problem is asking for the product of a number and a sum.
$(3 \times 10) + (3 \times 6) =$	The answer can be found by finding the sum of the products.
$30 + 18 = 48$	Finally, find the sum of these products.

The Distributive Property shows how multiplication affects an addition or subtraction problem. The Latin word *distribuere* means "assign separately." Look at the following examples.

$$3 \times (7 + 5) = (3 \times 7) + (3 \times 5) = 21 + 15 = 36$$

$$4 \times (6 - 3) = (4 \times 6) - (4 \times 3) = 24 - 12 = 12$$

The Distributive Property is useful in helping you compute mentally, and is important in algebra.

Examples

A. $6 \times 52 =$

$6 \times (50 + 2) =$

$(6 \times 50) + (6 \times 2) =$

$300 + 12 = 312$

You know that 52 equals 50 + 2, so you can use the Distributive Property to solve this problem mentally.

B. $4 \times 99 =$

$4 \times (100 - 1) =$

$(4 \times 100) - (4 \times 1) =$

$400 - 4 = 396$

You know that 99 equals 100 – 1, so you can use the Distributive Property to solve this problem mentally.

Try THESE

Replace each ■ with a number to make a true equation.

1. $5 \times (3 + 4) = (5 \times 3) + (5 \times ■)$
2. $8 \times (4 - 2) = (8 \times 4) - (8 \times ■)$
3. $12 \times (15 + 5) = (12 \times 15) + (12 \times ■)$
4. $6 \times 14 = (6 \times 10) + (6 \times ■)$
5. $8 \times 77 = (8 \times 70) + (8 \times ■)$
6. $5 \times 99 = (■ \times ■) - (5 \times 1)$

Exercises

Use the Distributive Property to solve. Show your work.

1. $7 \times (5 + 4)$
2. $12 \times (8 - 5)$
3. $11 \times (6 + 3)$
4. $100 \times (14 - 3)$
5. $17 \times (5 + 6)$
6. $21 \times (10 - 7)$

Compute mentally.

7. 2×27
8. 3×46
9. 3×34
10. 5×62
11. 2×64
12. 4×93
13. 7×61
14. 3×85
15. 6×55
16. 4×59
17. 7×42
18. 5×76
19. 7×104
20. 5×95

★21. 2×156

Problem SOLVING

22. Suppose George has 7 nickels and 7 dimes in his pocket. How much money does he have? Solve this problem in two ways using the Distributive Property.
23. A kitchen that is 15 feet long and a connecting dining room that is 12 feet long are both 12 feet wide. Find the area of the kitchen floor and dining room floor combined. Solve this problem in two ways using the Distributive Property.
(*Remember:* Area = length × width)

Constructed RESPONSE

24. Mary Ellen and Barbara used the Distributive Property to simplify $4 \times (6 + 3)$. Look at their work. Who was correct? Explain.

Mary Ellen	Barbara
$4 \times (6 + 3) =$	$4 \times (6 + 3) =$
$(4 \times 6) + (4 \times 3) =$	$(4 + 6) \times (4 + 3) =$
$24 + 12 =$	$10 \times 7 =$
36	**70**

1.6 Equations and Mental Math

Objective: to solve equations using mental math

The dictionary defines the word **translate** as "changing or transferring from one set of symbols to another." You can translate "How are you?" from English to Spanish as "¿Cómo esta usted?" *How* translates to *cómo, are* translates to *esta,* and *you* translates to *usted.*

Not only can you translate from English to Spanish, you can also translate verbal sentences to **equations**. Instead of using words, an equation uses variables, numbers, operation symbols, and an equal sign.

Here are some examples of equations.

$3 + 2 = 5$ $\qquad$ $4 + 8 = 12$

In each example the value on both sides of the equal sign is the same. They balance each other, just like a scale.

Some equations contain variables.

$x + 8 = 10$ $\qquad$ $x + 3 = 7$

You can solve the equation for the variable by replacing the variable with a number that makes the equation true. Remember that a variable represents an unknown quantity. The number you substitute for the variable is the **solution**.

The solution for $x + 8 = 10$ is 2 because $2 + 8 = 10$.

The solution for $x + 3 = 7$ is 4 because $4 + 3 = 7$.

What is the solution of the equation $x + 4 = 9$?

Now you will investigate equations.

Examples

You can solve these equations for the variable using mental math.

A. I am thinking of a number. Add 3 to it. The sum is 7. What is my number?

n $+ 3$ $= 7$

Solve $n + 3 = 7$ using mental math.
7 equals 3 plus what number?
$7 = 3 + n$
$n = 4$

B. I am thinking of a number. Subtract 8 from it. The difference is 12. What is my number?

n $- 8$ $= 12$

Solve $n - 8 = 12$ using mental math.
12 equals what number minus 8?
$12 = n - 8$
$n = 20$

Try THESE

Translate the following to equations and use mental math to solve the equations.

1. I am thinking of a number.
   It is increased by 7.
   The sum is 21.
   What is my number?

2. I am thinking of a number.
   Multiply it by 9.
   The product is 27.
   What is my number?

Solve the following equations using mental math.

3. $n - 5 = 12$

4. $2x + 1 = 7$

Exercises

Translate each sentence into an equation.

1. The product of 9 and a number is 45.
2. A number increased by 3 is 19.
3. A number divided by 6 is 3.
4. The difference of a number and 7 is 21.
5. Look at the equations you wrote above. Solve each equation using mental math.

Choose the correct answer.

6. $x + 8 = 19$ a. 7 b. 10 c. 11
7. $7 - n = 3$ a. 2 b. 4 c. 3
8. Which equation is true when $x = 7$?
   a. $14 = x + 4$ b. $x + 5 = 12$ c. $21 = x + 7$

Solve using mental math.

9. $a + 6 = 20$
10. $4d = 28$
11. $27 - x = 19$
12. $h + 12 = 18$
13. $\frac{x}{5} = 9$

Problem SOLVING

14. Suppose you bought a pizza for \$8.00. You also bought a salad, but you cannot remember how much you paid and you cannot find the receipt. You know that the total amount you spent was \$10.00. The equation $8 + s = 10$ can be used to represent this situation. Solve the equation to find how much you spent on the salad using mental math.

15. How many more books do you need to read over the summer if you already read 3 books and you need to read a total of 8 books? Write an equation to represent this situation and solve the equation using mental math.

How Many Numtherians?

Using a table may help you solve this problem.

On the planet Numtherius, the inhabitants are called Biplors, Trimeeps, and Quadrogs. Biplors have 2 legs, Trimeeps have 3 legs, and Quadrogs have 4 legs. On a recent trip to Earth, the Numtherians met a girl named Pauline, who took their picture before they left for their home planet. All three types of Numtherians were in the picture. When Pauline had the picture developed, she found that she had placed her finger over the lens of the camera, and all she could see were the Numtherians' heads and legs. If there were 12 heads and 27 legs, how many of each kind of Numtherian were there?

Extension

Find numbers to replace each ■ in the problem below. Then give the problem to someone else to solve. You must know the solution before you give it to someone.

You paid ■ for movie tickets. The tickets cost ■ for an adult and ■ for a child. How many of each ticket were bought?

1.7 Problem-Solving Strategy: Guess and Check

Objective: to solve problems using the guess-and-check strategy

Kevin makes dolls out of apples to sell at craft fairs. He spent $2.59 on apples. He paid no tax. Apples cost less than 40¢ each. Kevin bought fewer than 10 apples. What is the price of 1 apple? How many apples did Kevin buy in all?

One way to solve this problem is to guess and check.

- Make a guess that makes sense.
- Check the guess to see if it is correct.
- Decide how to make a better guess.
- Make another guess and check to see if it is correct.
- Guess and check until you solve the problem.

You need to find the price of 1 apple and the number of apples bought. You know the total cost. You also know that Kevin spent 39¢ or less on each apple and that he bought 9 or less apples.

First guess that the apple costs 39¢.
Divide $2.59 (259¢) by 39¢.

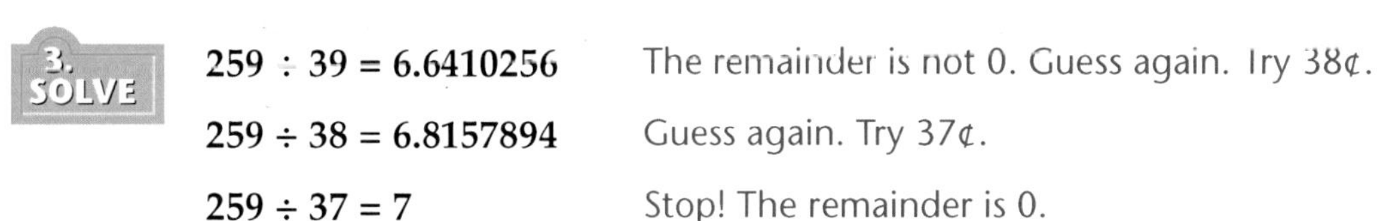

259 ÷ 39 = 6.6410256 The remainder is not 0. Guess again. Try 38¢.

259 ÷ 38 = 6.8157894 Guess again. Try 37¢.

259 ÷ 37 = 7 Stop! The remainder is 0.

Kevin bought 7 apples that cost 37¢ each.

How do you know that there is only one solution?

Since 7 is less than 10 and 37¢ is less than 40¢, the solution is reasonable.

Check:
$0.37 × 7 = 2.59 ⟶ $2.59 ✔

Try THESE

Use the guess-and-check strategy to solve the following problems.

1. The music teacher is three times as old as Walt. In 12 years, she will be twice as old as Walt. How old is Walt now?

2. Charlene is four times as old as her nephew. In 16 years, Charlene will be twice his age. How old is Charlene's nephew now?

3. When a certain number is multiplied by itself, the product is 2,304. Find the number.

Solve

1. Find the lowest prime number greater than 200. *Hint*: Use rules for divisiblity.

2. A certain number, when raised to the third power, equals 216. Find the number.

★3. Use each digit only once. Write two 2-digit whole numbers that have the greatest product possible.

★4. Use each digit only once. Write two 3-digit whole numbers that have the least product possible.

★5. A group of chickens and goats has a total of 6 heads and 18 feet. How many of each kind of animal is in the group? For each guess, explain why you made that guess before you check the answer.

Chapter 1 Review

Language and Concepts

Write *true* or *false*. If false, replace the underlined word or words to make a true statement.

1. In the expression 4^2, 4 is the exponent.
2. In the order of operations, you should multiply/divide, before you add/subtract.
3. The Distributive Property states that the product of a number and a sum is equal to the quotient of the products.
4. A variable is used to represent some quantity.
5. A mathematical sentence that uses numbers, variables, and an equal sign is called an equation.

Skills and Problem Solving

Write using exponents. (Section 1.2)

6. 5 cubed
7. $8 \times 8 \times 8 \times 8$
8. 7 squared
9. $3 \times 3 \times 3 \times 3 \times 3 \times 3$

Write in standard form. (Section 1.2)

10. 2^5
11. 21 squared
12. 5^5
13. If $x = 12$, what is x^2?

Solve. (Section 1.3)

14. $5 \times 2 + 6 - 5$
15. $10 \div 5 \times 2 - 1$
16. $5^2 + 3 \times 5$
17. $(3 \times 9) - 5^2 + 7$
18. $12(20 + 1)$
19. $14 - 4 \times 3 \div 2 + 5$

Evaluate each expression if $r = 5$, $s = 7$, and $t = 4.5$. (Section 1.4)

20. $r + 2s$
21. $8 + s - t$
22. $rs - t$

Replace each ■ with a number to make the equation true. (Section 1.5)

23. $6 \times (3 + 9) = (6 \times 3) + (6 \times$ ■$)$
24. $6 \times 37 = (6 \times$ ■$) + (6 \times 7)$

Compute mentally. (Section 1.5)

25. 9×41
26. 3×17

Translate each sentence into an equation. (Section 1.6)

27. The sum of 12 and a number is 30.

28. A number decreased by 7 is 20.

29. The product of 3 and a number is 12.

Solve using mental math. (Section 1.6)

30. $a + 5 = 25$ 31. $b - 3 = 12$ 32. $4c = 40$ 33. $\frac{x}{5} = 10$

Solve. (Sections 1.1, 1.7)

34. William is x years old. If I multiply his age by 6, I can find his father's age. His father is 30 years old. Write an equation and solve for William's age using mental math.

35. Sam earns $5 per hour baby-sitting. She works 3 hours on Friday, 5 hours on Saturday, and 2 hours on Sunday. How much did she earn baby-sitting these three days? Write an equation and solve using the Distributive Property.

36. I am thinking of a number. If I multiply the number by 5 and subtract 11, I get 124. What is my number?

37. Look at the expression below.

    $100 - 35 \div 5 + 3$

    a. What is the value of this expression?

    b. Explain why the value you determined is correct. Use what you know about order of operations in your explanation. Use words, numbers, and/or symbols in your explanation.

Chapter 1 Test

Write using exponents.

1. 10 squared
2. $5 \times 5 \times 5 \times 5 \times 5 \times 5 \times 5$
3. 3 cubed

Write in standard form.

4. 2^6
5. 15 squared
6. 3^4
7. 1 cubed

Solve.

8. $12 \times 2 + 7 - 5$
9. $18 \div 6 \times 2 - 1$
10. $4^3 + 2 \times 8$
11. $(8 \times 5) - 3^3 + 7$

Replace each ■ with a number to make the equation true.

12. $4 \times (7 + 11) = (4 \times 7) + (4 \times$ ■$)$
13. $5 \times 45 = (5 \times$ ■$) + (5 \times 5)$

Compute mentally.

14. 7×51
15. 2×19

Evaluate each expression if $a = 3$, $b = 10$, and $c = 1.5$.

16. $2a + c$
17. $a^2 + (b \times c)$

Translate each sentence into an equation.

18. The difference of 25 and a number is 5.
19. The product of 5 and a number is 50.

Solve using mental math.

20. $w + 10 = 22$
21. $x - 9 = 12$
22. $3y = 24$
23. $\frac{x}{6} = 5$

Solve.

24. Sue is x years old. If I multiply her age by 2, I can find her brother's age. Her brother is 6 years old. Write an equation and solve using mental math for Sue's age.
25. In a bag of red and green candy, there are 25 red pieces of candy. If the bag contains 42 pieces of candy total, how many pieces are green?
26. Joey went to buy a cup of coffee for $2 and a roll for $1. His friend wanted to order the same breakfast. Write an equation to find the total cost of breakfast for Joey and his friend. Then use the Distributive Property to solve.
27. Look at the expression $72 - 24 \div 3 + 12$.
    a. What is the value of this expression?
    b. Explain why the value you determined is correct. Use what you know about order of operations in your explanation. Use words, numbers, and/or symbols in your explanation.

Change of Pace

Base Five Numerals

Our base ten numeration system uses ten digits: 0, 1, 2, 3, 4, 5, 6, 7, 8, and 9. We group by ones, tens, hundreds, and so on. A **base five** system uses five digits: 0, 1, 2, 3, and 4. The grouping is by ones, fives, twenty-fives, and so on.

Base Ten

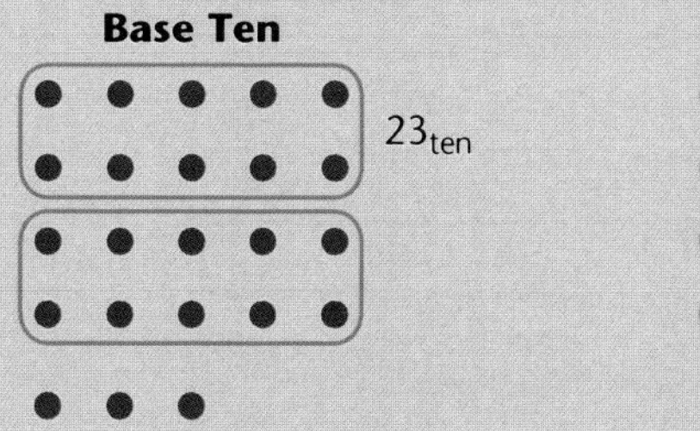

2 groups of ten and 3 ones

Base Five

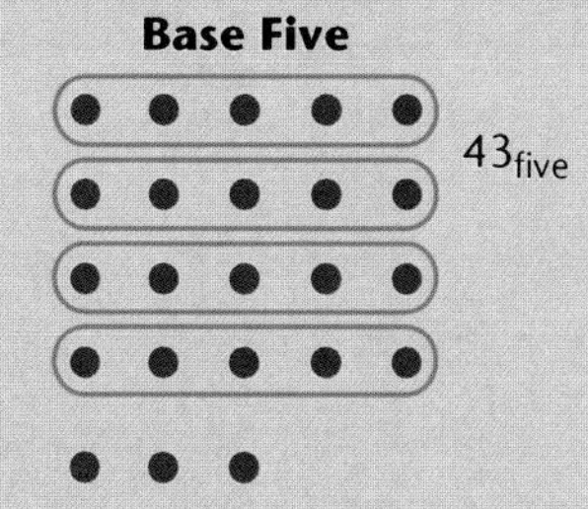

4 groups of five and 3 ones

Another Example

In base ten, the numeral 123 means 1 hundred plus 2 tens plus 3 ones.

$$100 + 20 + 3 = 123_{ten}$$

Base Ten

thousands	hundreds	tens	ones
	1	2	3

In base five, the numeral 123 means 1 twenty-five plus 2 fives plus 3 ones.

$$25 + 10 + 3 = 38_{five}$$

Base Five

one hundred twenty-fives	twenty-fives	fives	ones
	1	2	3

Group by fives to write each as a base five numeral.

1. 2. 3. 4.

5. 6. 7. 8.

Use expanded form to write each as a base ten numeral.

$14_{five} = 5 + 4$ or 9_{ten} $\quad$ $103_{five} = 25 + 3 = 28_{ten}$

9. 12_{five}
10. 10_{five}
11. 21_{five}
12. 23_{five}
13. 20_{five}
14. 31_{five}
15. 100_{five}
16. 111_{five}
17. 104_{five}
18. 444_{five}

★ 19. How tall are you in inches in the base five system?

Cumulative Test

1. What is 17,000 written in words?
   a. seventeen hundred
   b. seventy-five thousand
   c. sixteen thousand
   d. none of the above

2. $\begin{array}{r} 309 \\ +\ 288 \\ \hline \end{array}$
   a. 577
   b. 597
   c. 607
   d. none of the above

3. Which of the following distances is the greatest?
   a. 80,064 miles
   b. 80,564 miles
   c. 80,640 miles
   d. 80,656 miles

4. $594 – $299 = ____
   a. $275
   b. $295
   c. $305
   d. none of the above

5. Estimate.
   $\begin{array}{r} 626 \\ 272 \\ +\ 182 \\ \hline \end{array}$
   a. 800
   b. 900
   c. 1,000
   d. 1,100

6. What is 7,642 rounded to the nearest ten?
   a. 7,600
   b. 7,640
   c. 7,650
   d. none of the above

7. Ahmad buys two pairs of shorts that cost $7 each and three shirts that cost $9 each. What is his change from $50?
   a. $7
   b. $10
   c. $11
   d. none of the above

8. Which fact is not needed in order to solve this problem?

   Sue is 12 years old. She has collected 228 butterflies. Mary has collected 179 butterflies. How many more does Sue have than Mary?
   a. the number in Sue's collection
   b. the number in Mary's collection
   c. Sue's age
   d. none of the above

9. The Brock family is planning to drive to Yosemite National Park. What is one of the problems they will probably have to solve?
   a. How many people, on the average, attend the park each day?
   b. What is the area of Yosemite National Park?
   c. What is the average weight of each family member?
   d. What will they have to pay for gasoline, food, and lodging?

10. The Brocks drove 400 miles each day for 3 days. They drove 8 hours each day.

    From the data in the problem above, which question cannot be answered?
    a. Each day, how many hours were spent not driving?
    b. How many hours did they drive in the 3 days?
    c. How many miles did they drive in 3 days?
    d. How many miles per gallon of gas did they average?

CHAPTER 2

Adding and Subtracting Decimals

Enoch Porter
Alaska

2.1 Problem-Solving Strategy: Logic

Objective: to use logical reasoning to solve word problems

In a cross-country race, Shannon outran David. Joe beat Sue but lost to David. Who finished first?

A problem like the example requires **logical reasoning**. Logical reasoning allows you to make conclusions based on the given information. Sometimes you will need to find patterns, eliminate possible answers, or use information that you already know. Other times you may need to make **conjectures**, predictions about what may happen.

Examples

Use logical reasoning to answer the following questions.

A. In a cross-country race, Shannon outran David. Joe beat Sue but lost to David. Who finished first?

In the problem you are given four runners: Shannon, David, Joe, and Sue. You need to find out who finished first.

To answer the question, you need to determine the order of the racers. If you list them using the given information, you can determine the winner.

Shannon outran David, so list Shannon ahead of David. Joe beat Sue, so we know that Joe is ahead of Sue. You also know that Joe lost to David. Joe is behind David, and Sue is behind Joe. Your list is complete. Shannon finished first.

Shannon
David
Joe
Sue

Make sure that your list matches the problem. Shannon is ahead of David, and Joe is ahead of Sue but behind David. The solution works.

B. Reese, Royal, and Rhonda are each holding math books. One book is blue, one is green, and the other is red. Reese has a red book. Royal's book is neither red nor green. Which book is each person holding?

The three girls are each holding a book. One is blue, one is green, and other is red.

Match the books with each girl. You know that Reese has a red book, so you just need to determine which books the other girls are holding.

Reese has the red book. If Royal has a book that is neither red nor green, then it must be blue. That leaves Rhonda with the green book.

The solution checks with the problem.

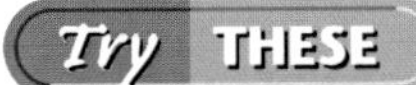

1. Matt has a favorite number. It is between 10 and 20. You can divide the number evenly by 7, but the number is even. What is Matt's number?

2. Joan, Alan, Maria, and Tom are standing in line. Maria is after Tom. Joan is before Alan and Tom. Alan is not second or fourth. Who is first, second, third, and fourth in line?

Exercises

1. Joshua, Kevin, Tammy, and Lisa are playing a board game. Use the clues to determine where each person is sitting.

   a. Joshua starts the game by passing the spinner to his right.

   b. Tammy gets the spinner first and Joshua gets the spinner last.

   c. Kevin and Joshua are partners.

   d. Lisa gets angry because Kevin bumps her left arm.

2. A red car, a green car, a blue car, and a brown car are parked one behind the other in a narrow driveway. Only the blue car can back out. The red car bumped the brown car as it drove in. The green car will be the last to leave. In what order are the cars parked in the driveway?

3. When the school nurse measured Mary, Carl, Ben, and Paul, she forgot to write their names on the chart. Use these clues to help her fill in her chart.

Name	Height
	5 ft 7 in.
	5 ft 5 in.
	5 ft 1 in.
	5 ft

Mary is taller than Ben, but shorter than Paul. Carl is not the tallest, but he is taller than Ben and Mary.

4. These are toggles: These are not toggles:

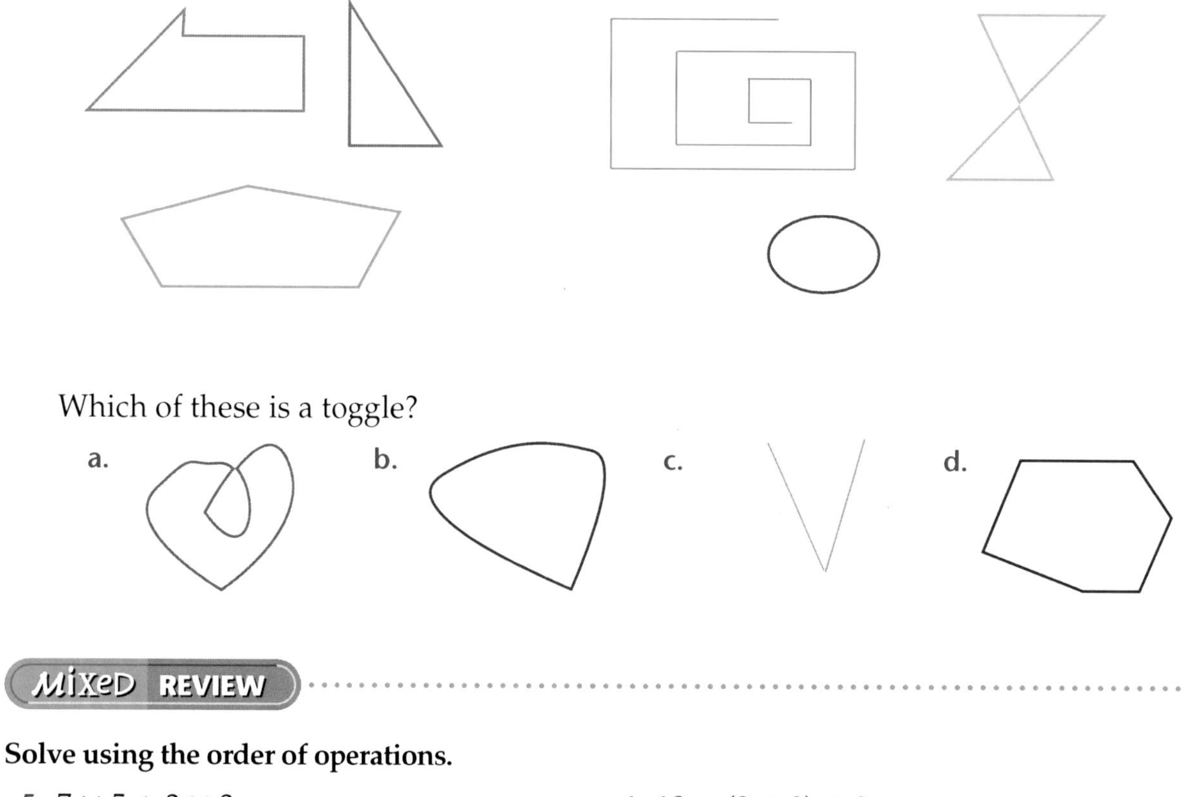

Which of these is a toggle?

a. b. c. d.

MIXED REVIEW

Solve using the order of operations.

5. $7 \times 5 + 2 \times 3$

6. $18 - (9 + 3) + 2$

7. $(45 + 21) \div 11$

8. $6 \times (38 - 12) + 4$

There's a Hole in the Bucket

Henry went to the well to get exactly 3 liters of water. However, Henry only had a 9-liter bucket with a hole in it, and a 4-liter bucket. The hole in the 9-liter bucket caused Henry to lose water at the rate of 1 liter per minute. How many ways can you find for Henry to bring back 3 liters, if it takes him 2 minutes to walk back to the house, and Henry does not have a watch?

Extension

Suppose you have a 9-pint pitcher and a 4-pint pitcher. What would you have to do to get exactly 6 pints of water?

2.2 Decimals and Place Value

Objective: to read and write decimals

Decimals are another way to write fractions and mixed numbers when the denominators are 10, 100, and so on.

$\frac{7}{10}$ written as a decimal is 0.7.

Write:	**Say:**
0.7	seven tenths

The mixed number $1\frac{3}{100}$ (one and three hundredths) is pictured at the right.

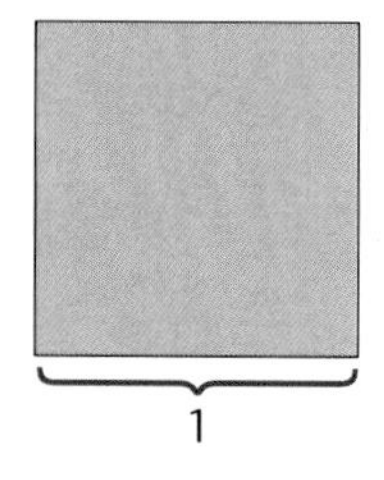

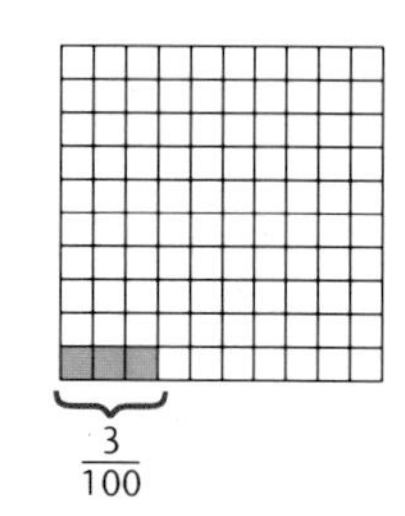

$1\frac{3}{100}$ written as a decimal is 1.03.

Write:	**Say:**
1.03	one and three hundredths

Read the decimal point as *and.*

One of the smallest bird eggs is that of the bee hummingbird. Its egg weighs about 0.0176 ounces. This number is shown in the place-value chart.

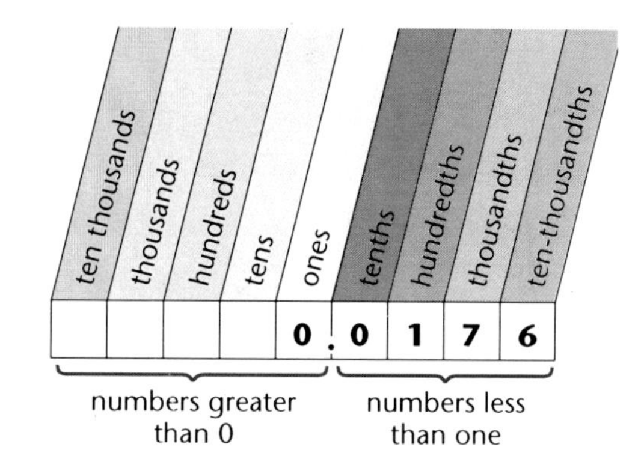

The digit 7 is in the thousandths position. This digit and its place-value position name the number seven thousandths, 0.007.

Write:	**Say:**
0.0176	one hundred seventy-six ten-thousandths

More Examples

Write:	**Say:**
A. 6.7	six and seven tenths
B. 10.08	ten and eight hundredths
C. 0.312	three hundred twelve thousandths
D. 7.2005	seven and two thousand five ten-thousandths

The word *and* is used to express the decimal point.

What pattern do you see in the names of the place-value positions to the left and right of the ones place?

Try THESE

Write the place-value position and number named for each of the following digits in 652.4198.

1. 4 2. 9 3. 5 4. 6 5. 1 6. 8

Exercises

Write as a decimal.

1. $\frac{6}{1,000}$ 2. $\frac{5}{10,000}$ 3. $\frac{143}{1,000}$ 4. $\frac{79}{10,000}$

5. five tenths
6. nine thousandths
7. seven hundredths
8. four and five hundredths
9. forty-eight thousandths
10. three hundred one thousandths
11. six hundred ten-thousandths
12. five and three hundred fifty-two thousandths
13. nine hundred and two hundred twenty-five ten-thousandths
14. What part of a dollar is a penny?
15. What part of a dollar is a quarter?

Write in words.

16. 0.2
17. 0.06
18. 16.7
19. 0.1004
20. 7.007
21. 5.0015
22. 40.035
23. 370.0411

Write each decimal in words.

24. The New York City subway system is 231.73 miles long.
25. The thinnest wristwatch is 0.039 inches thick.
26. The bee hummingbird weighs 0.056 ounces.

Problem SOLVING

27. Jonah ate $\frac{6}{10}$ of a pepperoni pizza. Write this as a decimal.
28. Susie ran one and eight hundred forty-two thousandth miles. Write this a decimal.

★ 29. A decimal is made by using the digits 6, 3, 9 only once.

a. What is the greatest possible decimal greater than 6 but less than 9?

b. Find the least possible decimal greater than 0 but less than 1.

2.3 Measuring Metric Length

Objective: to use decimals to express metric measurement

Things all around us come in different sizes. Most often you measure to find length, area, or volume. The metric system of measuring uses basic units with prefixes. The basic unit of length is the meter.

French mathematicians who developed the metric system wanted a system that was easy to use so they based the system on 10. First they measured the distance from the North Pole to the equator. Then they chose $\frac{1}{10,000,000}$ of that distance to be the base unit or meter.

Metric Units of Length			
Name of Unit	**Abbreviation**	**Equivalent in Meters**	**Measurable Objects**
1 millimeter	1 mm	0.001 meter	the thickness of a dime
1 centimeter	1 cm	0.01 meter	the radius of a penny
1 decimeter	1 dm	0.1 meter	the length of a new crayon
1 meter	1 m		about the length of a yardstick
1 dekameter	1 dkm	10 meters	the length of two parked cars
1 hectometer	1 hm	100 meters	the length of a football field
1 kilometer	1 km	1,000 meters	the length of a short airport runway

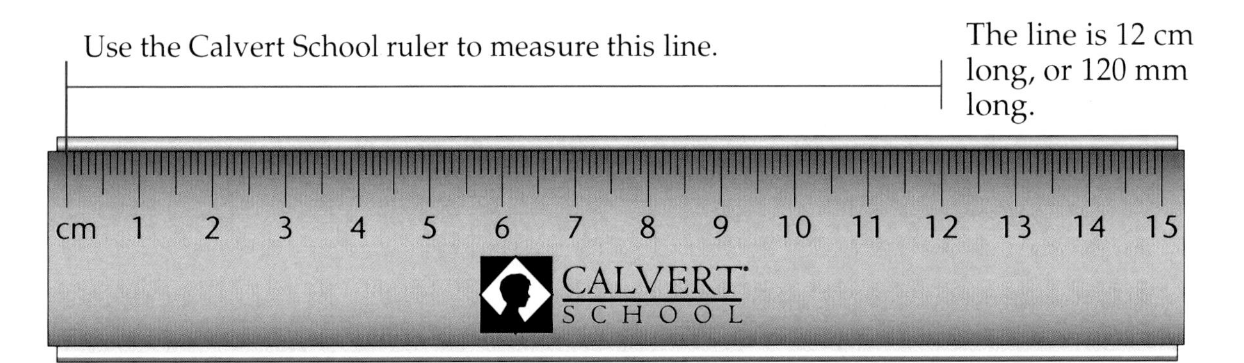

Another Example

Try THESE

Name the unit that is best to measure each.

1. the height of a flagpole
2. the width of your math book
3. the length of a pair of scissors
4. the length of a paper clip
5. your height

Exercises

Name the unit that is best to measure each.

1. the length of a car
2. the height of a classroom ceiling
3. the length of your desk
4. the width of your Calvert ruler
5. the length of this book

Using a ruler, measure each segment to the nearest centimeter.

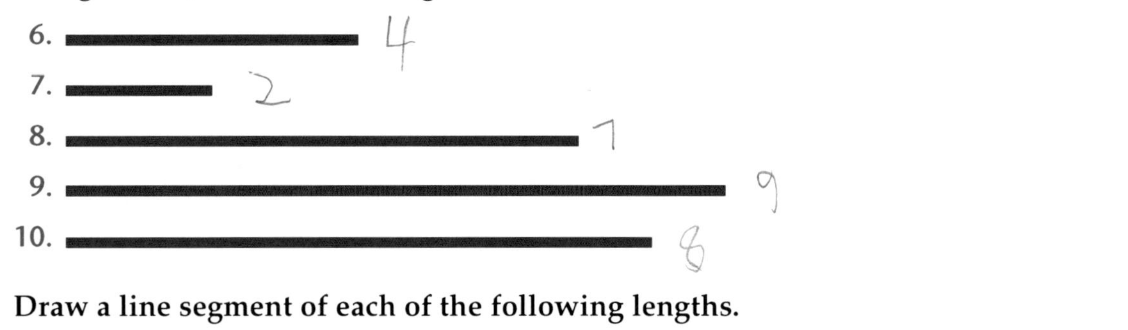

Draw a line segment of each of the following lengths.

11. 130 mm
12. 11 cm
13. 102 mm
14. 5 cm
15. 29 mm

Estimate the length of each side. Then check by measuring.

16.

17.

18.

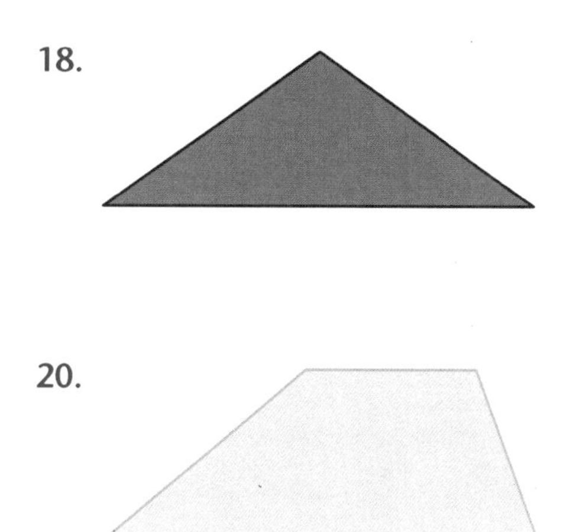

19.

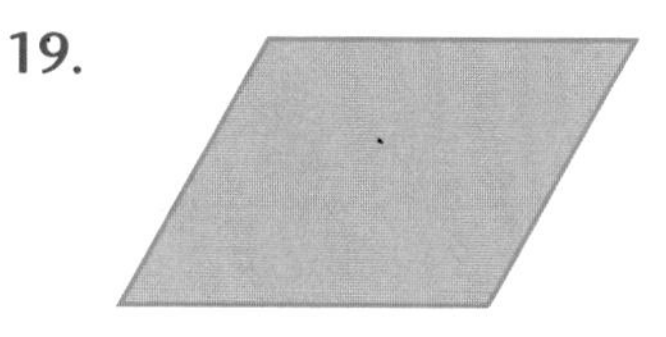

20.

Problem SOLVING

21. Which is greater, 25 millimeters or 5 centimeters? Explain.

Constructed RESPONSE

22. Susie measures a crayon to the nearest centimeter and the result is 6 centimeters. She then measures the same crayon to the nearest tenth of a centimeter.

   a. What results are possible? Explain.

   b. Susie thinks that measuring to the nearest tenth of a centimeter is the same as measuring to the nearest thousandth of a meter. Do you agree or disagree? Explain.

Mind BUILDER

The Soroban

The soroban is the Japanese abacus. The nine rods represent the columns of written figures. The beads on the top count as 5; the beads on the bottom count as 1. To use the soroban, the beads are moved closer to the bar. So, the array shown at the right would represent 13,987,401.

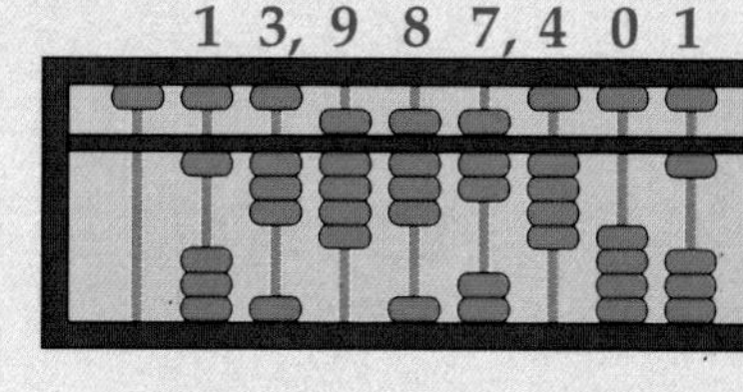

Write the steps you would take to do each operation on the soroban.

1. add 1,263
2. add 567
3. add 34
4. add 12,497

Cumulative Review

Write in words.

1. 45
2. 97
3. 1,400
4. 9,000,000,000
5. 0.3
6. 0.45
7. 9,325
8. 0.0003

Compare using <, >, or = .

9. 73 ● 37
10. 7,314 ● 7,514
11. 24,775 ● 24,775
12. 6 ● 6
13. 245 ● 243
14. 47,862 ● 7,682

Round to the underlined place-value position.

15. $\underline{3}76$
16. $\underline{8},249$
17. $2\underline{3}8,775$
18. $\underline{9}98,304$
19. $\underline{4}8$
20. $\underline{7}29$
21. $\underline{5}4$
22. $7,0\underline{2}9$

Estimate.

23. 723 + 681
24. 8,405 + 2,336
25. 2,316 × 594
26. 41)3,307

Compute.

27. $605 + 349
28. 7,306 + 4,784
29. 87,254 + 28,757
30. 477 + 23,356
31. 4,854 + 42,238
32. 6,576 − 384
33. 3,442 − 2,784
34. 14,000 − 5,326
35. 53,070 − 47,654
36. 26,010 − 19,094
37. 5,000 × 40
38. 528 × 47
39. 636 × 54
40. 719 × 421
41. 535 × 628
42. 4)816
43. 5)2,050
44. 23)234,991
45. 48)2,486
46. 87)24,994

Solve.

47. There are 54 tulips and 167 daisies. How many flowers are there in all?
48. Murray spends $24 more than Ted. Ted spends $19. How much does Murray spend?
49. A computer is purchased at $64 each month for 12 months. What is the total amount paid?
50. Each bus can carry 32 students. There are 224 students. How many buses are needed?

2.4 Comparing and Ordering Decimals

Objective: to compare and order decimals

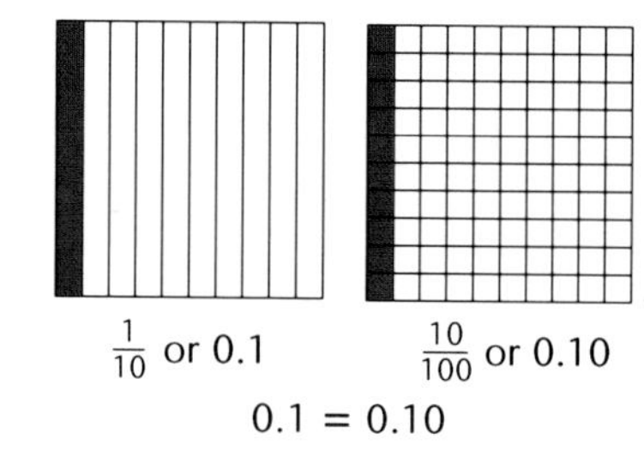

The men's 100 m butterfly is one event in the Summer Olympic Games. The length of the race, 100 meters, can be written as 0.1 kilometers or 0.10 kilometers.

The figures at the right show that 0.1 and 0.10 have the same value. Decimals that have the same value are **equivalent.**

You can attach, or place, zeros to the right of a decimal without changing its value. Why?

0.6 = 0.60 = 0.600 and so on

These values are read 6 tenths, 60 hundredths, 600 thousandths.

4 = 4.0 = 4.00 and so on

Use the following steps to compare decimals.

1. Write the decimals in a column, lining up the decimal points.
2. Attach a zero, if needed, so that each number has the same number of *decimal places.*
3. Start at the *left* and compare the digits in each place-value position.
4. At the first decimal place where the digits are different, you can determine which number is the greater and which is the lesser.

Examples

A. Compare 5.43 and 5.34.

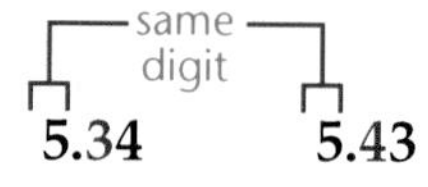

5.34 **5.43**

In the tenths place, $3 < 4$.

$5.34 < 5.43$

B. Compare 7.62 and 7.6.

same
digits

7.62 **7.60**

Attach a zero to compare the digits in the hundredths place.

In the hundredths place, $2 > 0$.

$7.62 > 7.6$

Compare using $<$, $>$, or $=$.

1. 3.4 ● 3.6
2. 0.09 ● 0.009
3. 17.639 ● 17.6391

4. 6.78 ● 6.78
5. 4.0 ● 0.4
6. 1.86 ● 1.859
7. 23.5 ● 23.4
8. 8.05 ● 8.15
9. 0.012 ● 0.021

Exercises

Compare using <, >, or = .

1. 0.51 ● 0.50
2. 0.29 ● 0.3
3. 1.0 ● 0.1
4. 3.001 ● 3.001
5. 0.04 ● 0.40
6. 0.73 ● 0.7300
7. 0.7 ● 0.6999
8. 0.4635 ● 0.37
9. 2.75 ● 2.7500

Write *true* or *false*.

10. $0.7 > 0.4$
11. $3.47 = 3.74$
12. $0.002 < 0.02$
13. $0.6 > 0.06$
14. $7.62 < 7.61$
15. $5.4 = 5.400$

Compare and then order from least to greatest.

16. 7.0, 0.07, 0.7, 0.077
17. 1.2, 0.12, 12, 0.012
18. 9.7, 9.71, 9.68, 9.90
19. 5.9, 5.89, 5.809, 5.8910

Problem SOLVING

20. The average price of a single scoop ice cream cone from various ice cream shops is given below. Order the costs from least to greatest.

Single Scoop Ice Cream Cones				
Shop	**Tasty Twist**	**Cone Caper**	**Flavor Factory**	**Softy Shop**
Price	$1.65	$1.74	$1.53	$1.44

★21. If the price of a cone in each ice cream shop goes up by the same amount, does the order of least expensive item to most expensive item change? Why or why not?

Solve using mental math.

22. $3x = 27$
23. $a + 5 = 16$
24. $\frac{x}{5} = 6$
25. $m - 4 = 9$

2.5 Rounding Decimals

Objective: to round decimals

The Beast at King's Island in Ohio is one of the fastest roller coasters in the world. Its top speed is 64.77 mph.

On the number line below, 64.77 is closer to 65 than 64. To the nearest whole number, 64.77 rounds to 65.

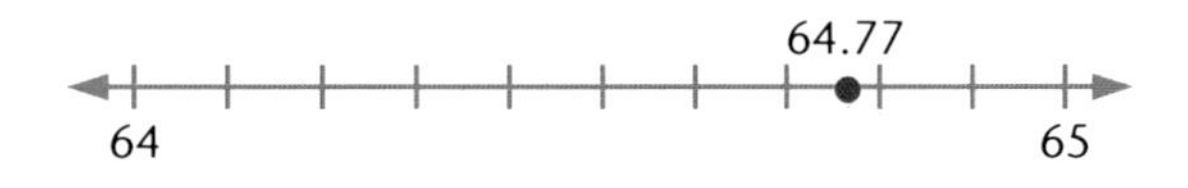

On the number line below, 64.77 is closer to 64.8 than 64.7. To the nearest tenth, 64.77 rounds to 64.8.

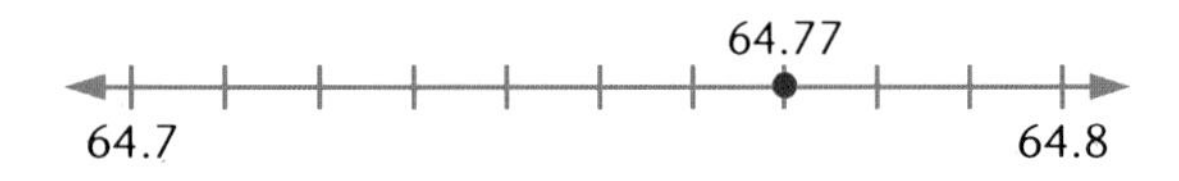

You can round to any place-value position without using a number line.

To round a decimal to a given place value, look at the digit in the place to the right:

- If the digit is 4 or less, the digit remains the same.
- If the digit is 5 or greater, round up.

More Examples

A. $1\underline{2}.5217$ rounded to the nearest whole number is 13.

THINK Round up since the digit to the right of the decimal point is 5.

B. $12.\underline{5}217$ rounded to the nearest tenth is 12.5.

THINK The nearest tenth is 0.5 since the digit to the right is 2.

Try THESE

Choose the correct answer for rounding to the underlined place-value position.

1. $\underline{7}.2$ a. 7 b. 8
2. $1\underline{3}.8$ a. 13 b. 14
3. $43\underline{6}.37$ a. 436 b. 437
4. $3.\underline{4}9$ a. 3.4 b. 3.5
5. $27.\underline{8}2$ a. 27.8 b. 27.9
6. $253.\underline{8}4$ a. 253.8 b. 253.9

Exercises

Round to the nearest whole number.

1. 9.3
2. 14.7
3. 28.31
4. 456.82
5. 72.73
6. 89.265
7. 604.68
8. 859.8913

Round to the nearest tenth.

9. 0.64
10. 0.751
11. 1.27
12. 4.31
13. 53.374
14. 81.29
15. 256.768
16. 539.9674

Round to the underlined place-value position.

17. $0.\underline{3}34$
18. $\underline{0}.871$
19. $2.\underline{6}43$
20. $\underline{5}.778$
21. $\underline{1}3.7602$
22. $9\underline{4}.445$
23. $7.\underline{6}005$
24. $9.\underline{3}863$
25. $4.\underline{2}054$
26. $\underline{6}.6049$
27. $6.\underline{1}432$
28. $\underline{0}.9871$
29. $0.\underline{0}655$
★30. $0.\underline{9}897$
★31. $4.\underline{9}999$

Problem SOLVING

32. The Thriller roller coaster reaches a speed of 61.09 miles per hour. Round the speed to the nearest whole number.
33. Kathryn buys a dress for $39.00. The sales tax is $2.145. To the nearest cent, what is the sales tax?
34. You and three friends go to Paul's Pizzeria where you split a pizza. You split the bill between the four of you and you each owe $4.2888. Explain how to round this to the nearest dime.

Constructed RESPONSE

35. Look at the problem below and find and correct the error. Explain.

> Round to the nearest tenth.
>
> 19.95 → 20

2.6 Problem-Solving Strategy: Make a List

Objective: to solve problems by making a list

Erika saves money to attend the State Fair. She saves 1¢ the first day, 2¢ the second day, 4¢ the third day, and so on. How much money does she save in one week?

1. READ

You need to find how much Erika saves in one week. You know how much she saves the first three days.

2. PLAN

Make a list to show what you know. Is there a pattern? How do you know? Then to continue the list, use the pattern.

3. SOLVE

Day	Amount Saved	Total Amount Saved
1	1¢	1¢
2	2¢	3¢ (1¢ + 2¢)
3	4¢	7¢ (1¢ + 2¢ + 4¢)
4	8¢	15¢ (1¢ + 2¢ + 4¢ + 8¢)
5	16¢	31¢ (1¢ + 2¢ + 4¢ + 8¢ + 16¢)
6	32¢	63¢ (1¢ + 2¢ + 4¢ + 8¢ + 16¢ + 32¢)
7	64¢	127¢ (1¢ + 2¢ + 4¢ + 8¢ + 16¢ + 32¢ + 64¢)

In one week she saves $1.27.

4. CHECK

You can check the solution by completing the list another way. Notice that the amount saved doubles each day. Then add the amount saved to the previous day's total amount saved.

Day 6 → $16 \times 2 = 32$ → 32¢ saved → $32 + 31 = 63$ → 63¢

Day 7 → $32 \times 2 = 64$ → 64¢ saved → $64 + 63 = 127$ → 127¢

Try THESE

Solve. Refer to the problem above.

1. How much money does Erika save in 9 days?
2. How many days does it take Erika to save $20.00?

Solve

1. Erika wants to save money to buy a gift. On the first day, Erika puts 1¢ into her piggy bank. On the second day, she puts in 3¢. Then, her uncle challenged her to double her deposits each day after the second day. How much will she save in a week if she takes the challenge?

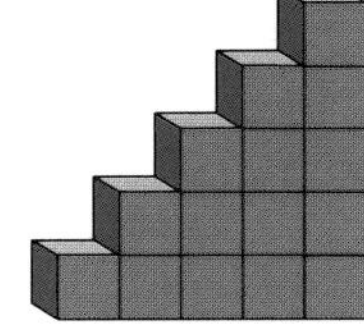

2. Refer to the staircase shown at the right. How many cubes would be needed for a staircase with 11 stairs?
3. Start with 1. What is the sum of the first ten odd numbers? What is the sum of the first ten even numbers?
4. A local frozen yogurt shop offers three flavors of frozen yogurt and four toppings. How many one flavor, one topping combinations are possible?
5. The science teacher is planning an experiment to determine the best growing conditions for her basil plants. She can keep the plants in high, medium, or low sunlight and she will give them a small, medium, or large amount of water. How many plants should the science teacher buy to test which conditions work best?
6. Joshua is preparing to run a marathon. He begins the first week by running 5 miles each day. Each week, he plans to increase his daily run by 2.5 miles. How far will he run daily during the eighth week of training?
7. Six students can sit in 6 chairs. How many ways can you arrange the students in the chairs?

Constructed RESPONSE

8. How many ways can you make 25 cents using pennies, nickels, and dimes? Explain how you would find the different ways to make 25 cents by using an organized list. Then make a list to answer the question.

Mid-Chapter REVIEW

Write as a decimal.

1. forty-one ten-thousandths
2. eighty-one and nine thousandths

Compare using <, >, or = .

3. 0.74 ● 0.73
4. 2.654 ● 2.6547
5. 56.9 ● 57.3

Round to the nearest tenth.

6. 0.37
7. 0.54
8. 2.86
9. 231.41

2.7 Decimal Estimation

Objective: to estimate sums and differences of decimals

The heaviest apple ever grown to date weighed 3.0625 pounds. The heaviest orange weighed 5.5 pounds. About how many pounds would they weigh together?

Since you only need to know *about* how many pounds, estimate. One way to estimate is to round each number to the same place-value position. Then add or subtract.

Estimate 3.0625 + 5.5.

$$\begin{array}{r} 5.5 \\ +\ 3.0625 \\ \hline \end{array} \rightarrow \begin{array}{r} 6 \\ +\ 3 \\ \hline 9 \end{array}$$

5.5 rounds to 6
3.0625 rounds to 3
estimate

When estimating with decimals, round so there are only whole numbers to add or subtract.

The apple and the orange would weigh about 9 pounds.

More Examples

A. $$\begin{array}{r} 61.39 \\ -\ 25.74 \\ \hline \end{array} \rightarrow \begin{array}{r} 60 \\ -\ 30 \\ \hline 30 \end{array}$$

Round to the nearest ten.

B. $$\begin{array}{r} 12.49 \\ 1.64 \\ +\ 0.967 \\ \hline \end{array} \rightarrow \begin{array}{r} 12 \\ 2 \\ +\ 1 \\ \hline 15 \end{array}$$

Round to the nearest one.

Try THESE

Choose the best estimate.

1. 18.4 − 9.7   **a.** 6   **b.** 8   **c.** 10
2. 7.51 + 11.6 + 0.77   **a.** 18   **b.** 21   **c.** 24
3. $22.36 − $13.89   **a.** $10   **b.** $15   **c.** $8

Exercises

Estimate.

1. $\begin{array}{r} 3.1 \\ +\ 4.8 \\ \hline \end{array}$

2. $\begin{array}{r} 9.67 \\ +\ 5.28 \\ \hline \end{array}$

3. $\begin{array}{r} \$3.25 \\ +\ 4.09 \\ \hline \end{array}$

4. $\begin{array}{r} 18.4 \\ 21.7 \\ +\ 19.3 \\ \hline \end{array}$

5. $\begin{array}{r} 276.12 \\ 308.06 \\ +\ 294.11 \\ \hline \end{array}$

6. $\begin{array}{r} 6.7 \\ -\ 4.3 \\ \hline \end{array}$

7. $\begin{array}{r} 22.92 \\ -\ 11.45 \\ \hline \end{array}$

8. $\begin{array}{r} \$7.54 \\ -\ 2.31 \\ \hline \end{array}$

9. $\begin{array}{r} 6.9311 \\ -\ 4.64 \\ \hline \end{array}$

10. $\begin{array}{r} 15.7 \\ -\ 7.980 \\ \hline \end{array}$

11. $2.89 + $6.28

12. 65.5 + 19.7

13. 75.05 − 69.775

Problem SOLVING

14. Carlos buys bacon, eggs, bread, and orange juice. About how much change does he receive from $10?

15. Elaine has $5. Does she have enough to buy bread, butter, and milk?

16. Jessie bought butter, two loaves of bread, and orange juice. About how much did she spend?

Food Mart

Item	Price
Bacon	$2.15
Bread	$0.89
Butter	$1.63
Eggs	$1.08
Milk	$1.59
Orange juice	$2.35

Constructed RESPONSE

17. You have $8 to spend on all of the groceries listed above. Do you have enough money? Explain your answer.

2.8 Adding and Subtracting Decimals

Objective: to add and subtract decimals

In the 1952 Olympic Games, Pat McCormick scored 79.37 points in the platform dive. In the 1956 Olympics, she won the gold medal with a score of 84.85 points.

To find the total points scored, add 79.37 and 84.85.

An estimate is 80 + 80 = 160 points.

Step 1	Step 2
Align the decimal points.	Add as with whole numbers.
79.37 + 84.85 → $\begin{array}{r} 79.37 \\ +\ 84.85 \\ \hline \end{array}$	$\begin{array}{r} ^{1\,1\ \,1} \\ 79.37 \\ +\ 84.85 \\ \hline 164.22 \end{array}$

She won with a total of 164.22 points.

To find the difference in scores, subtract 79.37 from 84.85.

An estimate is about 85 − 80 = 5.

Step 1	Step 2
Align the decimal points.	Subtract as with whole numbers.
84.85 − 79.37 → $\begin{array}{r} 84.85 \\ -\ 79.37 \\ \hline \end{array}$	$\begin{array}{r} ^{7\,14\ \,7\,15} \\ \not{8}\not{4}.\not{8}\not{5} \\ -\ 79.37 \\ \hline 5.48 \end{array}$

The difference in scores is 5.48 points.
You can use addition to check the answer.

Check: $\begin{array}{r} 84.85 \\ -\ 79.37 \\ \hline 5.48 \end{array}$ → $\begin{array}{r} 79.37 \\ +\ 5.48 \\ \hline 84.85 \end{array}$

More Examples

A. $\begin{array}{r} ^{1\,1} \\ 1.7893 \\ +\ 5.1162 \\ \hline 6.9055 \end{array}$

B. $\begin{array}{r} ^{8\ 10} \\ \not{9}.\not{0}875 \\ -\ 3.2154 \\ \hline 5.8721 \end{array}$ Rename 9 ones and 0 tenths as 8 ones and 10 tenths.

C. $\begin{array}{r} ^{6\ 9\ 10} \\ \not{7}.\not{0}\not{0} \\ -\ 3.46 \\ \hline 3.54 \end{array}$ Rename 7 ones as 69 tenths and 10 hundredths.

Try THESE

Add or subtract.

1. $\begin{array}{r} 0.5 \\ +\ 0.2 \\ \hline \end{array}$

2. $\begin{array}{r} \$0.76 \\ +\ \ 0.45 \\ \hline \end{array}$

3. $\begin{array}{r} \$67.54 \\ +\ 53.46 \\ \hline \end{array}$

4. $\begin{array}{r} 28.9 \\ -\ 13.8 \\ \hline \end{array}$

5. $\begin{array}{r} 67.94 \\ -\ \ 5.48 \\ \hline \end{array}$

Exercises

Add.

1. $\begin{array}{r} 1.0017 \\ +\ 4.8 \\ \hline \end{array}$

2. $\begin{array}{r} 16.2 \\ +\ 27.473 \\ \hline \end{array}$

3. $\begin{array}{r} \$8.86 \\ 9.43 \\ +\ 5.75 \\ \hline \end{array}$

4. $\begin{array}{r} \$26.82 \\ 47.23 \\ +\ 83.92 \\ \hline \end{array}$

5. $\begin{array}{r} 7.236 \\ 28.54 \\ +\ \ 3.7 \\ \hline \end{array}$

6. $\begin{array}{r} 18.056 \\ 5.498 \\ +\ 194.53 \\ \hline \end{array}$

7. $\begin{array}{r} 8.9 \\ 6.25 \\ +\ 0.672 \\ \hline \end{array}$

8. $\begin{array}{r} 78.5 \\ 9.06 \\ +\ 203.7 \\ \hline \end{array}$

9. $\begin{array}{r} \$638.24 \\ 53.18 \\ +\ \ 19 \\ \hline \end{array}$

10. $\begin{array}{r} 34.012 \\ 12 \\ +\ 321.15 \\ \hline \end{array}$

Subtract.

11. $\begin{array}{r} 0.9 \\ -\ 0.3 \\ \hline \end{array}$

12. 0.78 − 0.52

13. $\begin{array}{r} 0.897 \\ -\ 0.696 \\ \hline \end{array}$

14. 28.8 − 17.9

15. $\begin{array}{r} 0.42 \\ -\ 0.32 \\ \hline \end{array}$

16. $\begin{array}{r} 3.764 \\ -\ 0.437 \\ \hline \end{array}$

17. $\begin{array}{r} 46.63 \\ -\ \ 5.38 \\ \hline \end{array}$

18. 18.85 − 4.79

19. $\begin{array}{r} 2.946 \\ -\ 1.487 \\ \hline \end{array}$

20. $\begin{array}{r} \$5.03 \\ -\ \ 3.30 \\ \hline \end{array}$

Problem SOLVING

21. A world record for swimming 50 meters in 20.94 seconds was set by Fred Bousquet in 2009. The 50-meter record was 21.28 seconds in 2008. What is the difference in times?

22. In 2005 Dan Wheldon won the Indianapolis 500 with a top speed of 157.603 miles per hour. In 2006 Sam Hornish Jr. won with a top speed of 157.085 miles per hour. What is the difference in their speeds?

Constructed RESPONSE

23. At Burger Barn you have two options. You can buy a hamburger for $1.10, fries for $0.89, and a drink for $1.15 or you can buy a combo meal that includes a hamburger, fries, and a drink for $3.10. Which option is a better bargain? Explain.

Chapter 2 Review

Language and Concepts

Write *true* or *false*. If false, replace the underlined word or words to make a true sentence.

1. Logical reasoning allows us to make conclusions based on the given information.
2. Decimals are another way to write fractions and mixed numbers when the numerators are 10, 100, and so on.
3. When reading decimals, read the decimal point as and.
4. A number written four places to the right of the decimal point is in the ten-thousandths place-value position.
5. Decimals that have the same value are similar.
6. To compare decimals, start at the right and compare the digits in each place-value position.
7. To add or subtract decimals, line up the decimal points and compute as with whole numbers.

Skills and Problem Solving

Write as a fraction or mixed number. (Section 2.2)

8. nine tenths
9. eight and three hundredths

Write the place-value position for each of the following digits in 342.76. (Section 2.2)

10. 4 11. 3 12. 2 13. 6

Write as a decimal. (Section 2.2)

14. $\frac{4}{10}$ 15. $\frac{22}{1,000}$ 16. $\frac{15}{10,000}$

17. two hundred forty thousandths 18. thirty-six ten-thousandths

Compare using <, >, or = . (Section 2.4)

19. 0.07 ● 0.70 20. 6.43 ● 6.403 21. 1.8900 ● 1.89

Round to the underlined place-value position. (Section 2.5)

22. $\underline{4}.914$ 23. $0.\underline{0}566$ 24. $\underline{0}.8989$ 25. $2.\underline{4}949$

Estimate. (Section 2.6)

26. $8.6 + 3.1$

27. $5.09 - 2.6$

28. $4.049 + 3.17$

29. $0.775 - 0.233$

30. $\$16.99 + 11.69 + 2.07$

Compute. (Section 2.7)

31. $0.28 + 0.44$

32. $8.3 - 4.18$

33. $\$59.37 + 48.26$

34. $6.58 - 3.005$

35. $2.21 + 1.66 + 22.162$

36. 8.35 – 2.465

37. 82.542 + 66.03

38. 92.08 – 18.1360

39. 63.445 – 21.08

40. 5 + 4.064 + 1.4 + 10.3

41. 6.308 + 12.97 + 554.809

42. 47.16 – 28.4321

Solve. (Sections 2.1, 2.7–2.8)

43. The Lin family goes to a drive-in movie. Mr. Lin has $16. He spends $10.50 on tickets and $2.80 on drinks. Does he have enough money to buy popcorn for $2.55?

44. Rory has more money than Susie but less money than Erin. Howie has 20¢ more than Bailey. The amounts are $0.85, $1.90, $1.18, $0.65, and $1.09. How much does each person have?

45. Discount Department Store is having a clearance sale. The original and sale prices of the items are listed in the table below. Answer the following questions showing all of your work.

Item	Original Price ($)	Discounted Price ($)
Sofa	315.45	250
Table	84.70	45
Picture frame	15.35	3
Lamp	45.80	20

a. If you bought one of each item, how much would you spend at Discount Department Store?

b. How much money is Discount Department Store losing by selling all of the items at a discounted price?

c. Explain how you calculated the loss.

Chapter 2 Test

Name the fraction or mixed number.

1. thirteen thousandths
2. fourteen and seven tenths
3. eight and nine hundredths
4. twenty-seven ten-thousandths

Write as a decimal.

5. $\frac{1}{10}$
6. $\frac{12}{1,000}$
7. $\frac{21}{100}$
8. $\frac{28}{10,000}$
9. eleven ten-thousandths
10. fifty-five and fifty hundredths
11. thirty-six hundredths
12. sixteen and one thousandth

Order from least to greatest.

13. 0.6, 7.0, 0.68, 0.068
14. 3.6, 0.36, 36, 0.036
15. 4.5, 4.56, 4.32, 4.40

Round to the underlined place-value position.

16. $0.\underline{1}42$
17. $2.\underline{0}33$
18. $1.\underline{5}438$
19. $\underline{3}.511$

Estimate.

20. $\begin{array}{r} \$2.85 \\ +\ 1.34 \\ \hline \end{array}$
21. $\begin{array}{r} 10.25 \\ -\ 3.12 \\ \hline \end{array}$
22. $\begin{array}{r} 62.08 \\ +\ 12.13 \\ \hline \end{array}$
23. $\begin{array}{r} 23.62 \\ -\ 10.81 \\ \hline \end{array}$
24. $\begin{array}{r} 8.888 \\ +\ 9.990 \\ \hline \end{array}$

Add or subtract.

25. $\begin{array}{r} \$15.45 \\ +\ 4.94 \\ \hline \end{array}$
26. $\begin{array}{r} 8.602 \\ -\ 2.001 \\ \hline \end{array}$
27. $\begin{array}{r} 20.002 \\ +\ 0.388 \\ \hline \end{array}$
28. $\begin{array}{r} 9.98 \\ -\ 9.070 \\ \hline \end{array}$
29. $\begin{array}{r} 1.2 \\ +\ 4.71 \\ \hline \end{array}$
30. \$5.88 + \$10.15
31. 9.205 − 1.801
32. 21.04 + 12.40
33. \$1.08 + \$10.88 + \$8.00
34. 48.067 − 14.01

Solve.

35. You are baby-sitting to earn extra money. You plan to save \$1 the first week, \$2 the second week, \$4 the third week, \$8 the fourth week and so on.

    a. What is the amount of money you will have saved during the sixth week?

    b. If you are planning to purchase an item that costs \$100, how many weeks will it take for you to save enough money continuing the pattern above?

Change of Pace

Logic

A statement is a collection of symbols that is either true or false, but *not* both. Study these expressions.

(a) Austin is in Texas.

(b) $2 + 2 = 5$

(c) What is your name?

(d) Please sit down.

The expressions (a) and (b) are statements.

(a) is true.

(b) is false.

Expressions (c) and (d) are not statements because neither is true or false.

More Examples

The **negation** of a statement can be formed by writing "It is false that . . ." before the statement. Sometimes inserting the word *not* in the statement forms the negation.

Statement	Negation
Austin is in Texas.	Austin is not in Texas.
$2 + 2 = 5$	$2 + 2 \neq 5$

Tell whether each expression is a *statement* or *negation*. For each statement, write whether it is *true* or *false* and write its negation.

1. $2 + 6 \neq 8$

2. $12 + 12$

3. $6 < 5$

4. $5 + 9 = 2 \times 7$

5. London is in France.

6. Please stand up.

7. It is true that $2^2 = 4$.

8. Helena is in Montana.

9. Is the negation of a true statement always false?

10. What is the negation of the negation of a true statement?

Cumulative Test

1. Estimate. $\begin{array}{r} 8.37 \\ +\ 7.6 \\ \hline \end{array}$

   a. 1
   b. 16
   c. 56
   d. none of the above

2. $\begin{array}{r} 36.09 \\ +\ \ 9.812 \\ \hline \end{array}$

   a. 44.821
   b. 45.902
   c. 45.92
   d. 134.21

3. $\begin{array}{r} 80 \\ -\ \ 0.5 \\ \hline \end{array}$

   a. 79.05
   b. 79.15
   c. 79.5
   d. 80.05

4. $\begin{array}{r} 378 \\ \times\ \ 26 \\ \hline \end{array}$

   a. 3,758
   b. 6,948
   c. 9,828
   d. 12,000

5. Which number is equal to 10,000 + 800 + 40 + 9?

   a. 10,049
   b. 10,490
   c. 10,849
   d. 18,049

6. What is 467.885 rounded to the nearest tenth?

   a. 467.89
   b. 467.9
   c. 468
   d. 470

7. A state park planted 600 trees in 20 locations. If an equal number of trees is planted in each location, how many trees were planted in each location?

   a. 20
   b. 30
   c. 600
   d. none of the above

8. Preston bought a tie for $4.99, a shirt for $6.49, and a pair of slacks for $10.00. How much did he have left from $25?

   a. $3.48
   b. $3.98
   c. $3.52
   d. $4.52

9. Roxanne earns $12, $13, $15, and $12 baby-sitting. What is the average amount she earns each time?

   a. $12
   b. $13
   c. $15
   d. none of the above

10. Which number is between 4.28 and 5.02?

    a. 4.058
    b. 5.13
    c. 4.09
    d. 4.412

CHAPTER

Multiplying and Dividing Decimals

Melanie Tate
Louisiana

3.1 Decimal Estimation with Multiplying and Dividing

Objective: to round any decimal up or down in order to deal with estimates in real situations

The Elliotts bought a 16.7-pound ham for Sunday dinner. If the ham sold for $2.29 a pound, about how much did the Elliotts pay?

To find *about* how much it cost, *estimate.* One way to estimate is to round each factor to its greatest place-value position. Then multiply.

$$\begin{array}{r} \$2.29 \\ \times\ 16.7 \\ \hline \end{array} \longrightarrow \begin{array}{r} \$\ 2 \\ \times\ 20 \\ \hline \$40 \end{array}$$

$2.29 rounds to $2
16.7 rounds to 20
estimate

The ham cost about $40.

More Examples

A.
$$\begin{array}{r} 24.56 \\ \times\ 8.92 \\ \hline \end{array} \longrightarrow \begin{array}{r} 20 \\ \times\ 9 \\ \hline 180 \end{array}$$

24.56 rounds to 20
8.92 rounds to 9
estimate

B.
$$\begin{array}{r} 1.732 \\ \times\ 9.81 \\ \hline \end{array} \longrightarrow \begin{array}{r} 2 \\ \times\ 10 \\ \hline 20 \end{array}$$

Since $9 \times 2 = 18$, it is easy to multiply 9×20.

One way to estimate quotients is to round the divisor to its greatest place-value position. Then find the multiple of the rounded divisor that is close to the dividend. Using **compatible numbers** makes it easy to divide since the numbers divide evenly.

C. $4.7\overline{)29.47} \longrightarrow 5\overline{)30}$ — quotient 6, estimate

D. $8\overline{)74.43} \longrightarrow 8\overline{)72}$ — quotient 9, estimate

Try THESE

Round to the greatest place-value position.

1. $1.19
2. 4.75
3. 3.08
4. 37.4
5. 68.79
6. 23.908
7. 2.644
8. $16.95
9. 304.98

Exercises

Estimate.

1. $\begin{array}{r} 7.1 \\ \times\ 3.4 \\ \hline \end{array}$ $\begin{array}{r} 7 \\ \times\ 3 \\ \hline \end{array}$

2. $\begin{array}{r} 6.07 \\ \times\ 8.99 \\ \hline \end{array}$

3. $\begin{array}{r} 4.8 \\ \times\ 2.9 \\ \hline \end{array}$

4. $\begin{array}{r} 18.9 \\ \times\ 11.3 \\ \hline \end{array}$

5. $\begin{array}{r} 12.81 \\ \times\ \ 6.33 \\ \hline \end{array}$

6. $\begin{array}{r} 6.4 \\ \times\ 10.11 \\ \hline \end{array}$

7. $\begin{array}{r} \$5.09 \\ \times\ \ 0.97 \\ \hline \end{array}$

8. $\begin{array}{r} 31.059 \\ \times\ \ \ 7.22 \\ \hline \end{array}$

9. $2.9\overline{)5.4}$ $3\overline{)6}$

10. $3.7\overline{)12.4}$

11. $6.9\overline{)\$7.08}$

12. $8\overline{)26.78}$

13. $6.1\overline{)35.41}$

14. $9.2\overline{)73.05}$

15. $4.05\overline{)21.27}$

16. $7.78\overline{)46.54}$

17. $\$6.29 \times 4.6$

18. $19.6 \div 2.37$

19. 8.75×3.2

20. $296.8 \div 3.1$

21. 28.4×11.631

22. $14.4 \div 4.08$

★23. Is 36 or 40 a better estimate for 3.8×9.2?

★24. Is 18 or 20 a better estimate for 18.439×0.89?

Problem SOLVING

Estimate. Use the chart for 25–27.

Employee	Hourly Pay
Maria Benitez	$4.25
Sam Kohen	$6.85
Ellen Hall	$7.80
Fred Martin	$8.12
Tony Solino	$5.10

25. Sam Kohen works 37.5 hours a week. About how much does he earn per week?

26. Maria Benitez works 41.2 hours a week. About how much does she earn per week?

27. About how much more than Tony Solino does Ellen Hall earn in 40 hours?

★28. If a calculator rounds down to the preceding millionth, what will it show for the 0.01234567?

29. There are exactly 30.48 centimeters in one foot.

    a. Round 30.48 to the nearest tenth.

    b. Round 30.48 to the nearest ten.

★30. A store sells apples at three for $2.00. You want one. So you divide $2.00 by 3 to get the cost. Your calculator shows 0.666666. How much will you probably pay for the apple?

3.2 Multiplying Decimals and Whole Numbers

Objective: to estimate and find the product of decimals and whole numbers

When multiplying a decimal by a whole number, multiply as you would with whole numbers. Then estimate to place the decimal point in the product. You can also count the number of decimal places.

Examples

Multiply a whole number by a decimal.

A. 19.4×7

Method 1: Use estimation.		Method 2: Count decimal places.	
19.4 × 7	Round 19.4 to 20.	**19.4** **× 7** **135 8**	There is one place to the right of the decimal point. Count the same number of decimal places from right to left.
20 × 7 = 140	140 is the estimate.	**135.8**	
19.4 **× 7**	Multiply the numbers.		
135 8 **135.8**	Since the estimate is 140, place the decimal point after the 5.		

If there are not enough decimal places in the product, you need to add zeros to the left.

B. 3×0.015

0.015 **× 3**	There are three decimal places.
45 **0.045**	You need to add a zero in front of the four to make three decimal places.

3 × 0.015 = 0.045

Try THESE

Multiply.

1. 9×0.73
2. 3.4×6
3. 4×0.003
4. 12×0.0034

Exercises

Multiply.

1. 0.4×7
2. 0.18×2
3. 2.1×8
4. 6.32×8
5. $\begin{array}{r} 8.5 \\ \times\ 4 \\ \hline \end{array}$
6. $\begin{array}{r} 16.8 \\ \times\ 2 \\ \hline \end{array}$
7. $\begin{array}{r} 35.3 \\ \times\ 8 \\ \hline \end{array}$
8. $\begin{array}{r} \$36.74 \\ \times\ 5 \\ \hline \end{array}$
9. $\begin{array}{r} 82.69 \\ \times\ \ 9 \\ \hline \end{array}$
10. $\begin{array}{r} 7.9 \\ \times\ 15 \\ \hline \end{array}$
11. $\begin{array}{r} 0.39 \\ \times\ 9 \\ \hline \end{array}$
12. $\begin{array}{r} 0.007 \\ \times\ 7 \\ \hline \end{array}$

Problem SOLVING

13. If gasoline sells for $2.52 a gallon, how much will 8 gallons cost?

Use the table to answer questions 14–18.

United States Minimum Wage	
1938	$0.25
1950	$0.75
1968	$1.60
1978	$2.65
1980	$3.10
1991	$4.25
2005	$5.15

14. At the minimum wage, how much did a person earn for a 35-hour workweek in 1968?

15. If you had a job in 1980, and worked a 15-hour week for minimum wage, how much would you have earned?

16. Tyrone's grandmother worked at a bakery in 1950 for minimum wage. If she worked 40 hours each week for an entire year, how much money did she make before taxes were deducted that year? Explain your answer.

17. Compare the difference in a 30-hour week in 1978 to a 30-hour week in 2005 at minimum wage.

18. About how many times higher was a minimum wage in 1968 to a minimum wage in 1950? Justify your answer.

★19. Write a multiplication problem using one whole number and a decimal. The product should be between 3 and 4.

3.3 Multiplying Decimals

Objective: to multiply decimals

Philip buys 2.7 pounds of dried fruit. The fruit costs $4.42 per pound. To find the total cost of the fruit, multiply $4.42 by 2.7. This is similar to multiplying a decimal by a whole number except you have more decimals. You can still estimate to see if your answer makes sense, but when you want the exact number, you need to follow the rule as to where to place the decimal point.

```
 $4.42          $4
×  2.7  ──►   ×  3
              $12   estimate
```

To find the exact cost, multiply as with a whole number. Then place the decimal point in the product.

```
  $4.42      two decimal places
×   2.7      one decimal place
  3 094
  8 84
$11.934      three decimal places
```

RULE

Add the decimal places in the factors to find the number of decimal places in the product.

Money is usually rounded up to the next cent. Therefore, the total cost of the fruit is $11.94.

How does this compare to the estimate?

```
A.   0.715         3 decimal places
   ×   0.5         1 decimal place
     0.3575  ◄──   4 decimal places

B.    12.3         1 decimal place
   ×  0.02         2 decimal places
     0.246   ◄──   3 decimal places
```

Try THESE

Use estimation to choose the correct product.

1. 6.8 × 7 — a. 4.76 — b. 47.6 — c. 476
2. 12 × 4.3 — a. 5.16 — b. 51.6 — c. 516
3. 6.1 × 0.4 — a. 2.44 — b. 24.4 — c. 244
4. 10.2 × 7.8 — a. 7.956 — b. 79.56 — c. 795.6
5. 5.5 × 0.05 — a. 0.275 — b. 2.75 — c. 27.5
6. 6.38 × 1.24 — a. 791.12 — b. 79.112 — c. 7.9112

Exercises

Multiply.

1. $\begin{array}{r} 8.5 \\ \times\ 4 \\ \hline \end{array}$

2. $\begin{array}{r} 16.8 \\ \times\ 7 \\ \hline \end{array}$

3. $\begin{array}{r} 7.83 \\ \times\ 9.4 \\ \hline \end{array}$

4. $\begin{array}{r} 30.52 \\ \times 0.95 \\ \hline \end{array}$

5. $\begin{array}{r} 0.8 \\ \times\ 0.8 \\ \hline \end{array}$

6. $\begin{array}{r} 6.4 \\ \times\ 0.8 \\ \hline \end{array}$

7. $\begin{array}{r} 3.3 \\ \times\ 0.5 \\ \hline \end{array}$

8. $\begin{array}{r} 9.503 \\ \times\ 0.7 \\ \hline \end{array}$

9. $\begin{array}{r} 15.3 \\ \times\ 0.28 \\ \hline \end{array}$

10. 0.81×0.4

11. 0.67×0.3

12. 9.8×0.02

13. 1.25×0.951

14. 1.256×0.9

15. 14.56×12

16. 25.4×0.781

17. 37.6×0.86

18. 5.83×0.27

19. 1.89×0.65

20. 2.376×3.4

21. 0.24×7.2

22. 0.8×0.46

23. 5.34×15.3

24. 3.17×0.56

25. 8.32×2.33

26. 17.54×1.14

27. 1.2×0.005

28. Find the product of $2.59 and 3.2. Round to the nearest cent.

Fill in each _____ with <, >, or =.

29. 4.3×5.6 _____ 21

30. 33.5×0.01 _____ 3.35

31. 31.58×2.96 _____ 80

32. 2.2×0.11 _____ 1

Problem SOLVING

33. Pears cost $0.65 per pound. What is the cost of 2.4 pounds of pears?

34. Jaime earns $5.28 per hour. How much does she earn in a 30-hour week?

35. Suppose you earn $4.10 per hour. For each hour over 40, you earn $4.10 plus half of $4.10. How much do you earn in a 42-hour week?

36. A turtle is traveling at a speed of 3.028 meters per minute. How far does the turtle travel in 7.5 minutes? Round your answer to the nearest meter.

Constructed RESPONSE

37. When you multiply two decimals between 0 and 1, is the product less than or greater than both factors? Explain.

3.4 Dividing by a Whole Number

Objective: to divide decimals by whole numbers

Frozen grape juice costs $0.96 for 8 ounces. To find the cost per ounce, you will need to divide a decimal by a whole number. Dividing decimals is similar to dividing whole numbers. To find the cost per ounce, divide $0.96 by 8. An estimate is $0.80 \div 8 = 0.10$.

Step 1	Step 2	Step 3
Find the decimal point in the quotient. Place the decimal point in the quotient directly above the decimal point in the dividend as shown. $0. 8)$0.96 ↑	Divide as with whole numbers. $0.1 8)$0.96 − 8 16	Divide the hundredths. $0.12 8)$0.96 − 8 16 − 16 0

The cost per ounce is $0.12.

To check the answer, multiply.

Check:

```
    1
 $0.12
×    8
 $0.96 ✓
```

Compared to the estimate, the answer is reasonable.

Examples

A.
```
   3.08
6)18.48
 − 18
    48
  − 48
     0
```

Sometimes you have to write zeros in the quotient. Why?

B.
```
  0.08
9)0.72
 − 72
     0
```

C. To round a quotient to a certain place, divide to one extra place. Then round.

```
  5.7
6)34.2
− 30
   4 2
 − 4 2
     0
```

To the nearest whole number, the quotient is 6.

5.7 rounds to 6.

A **terminating decimal** is a decimal that terminates or stops. Some examples are 0.25 and 0.125. A **repeating decimal** repeats the same digit or group of digits. You show repeating digits with a bar drawn over the repeating decimals. You write $1.\overline{91}$ for the number 1.9191919191...

D. 0.3 ÷ 8

$$\begin{array}{r} 0.0375 \\ 8\overline{)0.3000} \\ -\underline{24} \\ 60 \\ -\underline{56} \\ 40 \\ -\underline{40} \\ 0 \end{array}$$

← Insert zeros when needed.

E. 0.05 ÷ 6

$$\begin{array}{r} 0.00833 = 0.008\overline{3} \\ 6\overline{)0.05000} \\ -\underline{48} \\ 20 \\ -\underline{18} \\ 20 \\ -\underline{18} \\ 2 \end{array}$$

← Insert zeros when needed.

Where did the zeros come from?

When dividing decimals, remainders are no longer used. If there is a remainder, add zeros to the dividend until the quotient terminates or repeats.

Try THESE

Copy. Then place the decimal point and any necessary zeros in the quotient.

1. $\begin{array}{r}1\,3\,9\\2\overline{)27.8}\end{array}$
2. $\begin{array}{r}1\,4\,9\\5\overline{)74.5}\end{array}$
3. $\begin{array}{r}2\,4\\4\overline{)0.96}\end{array}$
4. $\begin{array}{r}1\,2\,7\\7\overline{)8.89}\end{array}$
5. $\begin{array}{r}1\,5\\15\overline{)0.225}\end{array}$
6. $\begin{array}{r}1\,0\,5\\12\overline{)1.260}\end{array}$
7. $\begin{array}{r}8\,3\\6\overline{)0.498}\end{array}$
8. $\begin{array}{r}1\,8\\3\overline{)5.4}\end{array}$
9. $\begin{array}{r}1\,0\,8\\8\overline{)86.4}\end{array}$
10. $\begin{array}{r}3\,9\\11\overline{)0.429}\end{array}$

Exercises

Divide.

1. $3\overline{)8.58}$
2. $4\overline{)99.2}$
3. $8\overline{)25.6}$
4. $6\overline{)7.68}$
5. $5\overline{)\$1.65}$
6. $7\overline{)3.57}$
7. $9\overline{)0.54}$
8. $5\overline{)\$15.95}$
9. $7\overline{)28.42}$
10. $8\overline{)24.64}$
11. $6\overline{)42.54}$
12. $7\overline{)35.35}$
13. $11\overline{)7.7}$
14. $23\overline{)11.5}$
15. $36\overline{)32.4}$
16. $12\overline{)\$96.84}$
17. 29.4 ÷ 2
18. 0.42 ÷ 6
19. $18.48 ÷ 6
20. 42.28 ÷ 7
21. 36.4 ÷ 52
22. 275.04 ÷ 16

Divide. Round to the nearest whole number.

23. $7\overline{)23.8}$
24. $9\overline{)367.2}$
25. $22\overline{)409.2}$
26. $32\overline{)1{,}596.8}$

Find each quotient. Identify each as *terminating* or *repeating*.

27. 0.9 ÷ 4
28. 1.2 ÷ 30
29. 2.2 ÷ 7

Problem SOLVING

30. Three pounds of pears cost $1.95. How much does one pound of pears cost?

31. A class set of 30 calculators would have cost $3,300.30 in 1975. In 2004, 30 calculators could be purchased for $321.30. How much less was the average price of one calculator in 2004 than in 1975?

Constructed RESPONSE

32. When shopping in a grocery store, it is good to look at the unit price for items. To find the unit price, divide the cost of the item by its size. Fill in the table to find the unit price of each size orange juice. Round to the nearest cent. Which size of orange juice is a better buy? Explain.

Orange Juice Prices

Cost	Size	Unit Price
$5.80	128 ounces	
$2.50	48 ounces	
$2.25	32 ounces	
$0.95	8 ounces	

Mid-Chapter REVIEW

Estimate.

1. 6.3×2.4
2. 2.56×5.74
3. 14.86×8.41
4. $16.8 \div 7$
5. $24.8 \div 4$
6. $38.4 \div 6$

Multiply.

7. 3.7×2
8. 0.43×5
9. 0.7×0.3
10. 5.9×6.3
11. 3.5×0.05
12. 1.25×0.7
13. 0.12×0.4

King Omar's Necklace

When the tomb of the great King Omar was opened, a painting was found of the king with a single strand necklace looped around his neck. The king's people were very organized, and every seven beads on the king's necklace went in the following repeating sequence.

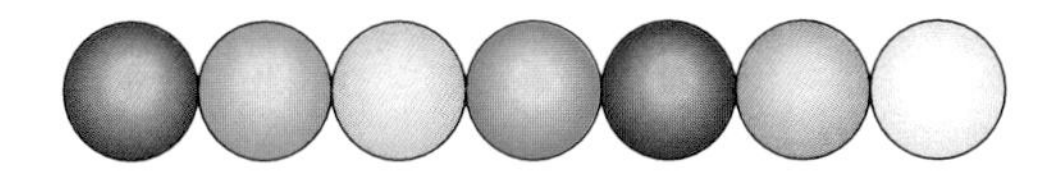

There are at least five beads of each color showing. If you know there is a red bead in the upper left-hand corner, and no color is repeated in any column, or diagonal, can you write the first letter of the color of each bead?

Extension

Study the figures below to determine the pattern. Then draw the next four figures in the pattern.

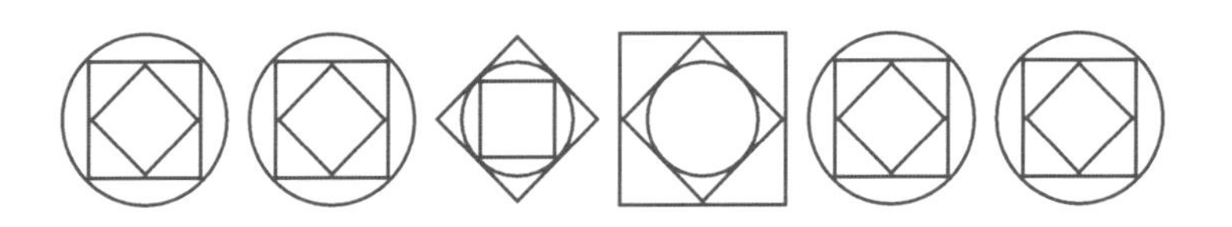

3.5 Multiplying and Dividing by Powers of Ten

Objective: to multiply and divide by powers of 10 using mental math

Exploration Exercise

What happens when you multiply by a power of 10?

1. Complete the table.
2. What do you notice about the movement of the decimal point when you multiply by whole number powers of 10?
3. What happens to the decimal when you multiply by decimal powers of 10?

Whole Number Powers of Ten	Decimal Powers of Ten
10 × 3.4 = ___	0.1 × 3.4 = ___
100 × 3.4 = ___	0.01 × 3.4 = ___
1,000 × 3.4 = ___	0.001 × 3.4 = ___
10,000 × 3.4 = ___	0.0001 × 3.4 = ___

Multiplying by Powers of Ten

Multiplying by Whole Number Powers of 10
Move the decimal point one place *to the right* for each zero in the whole number power of 10.
Ex. 5.594 × 100 = 559.4

Multiplying by Decimal Powers of 10
Move the decimal point one place *to the left* for each decimal place in the power of 10.
Ex. 559.4 × 0.001 = 0.5594

Dividing by Powers of Ten

Dividing by Whole Number Powers of 10
Move the decimal point one place *to the left* for each zero in the whole number power of 10.
Ex. 52 ÷ 100 = 0.52

Dividing by Decimal Powers of 10
Move the decimal point one place *to the right* for each decimal place in the power of 10.
Ex. 52 ÷ 0.001 = 52,000

Examples

Multiply decimals using mental math.

A. 4.25 • 10 = 42.5 — Move 1 place to the right.
B. 4.23 × 1,000 = 4,230 — Move 3 places to the right.
C. 215.3(0.001) = 0.2153 — Move 3 places to the left.

Did you know?
There are three ways to show multiplication: 2 × 5, 2 • 5, and 2(5) all mean 2 times 5!

Divide decimals using mental math.

D. 21.6 ÷ 100 = 0.216 — Move 2 places to the left.
E. 715.78 ÷ 0.01 = 71,578 — Move 2 places to the right.

Try THESE

Copy. Then place the decimal point and any necessary zeros in each product or quotient.

1. 7.3 × 10 = 73
2. 23.6 × 100 = 236
3. 0.56 × 1,000 = 56
4. 89 × 10 = 89
5. 6.54 × 100 = 654
6. 7.33 × 1,000 = 733

Exercises

Multiply or divide.

1. 97.96 × 10
2. 74.3 ÷ 100
3. 5.7 × 1,000
4. 56 ÷ 100
5. 6.92 • 10
6. 406 ÷ 10
7. 0.789 ÷ 10
8. 0.51 × 1,000
9. 0.17 (100)
10. 37.46 • 1,000
11. 4 ÷ 1,000
12. 46.34 ÷ 100
13. 73.8 ÷ 1,000
14. 9.8 (100)
15. 3.26 ÷ 1,000
16. In 173.6 ÷ 100, how many places do you move the decimal point?
17. Which is greater, 43.1 ÷ 10 or 0.67 × 10?
18. Copy and complete each statement using <, >, or =.

    a. 348 ÷ 10__3.4

    b. 14.4 ÷ 0.1__3.4

Problem SOLVING

19. Sally bought 10 tickets to a concert. Each ticket cost $8.50. How much did Sally spend on tickets?
20. Bob collected 100 pop bottles. He got $10 for all the bottles. How much did Bob get for one bottle?

Constructed RESPONSE

21. Tell whether the statement is *true* or *false* and explain your reasoning. When you divide a whole number by a whole number power of 10, the quotient is less than or equal to the dividend.

3.6 Dividing Decimals

Objective: to divide decimals by decimals

Mary Stollen buys 2 pounds of ground beef for a cookout. She makes 0.25-pound patties. To find how many patties she makes, divide 2 by 0.25.

The model shows that there are 8 groups of 25 hundredths in 2. The model also shows that there are 8 groups of 25 in 200.

2

0.25

Step 1	Step 2	Step 3
Move the decimal point in the divisor to the right until the divisor is a whole number. $0.25\overline{)2.}$	Move the decimal point in the dividend the same number of places. Fill in with zeros if necessary. $0.25\overline{)2.00}$	Divide as with whole numbers. $0.25\overline{)2.00}$ = 8; $-\,2\,00$; 0

More Examples

A. $1.5\overline{)7.35}$ = 4.9

$-\,6\,0$
$1\,35$
$-\,1\,35$
0

B. $0.83\overline{)6.225}$ = 7.5

$-\,5\,81$
415
$-\,415$
0

C. $0.047\overline{)2.632}$ = 56

$-\,2\,35$
282
$-\,282$
0

By what numbers were each divisor and dividend multiplied? Why?

Try THESE

Rewrite each so the divisor is a whole number.

1. $0.9\overline{)8.1}$
2. $2.8\overline{)4.48}$
3. $0.06\overline{)0.048}$
4. $2.06\overline{)14.42}$
5. $0.45\overline{)0.945}$
6. $0.004\overline{)1{,}284}$
7. $0.002\overline{)0.032}$
8. $0.039\overline{)0.2496}$

Exercises

Divide. Attach zeros if necessary.

1. $0.4\overline{)3.6}$
2. $0.6\overline{)0.48}$
3. $0.7\overline{)1.61}$
4. $0.5\overline{)20.15}$
5. $2.5\overline{)6.25}$
6. $6.1\overline{)1.342}$
7. $0.03\overline{)0.27}$
8. $0.05\overline{)0.515}$
9. $0.16\overline{)1.296}$
10. $0.06\overline{)0.246}$
11. $0.004\overline{)0.032}$
12. $0.009\overline{)1.107}$
13. $0.045\overline{)0.1035}$
14. $0.083\overline{)0.6225}$
15. $115 \div 0.5$
16. $0.924 \div 3.3$
17. $908.8 \div 1.6$
18. $6.12 \div 0.68$
19. $8.5 \div 0.34$
20. $7.7 \div 0.55$
21. $0.0371 \div 0.053$
22. $4.35 \div 0.075$
23. $96.6 \div 0.092$

Problem SOLVING

24. Find the error. Describe and correct the error in rewriting the problem.

    ✗ $0.24\overline{)15} \longrightarrow 24\overline{)150}$

25. Janet earns $5.90 per hour. If her weekly check totaled $224.20, how many hours did she work?

26. Suppose you have 1.7 pounds of hamburger, how many 0.1-pound meatballs can you make?

Test PREP

	a.	b.	c.	d.
27. $21.48 \div 12$	a. 17.9	b. 1.79	c. 179	d. 1,790
28. $1.08 \div 2.7$	a. 0.04	b. 4	c. 40	d. 0.4
29. $2.07 \div 0.9$	a. 23	b. 0.23	c. 2.3	d. 230
30. $0.0242 \div 0.4$	a. 6.05	b. 0.0605	c. 60.5	d. 0.605

3.7 Scientific Notation

Objective: to use scientific notation to represent large numbers

Jupiter is about 778,000,000 kilometers from the Sun.

Numbers like 778,000,000 can be written in **scientific notation**. A number in scientific notation is written as the product of a factor and a power of 10. The factor has to be from 1 through 9.

$778{,}000{,}000 = 7.78 \times 100{,}000{,}000 = 7.78 \times 10^8$

Move the decimal point to the left, so that you have a factor between 1 and 10.

The exponent shows how many places you moved the decimal point.

You can write a number in scientific notation in standard form as follows.

$3.49 \times 10^4 = 3.49 \times 10{,}000$
$= 3.4900$
$= 34{,}900$

To multiply by 10^4 or 10,000, move the decimal point four places to the right.

The exponent and the number of places the decimal point is moved are the same.

More Examples

Write in scientific notation.

A. $4{,}329 = 4.329 \times 1{,}000$
$= 4.329 \times 10^3$

Write in standard form.

B. $1.657 \times 10^3 = 1.657 \times 1{,}000$
$= 1.657$
$= 1{,}657$

Try THESE

Express the scientific notation in standard form.

1. 5.2×10^2
2. 2.07×10^3
3. 1.4×10^4

Express the standard form in scientific notation.

4. 163
5. 2,270
6. 12,000

Exercises

Express these standard numerals in scientific notation.

1. 200
2. 5,000
3. 1,500
4. 700,000
5. 63,900
6. 9,200,000
7. 607,300
8. 25,800,000
9. 458,321
10. 5,243,000,000

Express the scientific notation in standard form.

11. 4×10^2
12. 8×10^4
13. 5×10^6
14. 6×10^0
15. 4.3×10^2
16. 5.8×10^4
17. 9.07×10^1
18. 3.706×10^2
19. 6.2×10^7
20. 5.2608×10^9

Problem SOLVING

21. Saturn's orbit reaches a maximum distance from the Sun of 9.36×10^8 miles. Write this figure in standard form.
22. Earth's orbit is about 93,000,000 miles from the Sun. Express this in scientific notation.
23. A septillion is 1 followed by 24 zeros. Express this in scientific notation.

Mind BUILDER

Add and Subtract in Scientific Notation

It may be necessary in science to add or subtract numbers in scientific notation. To do this when both values have the same exponents, simply add or subtract the figures.

$(7.4 \times 10^6) + (1.8 \times 10^6) = (7.4 + 1.8) \times 10^6 = 9.2 \times 10^6$

$(8.8 \times 10^2) - (2.7 \times 10^2) = (8.8 - 2.7) \times 10^2 = 6.1 \times 10^2$

Compute these scientific notations.

1. $(8.1 \times 10^4) + (1.2 \times 10^4)$
2. $(4.1 \times 10^2) + (1.86 \times 10^2)$
3. $(3.281 \times 10^4) - (1.601 \times 10^4)$
4. $(9.217 \times 10^5) - (6.02 \times 10^5)$

3.8 Mass and Capacity in the Metric System

Objective: to use metric units of mass and capacity

The metric system is used over most of the world. It is based on decimals instead of fractions. If you had to find the dimensions of a room, would you rather add $2\frac{9}{16}$ inches or 6.161 meters?

The prefix *kilo-* means 1,000. There are 1,000 grams in a kilogram. The prefix *milli-* means $\frac{1}{1{,}000}$. There are 1,000 milligrams in a gram and 1,000 milliliters in a liter.

The most commonly used metric units of mass are milligrams, grams, and kilograms.

Unit	Abbreviation	Model	Benchmark
1 milligram	mg	grain of salt	1 mg ~ 0.00004 ounce
1 gram	g	small paper clip	1 g ≈ 0.04 ounce
1 kilogram	kg	six medium apples	1 kg ~ 2 pounds

Examples

Write the metric unit of mass that you would use to measure each of the following. Explain why you chose the unit, and then estimate the mass.

A. a sheet of notebook paper

A sheet of paper has a mass greater than a paper clip, but much less than six apples. The gram is the appropriate unit. An estimate for the mass of a sheet of paper is 5 grams.

B. a box of ten books

A box of ten books weighs more than six apples, so the kilogram is the appropriate unit. One book probably weighs about the same as six apples, so 10 kilograms is a good estimate.

The most commonly used metric units of capacity are milliliters and liters.

Unit	Abbreviation	Model	Benchmark
1 milliliter	mL	half an eyedropper	1 mL ≈ 0.03 ounce
1 liter	L	small pitcher	1 L ≈ 1 quart

Write the metric unit of capacity that you would use to measure the following. Explain why you chose the unit, and then estimate the capacity.

C. a can of soda
A can of soda is between the eyedropper and the small pitcher, so the milliliter should be used. A can of soda is a little more than a cup, and there are 4 cups in a quart. A cup would be around 250 mL, so a good estimate for a can of soda would be 350 mL.

Exercises

Write the metric unit of mass or capacity that you would use to measure each of the following. Then estimate the mass or capacity.

1. dollar bill
2. crayon
3. football player
4. bowl of punch
5. postage stamp
6. horse
7. small glass of juice
8. pencil
9. thumb tack
10. raindrop
11. bathtub
12. small aquarium

Choose the most reasonable unit for measuring mass or capacity.

13. brick a. gram b. kilogram c. liter
14. feather a. gram b. kilogram c. liter
15. baseball bat a. gram b. kilogram c. liter
16. pair of socks a. gram b. kilogram c. liter

Choose the most reasonable measurement of mass or capacity.

17. small aquarium a. 7 L b. 300 mL c. 200 L
18. bag of flour a. 2 kg b. 2 g c. 20 kg
19. inflated balloon a. 40 L b. 200 mL c. 1 L
20. cup of coffee a. 2 mL b. 20 mL c. 200 mL

Problem SOLVING

For problems 21–24, use the recipe on the right.

21. Is the total amount of butter, flour, sugar, and brown sugar more or less than 1 kilogram?

22. Write the quantities of ingredients needed for two batches of cookies.

23. Is the total amount of butter, flour, sugar, brown sugar, and chocolate chips for two batches of cookies more or less than 3 kilograms?

24. Liquid soap comes in different sizes. Which is larger, a 1.89-liter container or a 332-milliliter container? Explain.

Chocolate Chip Cookies
2 eggs
250 grams of butter
300 grams of flour
200 grams of sugar
225 grams of brown sugar
500 grams of chocolate chips
5 milliliters of vanilla extract

25. Your doctor told you to take 2 tablespoons of medicine every day. She gave you a bottle with a capacity of 500 mL. If one tablespoon is 15 mL, how many days can you take a full dose of the medicine?

26. You and a partner completed a measurement activity. Your partner recorded the mass of a dime as 2 but did not write the units. What unit of mass did she use?

MIXED REVIEW

Add or subtract.

27. $\begin{array}{r} 48.54 \\ +\ 10.01 \\ \hline \end{array}$

28. $\begin{array}{r} 517.421 \\ +\ 2.474 \\ \hline \end{array}$

29. $\begin{array}{r} 8.16 \\ -\ 0.83 \\ \hline \end{array}$

30. $\begin{array}{r} 7.2 \\ -\ 2 \\ \hline \end{array}$

31. $4.431 + 2.4 + 76.24$

32. $345.1 - 4.88$

Cumulative Review

Compare using <, >, or = .

1. 48 ● 84
2. 8,396 ● 949
3. 16,151 ● 16,154
4. 0.67 ● 6.7
5. 8.5 ● 8.05
6. 4.1 ● 4.10

Round to the underlined place-value position.

7. $\underline{8}8$
8. $2\underline{6}4$
9. $4\underline{8},150$
10. $\underline{9},850$
11. $\underline{2}.6$
12. $0.\underline{8}5$
13. $0.0\underline{6}6$
14. $7.0\underline{9}8$

Estimate.

15. 651 + 908
16. 42,100 − 21,496
17. 285 × 6
18. 57 × 31
19. $21\overline{)8,051}$

Compute.

20. 86 + 18
21. \$971 + 48
22. 2.051 + 0.889
23. 4.85 + 26.1 + 8.2
24. 15,040 + 8,973 + 22,641
25. 43 − 8
26. 615 − 224
27. 9.1 − 6.28
28. 4,055 − 890
29. 45.520 − 29.040
30. 53.6 + 28.8
31. 6,507 − 2,320
32. \$9.65 − \$2.20
33. 4.6 + 2.1 + 8.9
34. 5.013 + 0.26 + 7.6
35. 26 × 5
36. 462 × 12
37. 88 × 72
38. 0.9 × 0.3
39. 0.014 × 5
40. $4\overline{)252}$
41. $38\overline{)1,026}$
42. $7\overline{)13.3}$
43. $2\overline{)35}$
44. 15 × 15
45. 40 × 41
46. 0.6 × 2.1
47. 400 ÷ 4
48. 1,000 ÷ 10
49. 4,000 ÷ 40

Solve.

50. Damon paid \$2.25 for a burger, \$1.63 for fries, and \$1.18 for a soda. How much did he pay altogether?
51. A case of cat food costs a grocer \$21.12. If there are 48 cans in a case, and the grocer sells each can for \$0.05 more than he paid for them, what is the profit?

3.9 Metric Measurement Conversions

Objective: to change units within the metric system

Bart Jackson works in a cafeteria. He records how many kilograms of meat he uses each day. Bart used a recipe that needed 750 grams of chicken. How many kilograms of chicken did he use?

Use the chart below to help you change from one metric unit to the other.

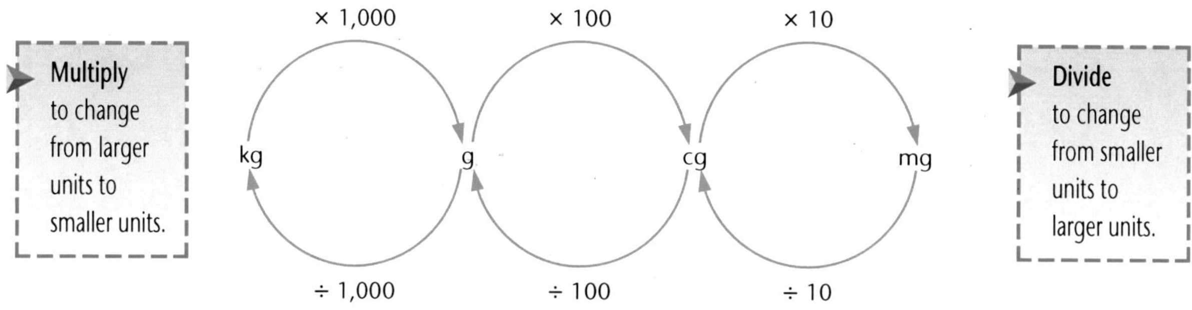

750 grams = ■ kilograms

You are changing from a smaller unit to a larger unit, so *divide.*

Since 1 kg = 1,000 g, divide by 1,000.

$$1{,}000\overline{)750.00} = 0.75$$

750 g = 0.75 kg

Bart used 0.75 kilograms of chicken.

A board is 15 meters long. How many centimeters is this?

15 meters = ■ centimeters

You are changing from a larger unit to a smaller unit, so *multiply.*

Since 1 m = 100 cm, multiply by 100.

15 × 100 = 1,500

15 m = 1,500 cm

The board is 1,500 centimeters long.

Examples

A. **6,225 grams = __?__ kilograms**

Kilograms are larger than grams, so you need to divide by 1,000.

6,225 ÷ 1,000 = 6.225

6,225 grams = 6.225 kilograms

B. **252 meters = __?__ millimeters**

Millimeters are smaller than meters, so you need to multiply by 1,000.

252 × 1,000 = 252,000

252 meters = 252,000 millimeters

Exercises

Solve.

1. 12 L = ____ mL
2. 1.35 kg = ____ g
3. 50 cm = ____ mm
4. ____ L = 95 mL
5. ____ mg = 8.2 g
6. 575 m = ____ km
7. 0.5 L = ____ mL
8. 1,500 cm = ____ km
9. 500 mg = ____ kg
10. 25.9 km = ____ mm

Problem SOLVING

11. Josie has 350 mL of orange juice. How many liters does she have?
12. A dog has a brain with a mass of 70 grams, and a cat has a brain with a mass of 30,000 milligrams. Which animal has the larger brain?
13. A 5-kg ham is sliced for sandwiches. There are 125 slices. How many grams are in each slice?
14. A can of soda costs 75 cents and has a capacity of 355 mL. A bottle of soda costs $1.25 and has a capacity of 0.59 L. Which is the better buy? Find the cost per milliliter and compare.

3.10 Problem-Solving Strategy: Write an Equation

Objective: to solve a word problem by writing an equation and using mental math

William made 7 caramel apples. He gave some to his brother and had 4 left. How many did he give to his brother?

You need to find the number of caramel apples William gave to his brother. You know how many caramel apples William made and how many he has left.

Let a represent the number of caramel apples William gave to his brother. William made 7 caramel apples. After he gave some to his brother he had 4 left. Therefore, the equation is $7 - a = 4$.

Solve the equation using mental math.

$7 - a = 4$ What number when subtracted from 7 is equal to 4?

$7 - 3 = 4$ The solution is 3.

William gave his brother 3 caramel apples.

If William made 7 caramel apples and he gave 3 to his brother, he had 4 left. The solution is correct.

Choose the equation that can be used to find the answer.

1. Samantha planted 24 marigolds in her backyard. If she planted 6 in each row, how many rows of marigolds did she plant?

   a. $m + 6 = 24$ b. $6m = 24$ c. $m - 6 = 24$ d. $m \div 6 = 24$

2. Jamal spends $15 on a pair of soccer shorts. He has $10 left. How much money did he have before buying the shorts?

   a. $10s = 15$ b. $s \div 10 = 15$ c. $s + 10 = 15$ d. $s - 15 = 10$

Solve

Write an equation and solve using mental math.

1. You bought a set of books for \$25. The set contained 5 books. What is the price of each book?

2. You and three friends plan to share the cost of a pizza. If the pizza costs \$16, determine the cost per person.

3. Mrs. Glub bakes 36 cookies. She gave some to her children and had 30 cookies left over. How many cookies did she give to her children?

4. The school conducted a fundraiser and collected \$100. If 50 students contributed the same amount to the fundraiser, how much did each student contribute?

5. Jane ran 8 miles in 2 days. If she ran 3 miles the first day, how many miles did she run the second day?

6. There are 70 lollipops in a bowl. If each student receives 7 lollipops, how many students are in the class?

7. Jamie counts 18 butterflies in her garden. After a few minutes, she counts again and there are 26 butterflies. How many new butterflies came into her garden?

8. Tim swims twice as many laps on Monday as he does on Friday. If he swims 30 laps on Monday, how many laps does he swim on Friday?

Mixed Review

Simplify.

9. $5 + 3 - 2^2 + 1$

10. $(5 - 3) + 9$

11. $18 \div 2 - 3^2$

12. $12 - 4 \times 2$

13. $5^2 + 4^2$

14. $8 \times 2 - (5 - 4)$

Chapter 3 Review

Language and Concepts

Choose the correct word to complete each sentence.

1. The number of decimal places in the product is the same as the sum of the number of decimal (points, places) in the factors.
2. If there are not enough decimal places in a product, zeros are attached to the (left, right) of the product to place the decimal point correctly.
3. To round a quotient to a certain place, divide to (one, two) extra place(s).
4. Sometimes you need to write zeros after the (first, last) digit in the dividend so that the problem will divide evenly.
5. When dividing by 100, move the decimal point (two, three) places to the left.
6. A number in scientific notation is written as the (product, quotient) of a factor and a power of 10.

Skills and Problem Solving

Estimate. (Section 3.1)

7. $\$6.19 \times 5$

8. 4.97×5.34

9. $9\overline{)18.3}$

10. $5.1\overline{)25.12}$

Copy. Then place the decimal point in the product or quotient. (Sections 3.2–3.5)

11. $3.9 \times 9 = 351$

12. $0.85 \times 0.22 = 1870$

13. $1.5\overline{)1.65}$ quotient: 11

14. $0.08\overline{)6.1}$ quotient: 7625

Compute. Sections (3.2–3.6)

15. 0.7×5

16. 0.4×6

17. 8.5×7

18. 11.6×9

19. 0.4×0.3

20. 0.5×0.9

21. 6.8×0.8

22. 60.7×0.3

23. 0.6×0.6

24. 0.42×0.7

25. $4\overline{)0.96}$ 26. $7\overline{)3.64}$ 27. $5\overline{)35.45}$ 28. $5\overline{)3.2}$ 29. $4\overline{)2.5}$

30. $1.5\overline{)70.5}$ 31. $0.6\overline{)5.4}$ 32. $0.8\overline{)0.64}$ 33. $4.3\overline{)34.4}$ 34. $1.6\overline{)92.48}$

35. 2.17×0.038 36. $0.79 • 3.7 → $\$0.79 \bullet 3.7$ 37. 0.3×0.2

38. $0.56 \div 0.07$ 39. $1.215 \div 0.15$ 40. $1.586 \div 0.26$

41. $9.34 \bullet 10$ 42. $8.63 \times 1{,}000$ 43. $3.27 \div 10$

44. $3.29 \div 100$ 45. $0.76 \bullet 100$ 46. $5.78 \div 1{,}000$

Express in scientific notation. (Section 3.7)

47. 3,800,000 48. 7,843 49. 57,600

Express in standard form. (Section 3.7)

50. 4.7×10^8 51. 6.15×10^3 52. 9.426×10^7

Solve. (Sections 3.8–3.9)

53. 5 L = ____ mL 54. 2.5 kg = ____ g

55. 80 cm = ____ mm 56. ____ L = 125 mL

57. ____ mg = 2.8 g 58. 2,550 m = ____ km

59. You are given two choices of soda. One contains 355 mL and the other contains 0.591 L. Which choice will give you more soda? Explain how you determined your answer.

Solve. (Sections 3.2–3.3, 3.10)

60. At a garden store a cherry tree costs five times as much as a pine tree. If a cherry tree costs $30, how much does a pine tree cost?

61. A 12-ounce bottle of shampoo costs $1.99. An 8-ounce bottle of shampoo costs $1.49. Which is the better buy?

62. Jeff earns $5.56 per hour. Erica earns $6.18 per hour. Each works 38 hours a week. How much more does Erica earn in one week?

63. What is the value of a roll of quarters if a roll contains 40 quarters?

Chapter 3 Test

Estimate.

1. $\$7.29 \times 4$

2. $\$5.99 \times 3.2$

3. $5.9\overline{)307.4}$

4. $4.89\overline{)33.6}$

Compute.

5. 0.5×9

6. 0.4×0.6

7. 0.8×0.9

8. 43.6×0.8

9. 0.7×0.7

10. $6\overline{)13.8}$

11. $7\overline{)150.5}$

12. $16\overline{)406.4}$

13. $0.7\overline{)4.2}$

14. $1.3\overline{)27.82}$

15. 5.27×1.45

16. 17.8×0.39

17. $0.07 \bullet 0.1$

18. $0.43 \bullet 100$

19. $0.42 \bullet 0.21$

20. 0.005×0.03

21. $3.64 \div 0.13$

22. $43.3 \div 1{,}000$

Write in scientific notation.

23. 12,000

24. 483

25. 6,000,000

Write in standard form.

26. 4×10^5

27. 3.2×10^3

28. 4.23×10^1

Solve.

29. 15 mL = __________ L

30. 25 km = __________ m

31. Madeline runs 2.5 kilometers every day. Each day her brother Max runs eight 200-meter sprints, and then runs 800 meters. How far does each sibling run in meters? Who runs the farthest?

32. A 1-pound can of coffee costs \$2.59. A 2-pound can costs \$4.98, and a 3-pound can costs \$7.69. Which is the best buy?

33. June and Jen were trying to decide which contains more lemonade, a 1-gallon jug or two 2-liter bottles. June discovered that a gallon contains 3,780 milliliters.

   a. Determine which contains more lemonade, a 1-gallon jug or two 2-liter bottles.

   b. Explain why the answer you determined is correct. Use what you know about metric units in your explanation. Use words, numbers, and/or symbols in your explanation.

What's Missing?

What is missing from each problem? To find out use your computation and number sense skills.

1. $$\begin{array}{r} 8.5 \\ +\ \blacksquare.2 \\ \hline 15.\blacksquare \end{array}$$

2. $$\begin{array}{r} 31.4 \\ -\ \blacksquare.\blacksquare \\ \hline 31.2 \end{array}$$

3. $$\begin{array}{r} 7.\blacksquare \\ \blacksquare.8 \\ +\ 12.9 \\ \hline 28.9 \end{array}$$

4. $$\begin{array}{r} 2.\blacksquare \\ -\ 1.3 \\ \hline 1.4 \end{array}$$

5. $$\begin{array}{r} 18.1 \\ -\ 12.\blacksquare \\ \hline 5.8 \end{array}$$

6. $$\begin{array}{r} 3\blacksquare.7 \\ -\ 18.8 \\ \hline 14.9 \end{array}$$

7. $$\begin{array}{r} 0.3\blacksquare \\ \times\ 1.7 \\ \hline 0.544 \end{array}$$

8. $$\begin{array}{r} 3.26 \\ \times\ 1.\blacksquare \\ \hline 5.216 \end{array}$$

9. $$\begin{array}{r} 123 \\ 56.7\overline{)6\blacksquare 74.1} \end{array}$$

10. $$\begin{array}{r} 887.2 \\ 8\overline{)70\blacksquare 7.6} \end{array}$$

Cumulative Test

1. $0.33 \times 0.63 =$ ____
   a. 0.02079
   b. 0.2079
   c. 0.279
   d. 2.79

2. Estimate. $5.4\overline{)55.292}$
   a. 1.1
   b. 11
   c. 110
   d. 1,100

3. $3.52 \div 0.044 =$ ____
   a. 0.08
   b. 0.8
   c. 8
   d. 80

4. $\begin{array}{r} 4.96 \\ \times \ 21 \\ \hline \end{array}$
   a. 1.0416
   b. 10.416
   c. 104.16
   d. 1,041.6

5. $\begin{array}{r} 36.09 \\ +\ 9.812 \\ \hline \end{array}$
   a. 44.821
   b. 45.902
   c. 135.21
   d. none of the above

6. Which statement is true?
   a. $0.0713 > 0.071$
   b. $0.0713 < 0.071$
   c. $0.0713 = 0.071$
   d. none of the above

7. Find the value of $6 \times 5 - 9 \div 3$.
   a. 21
   b. 12
   c. 27
   d. 30

8. Rochelle bought a dress for $40 and a blouse for $20. If the sales tax is $0.06 on each dollar spent, what was the total amount she paid?
   a. $60 c. $63.60
   b. $60.06 d. $66.60

9. Three friends, Mark, Mike, and Mo, have last names Krantz, Kelly, and Kuan, not necessarily in that order. Who is who if:
   1. Mark beat Kelly at handball.
   2. Krantz drove Mo and Kuan to work.
   3. Mike is married to Krantz's sister.

   a. Mark Kuan, Mike Krantz, Mo Kelly
   b. Mark Kelly, Mike Krantz, Mo Kuan
   c. Mark Krantz, Mike Kelly, Mo Kuan
   d. Mark Krantz, Mike Kuan, Mo Kelly

10. Ken noticed a pattern in the base of a pine cone. He counted the scales and found this sequence:

    1, 1, 2, 3, 5, 8, . . .

    What is the next number?
    a. 12
    b. 13
    c. 15
    d. 16

CHAPTER

Fractions and Decimals

Ashley Paquet
Maryland

4.1 Divisibility Patterns

Objective: to use divisibility patterns for 2, 3, 4, 5, 6, 9, and 10

At the county fair, 3 tickets are needed to ride the Ferris wheel. Tiffany and her friends have 21 tickets. Can they use them all by riding only the Ferris wheel?

To answer the question, we need to know if 3 is a **factor** of 21. A factor is a whole number that divides into another whole number with a remainder of zero. A whole number is **divisible** by its factors.

Since $21 \div 3 = 7$, 3 is a factor of 21, and 21 is divisible by 3. Sometimes we can use patterns to find factors. Look at the chart below.

Factor	Rule	Example
2	The ones digit is divisible by 2. The number is *even*.	2, 4, 6, 8, 10, 12, . . .
3	The sum of the digits is divisible by 3.	567 $5 + 6 + 7 = 18$ $18 \div 3 = 6$ Therefore 567 is divisible by 3.
5	The ones digit is 0 or 5.	5, 10, 15, 20, 25, . . .
10	The ones digit is 0.	10, 20, 30, 40, . . .

A whole number is **even** if it is divisible by 2. A whole number is **odd** if it is not divisible by 2.

Example

Tell whether 2,320 is divisible by 2, 3, 5, or 10. Then classify the number as *even* or *odd*.

2: Yes; the ones digit is 0, so the number is divisible by 2.

3: No; $2 + 3 + 2 + 0 = 7$, and 7 is not divisible by 3.

5: Yes; the ones digit is 0.

10: Yes; the ones digit is 0.

The number 2,320 is even because it is divisible by 2.

The rules for 4, 6, and 9 are related to the rules for 2 and 3.

Factor	Rule	Example
4	The number formed by the last two digits is divisible by 4.	4, 8, 12, 112 112: 12 ÷ 4 = 3
6	The number is divisible by both 2 and 3.	564 2: Last digit is 4 3: 5 + 6 + 4 = 15, 15 ÷ 3 = 5 Therefore 564 is divisible by 6.
9	The sum of the digits is divisible by 9.	459: 4 + 5 + 9 = 18 18 ÷ 9 = 2 Therefore 459 is divisible by 9.

Try THESE

Tell whether each number is divisible by 2, 3, 4, 5, 6, 9, or 10. Then classify the number as *even* or *odd*.

1. 126 **2.** 684 **3.** 2,835

Exercises

Tell whether each number is divisible by 2, 3, 4, 5, 6, 9, or 10. Then classify the number as *even* or *odd*.

1. 60 **2.** 489 **3.** 8,505 **4.** 45 **5.** 605

6. 9,948 **7.** 80 **8.** 900 **9.** 14,980 **10.** 78

11. 3,135 **12.** 18,321 **13.** 138 **14.** 6,950 **15.** 5,203,570

16. Find a number that is divisible by both 3 and 5.

Problem SOLVING

17. Find a number that is divisible by 2, 9, and 10.

18. Maxine is packaging brownies for a bake sale. She has 192 brownies.

a. Can Maxine package the brownies in groups of 4 with none left over?

b. What are the other ways she can package all of the brownies in equal-sized packages with 10 or less in a package?

Constructed RESPONSE

19. If all even numbers are divisible by 2, is it true that all odd numbers are divisible by 3, 5, or 7? If so, explain your reasoning. If not, explain by giving numbers for which this is not true.

4.2 Prime and Composite Numbers

Objective: to recognize the difference between prime and composite numbers

Exploration Exercise

When two or more numbers are multiplied, each number is called a **factor**. The answer when multiplying is called the *product*.

A whole number that has exactly two unique factors, 1 and itself, is called a **prime number**. A number greater than 1 with more than two factors is called a **composite number**.

Copy the numbers shown in the chart on grid paper.

The Sieve of Eratosthenes

1	2	3	4	5	6	7	8	9	10
11	12	13	14	15	16	17	18	19	20
21	22	23	24	25	26	27	28	29	30
31	32	33	34	35	36	37	38	39	40
41	42	43	44	45	46	47	48	49	50
51	52	53	54	55	56	57	58	59	60
61	62	63	64	65	66	67	68	69	70
71	72	73	74	75	76	77	78	79	80
81	82	83	84	85	86	87	88	89	90
91	92	93	94	95	96	97	98	99	100

1. Cross out the number 1. One is neither prime nor composite; the only factor of 1 is 1.
2. Circle the number 2. Then, using a yellow marker, shade all of the numbers that are divisible by 2. Do you see a pattern? Describe the pattern.
3. Circle the number 3. Then, using a pink marker, underline all of the numbers that are divisible by 3. Do you see a pattern? Describe the pattern.
4. Circle the number 5. Then, using a blue marker, put an X on all of the numbers that are divisible by 5. Do you see a pattern? Describe the pattern.
5. Circle the number 7. Then, using a red marker, draw a square around all of the numbers that are divisible by 7. Do you see a pattern? Describe the pattern.
6. Circle all remaining numbers that have not been marked. All of the circled numbers are prime numbers.

Try THESE

Tell whether each given number is *prime*, *composite*, or *neither*.

1. 28
2. 11
3. 103

Exercises

Tell whether each given number is *prime*, *composite*, or *neither*.

1. 17
2. 12
3. 53
4. 114
5. 15
6. 45
7. 93
8. 179
9. 0
10. 29
11. 125
12. 291
13. Write 38 as a product of two prime numbers.
14. Find the least prime number that is greater than 60.
15. How many prime numbers are less than 100?
16. Name five prime numbers between 100 and 200.

Problem SOLVING

17. A **perfect number** is a number in which all the factors of a number, except the number itself, add up to the number. There are only two perfect numbers between 1 and 100. They are both less than 30. Which numbers are they?
18. **Twin primes** are two prime numbers that are consecutive odd integers such as 3 and 5, 5 and 7, and 11 and 13. Find all of the twin primes that are less than 100.
19. **Reversal primes** are prime numbers that make another prime number when reversed, like 17. List all reversal primes less than 100.

State whether the statement is *true* or *false*.

20. The sum of two prime numbers is always an even number.
21. There is a smallest prime number, but not a largest prime number.
22. Every whole number is either prime or composite.
23. All odd numbers greater than 7 can be expressed as the sum of three prime numbers. Which three prime numbers have a sum of 37?

★ 24. *True* or *False*: Any even composite number greater than 2 is the sum of two prime numbers. Show examples to support your answer.

4.3 Prime Factorization

Objective: to find the prime factorization of a composite number

Exploration Exercise

Any given number of squares can be arranged into one or more different rectangles. Look at the table below. The table shows the different rectangles that can be made using 2, 3, 4, 5, or 6 squares. A 1×5 rectangle is the same as a 5×1 rectangle.

Number of Squares	Sketch of Rectangles Formed	Dimensions of Each Rectangle
2		1×2
3		1×3
4		1×4; 2×2
5		1×5
6		1×6; 2×3

Your turn! Use square tiles or sketch squares in order to complete the table down to 20.

1. For what numbers can more than one rectangle be formed?
2. For what numbers can only one rectangle be formed?
3. What kind of numbers can only form one rectangle?
4. What kind of numbers formed more than one rectangle?

Every composite number can be expressed as a product of more than two factors. Every composite number can also be expressed as a product of prime numbers. This is called **prime factorization**. A factor tree can be used to find the prime factorization of a number.

To find the prime factorization, choose any pair of whole number factors and continue to factor any number that is not prime.

Example

Find the prime factorization of 28.

28
2 × 14
2 × 2 × 7

Write the number that is being factored at the top.
Choose any pair of whole number factors of 28.
Continue to factor any number that is not prime until all factors are prime.

28
4 × 7
2 × 2 × 7

The prime factorization of 28 is $2 \times 2 \times 7$ or $2^2 \times 7$.

Exercises

List all of the factors of each number.

1. 12
2. 24
3. 48
4. 36
5. 50

Find the prime factorization of each number.

6. 24
7. 18
8. 75
9. 49
10. 104
11. 42
12. 126
13. 225
14. 1050
15. 72

Write the prime factorization for the following numbers using exponents. (For example, $54 = 2 \times 3 \times 3 \times 3 = 2 \times 3^3$)

16. 40
17. 32
18. 140
19. 56
20. 125

Problem SOLVING

21. Jennifer bought bags of snacks that each cost the same. She spent a total of $40. Find three possibilities for possible costs per bag and the number of bags that she could have purchased.

22. To find the volume of a box, you can multiply its length, width, and height. The measure of the volume of a box is 429. Find its possible dimensions.

Test PREP

23. Which of the following is the prime factorization of 72?

a. $2^2 \times 3^3$ b. $2^2 \times 3 \times 6$ c. $2^3 \times 3^2$ d. $2^3 + 3^2$

4.4 Problem-Solving Strategy: Use a Venn Diagram

Objective: to make and use Venn diagrams to organize information

A **Venn diagram** uses overlapping circles to show the similarities and differences of two or more groups of items. Any item that is located where the circles overlap has a characteristic of both circles.

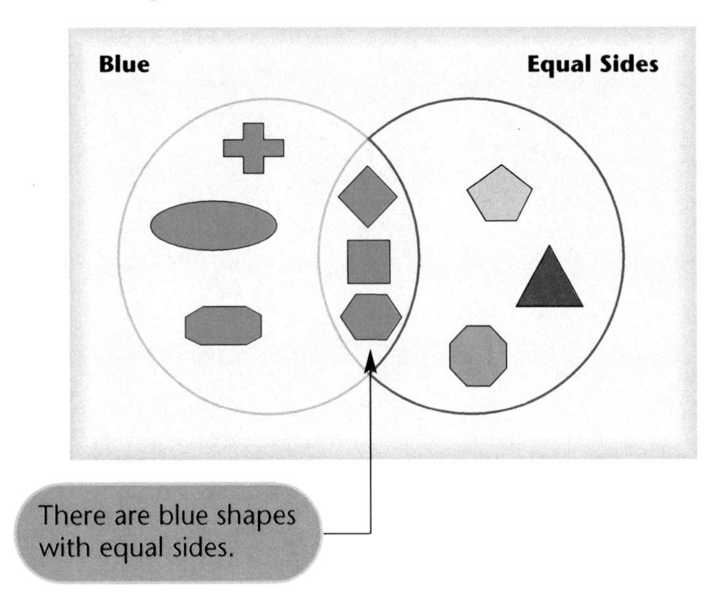

You can also make a Venn diagram using numbers. The Venn diagram below shows the factors of 12 in one circle and the factors of 18 in the second circle.

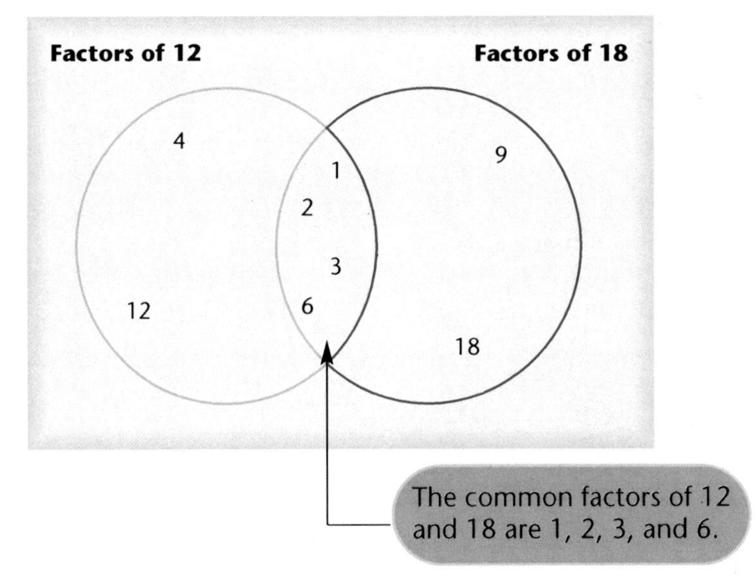

Try THESE

1. Make a Venn diagram to show all of the factors of 20 and 30.

Exercises

Make a Venn diagram that shows the factors for each pair of numbers.

1. 8, 12
2. 25, 28
3. 42, 56

Use the Venn diagram below to answer questions 4–8.

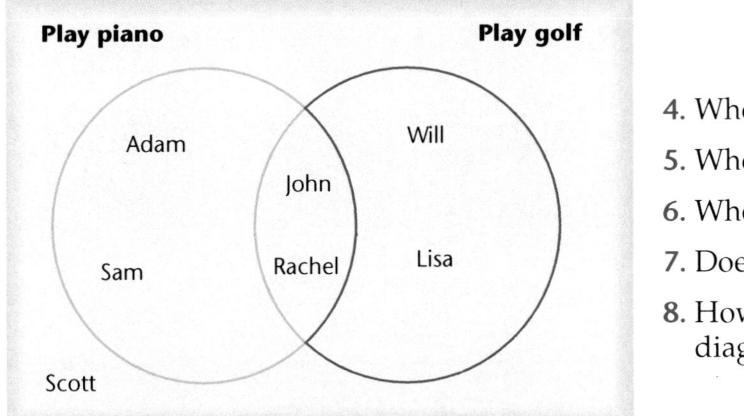

4. Who plays golf?
5. Who plays piano?
6. Who plays both golf and piano?
7. Does Scott play golf or piano?
8. How many people are listed in the diagram?

Use the Venn diagram below to answer questions 9–12.

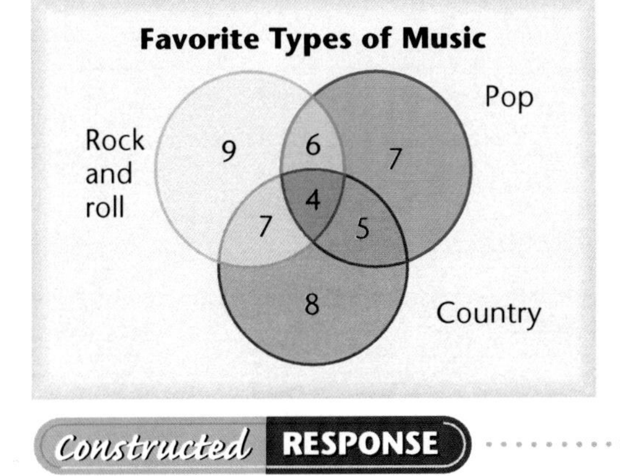

9. How many liked only rock and roll?
10. How many liked only country and pop?
11. How many liked only country or pop?
12. How many were asked altogether?

Constructed RESPONSE

13. Organize the numbers 2, 5, 9, 27, 29, 35, and 43 into a Venn diagram comparing prime numbers and composite numbers. What numbers are in the overlapping circles? Explain.

★14. Make a Venn diagram to show the factors of 24, 42, 56, and 80.

4.5 Greatest Common Factor

Objective: to find the greatest common factor of two or more numbers

Numbers often have common factors. The greatest of the common factors of two or more numbers is the **greatest common factor (GCF)** of the numbers. There are different ways to find the GCF.

What is the GCF of 12 and 18?

■ Method 1: Make a list of the factors.

Factors of 12
1, 12
2, 6
3, 4

Factors of 18
1, 18
2, 9
3, 6

The common factors are 1, 2, 3, and 6. The greatest common factor (GCF) is 6.

What is the GCF of 18 and 30?

■ Method 2: Use prime factorization.

$18 = \boxed{2} \times \boxed{3} \times 3$
$30 = \boxed{2} \times \boxed{3} \times 5$

2 and 3 are common prime factors of 18 and 30, and $2 \times 3 = 6$. Therefore 6 is the GCF of 18 and 30.

Try THESE

Find the common factors of both numbers. Circle the GCF.

1. 4, 8
2. 18, 20
3. 18, 45

Exercises

Find the common factors of both numbers. Circle the GCF.

1. 10, 35
2. 28, 70
3. 36, 72
4. 15, 24

Find the GCF of each set of numbers.

5. 12, 60
6. 24, 48, 84
7. 25, 45, 90

Use prime factorization to find the GCF of each set of numbers.

8. 630, 712
9. 90, 135
★ 10. 16, 24, 30, 42

Problem SOLVING

Mrs. Kirby's Class Field Trip Money	
Wednesday	$54
Thursday	$45
Friday	$63

11. Mrs. Kirby recorded the amount of money collected from sixth grade classes for a field trip to the orchestra. Each student paid the same amount. What is the most the field trip could cost per student?

12. How many students have paid to attend the field trip to the orchestra using the ticket cost from problem 11?

★ 13. Write three numbers whose GCF is 14 with one of the numbers being larger than 100.

Constructed RESPONSE

14. Write a rule for determining the GCF of two prime numbers. Then develop a rule for finding the GCF of one prime and one composite number.

15. Is the GCF of any two even numbers always even? Is the GCF of any two odd numbers always odd? Explain, and support your reasoning with examples.

MixeD REVIEW

Compute.

16. 15.2×2.3
17. $3.4\overline{)23.12}$
18. 6.341×5.9
19. $1.5\overline{)9.345}$

4.6 Equivalent Fractions

Objective: to use models to recognize equivalent fractions

Tim, Tom, and Tammy each have individual pizzas for lunch. Tim cuts his in two slices, Tom cuts his in four slices, and Tammy cuts her pizza into eight slices. Tim eats one slice, Tom eats two, but Tammy eats four slices. Did they eat the same amount?

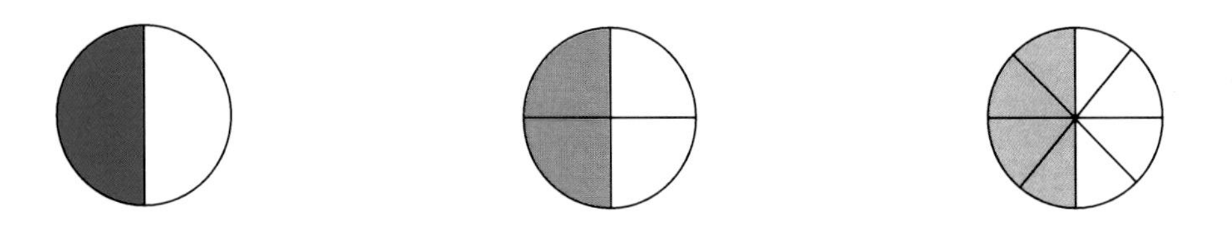

Look at the pizzas. Even though each person ate a different number of slices, they ate the same amount of pizza. In fact, each person ate the same *fraction* of his or her individual pizza. Fractions that represent the same value are called **equivalent fractions**.

You can use multiplication and division to write a fraction as an equivalent fraction. Study the examples below.

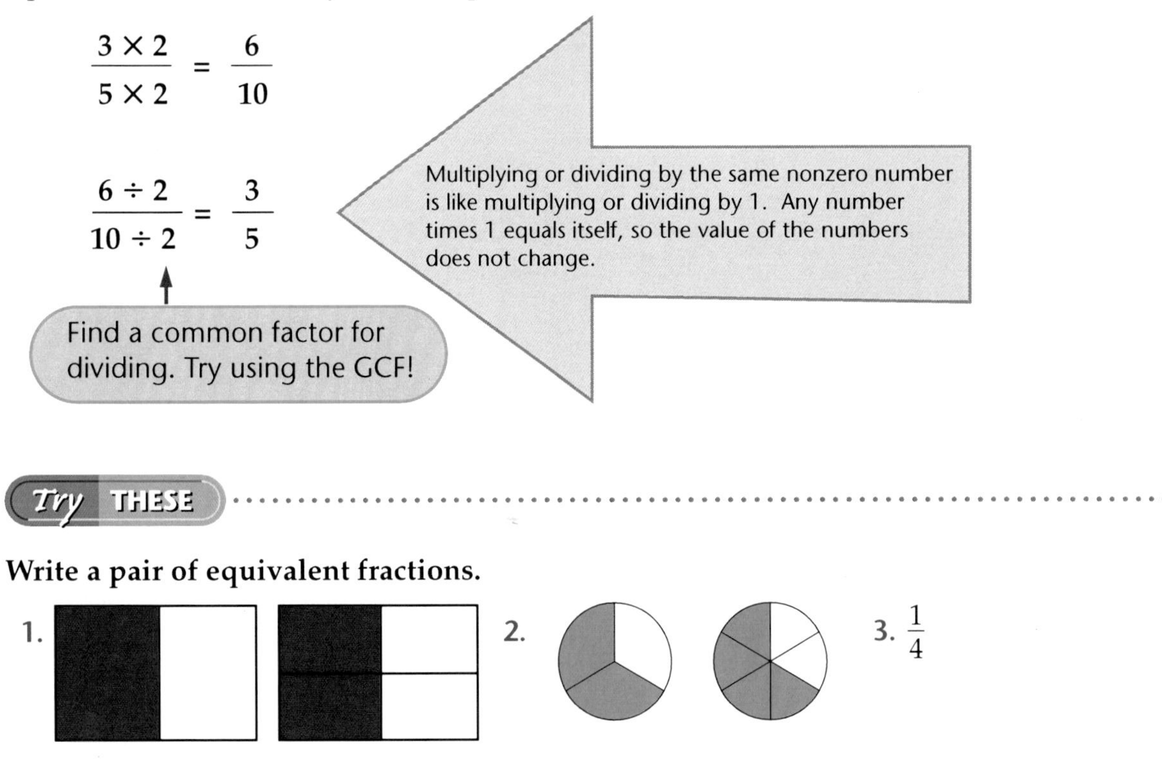

$$\frac{3 \times 2}{5 \times 2} = \frac{6}{10}$$

$$\frac{6 \div 2}{10 \div 2} = \frac{3}{5}$$

Try THESE

Write a pair of equivalent fractions.

1.

2.

3. $\frac{1}{4}$

Exercises

Find two fractions that are equivalent to each number.

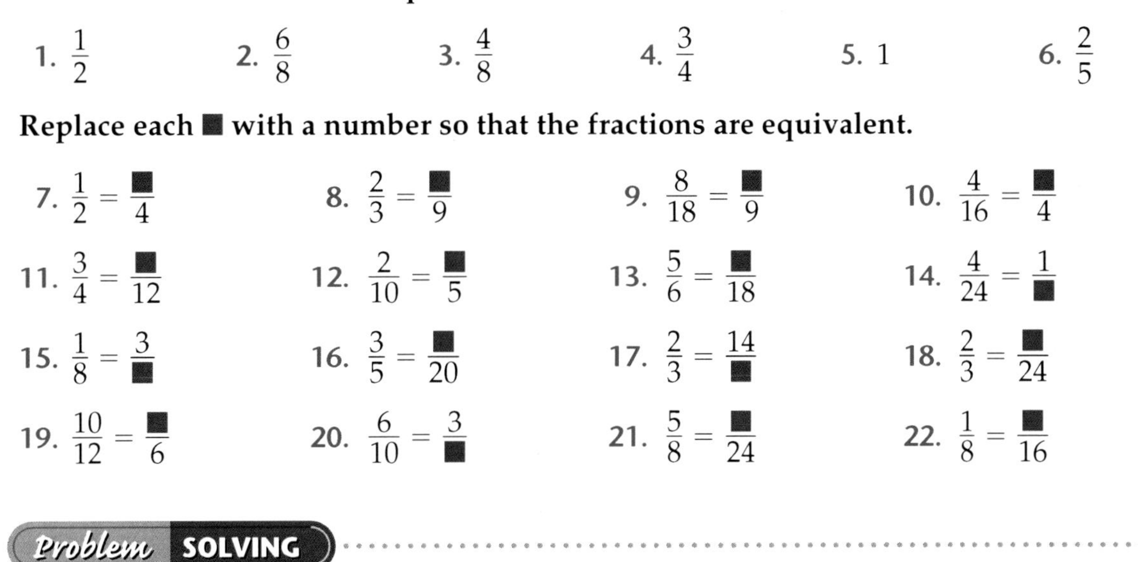

1. $\frac{1}{2}$ 2. $\frac{6}{8}$ 3. $\frac{4}{8}$ 4. $\frac{3}{4}$ 5. 1 6. $\frac{2}{5}$

Replace each ■ with a number so that the fractions are equivalent.

7. $\frac{1}{2} = \frac{\blacksquare}{4}$ 8. $\frac{2}{3} = \frac{\blacksquare}{9}$ 9. $\frac{8}{18} = \frac{\blacksquare}{9}$ 10. $\frac{4}{16} = \frac{\blacksquare}{4}$

11. $\frac{3}{4} = \frac{\blacksquare}{12}$ 12. $\frac{2}{10} = \frac{\blacksquare}{5}$ 13. $\frac{5}{6} = \frac{\blacksquare}{18}$ 14. $\frac{4}{24} = \frac{1}{\blacksquare}$

15. $\frac{1}{8} = \frac{3}{\blacksquare}$ 16. $\frac{3}{5} = \frac{\blacksquare}{20}$ 17. $\frac{2}{3} = \frac{14}{\blacksquare}$ 18. $\frac{2}{3} = \frac{\blacksquare}{24}$

19. $\frac{10}{12} = \frac{\blacksquare}{6}$ 20. $\frac{6}{10} = \frac{3}{\blacksquare}$ 21. $\frac{5}{8} = \frac{\blacksquare}{24}$ 22. $\frac{1}{8} = \frac{\blacksquare}{16}$

Problem SOLVING

23. Shana wanted to make a birthday cake for her mother. Her old-fashioned recipe called for $\frac{3}{12}$ of a teaspoon of salt, but her measuring spoons only measured in fourths. How many fourths of a teaspoon should she use?

24. On the middle school football team 20 out of 32 players are eighth graders. Write two equivalent fractions for the fraction of the football team that are eighth graders.

Mid-Chapter REVIEW

Tell whether each number is divisible by 2, 3, 4, 5, 6, 9, or 10.

1. 81 **2.** 123 **3.** 600 **4.** 1,228

Find the prime factorization of each number.

5. 63 **6.** 200 **7.** 350 **8.** 42

Find the greatest common factor of each pair of numbers.

9. 12 and 15 **10.** 10 and 35 **11.** 26 and 36

4.7 Simplifying Fractions

Objective: to express fractions in simplest form

Jackson surveyed his grade. He asked the students to name their favorite sports. He created a bar graph of the results.

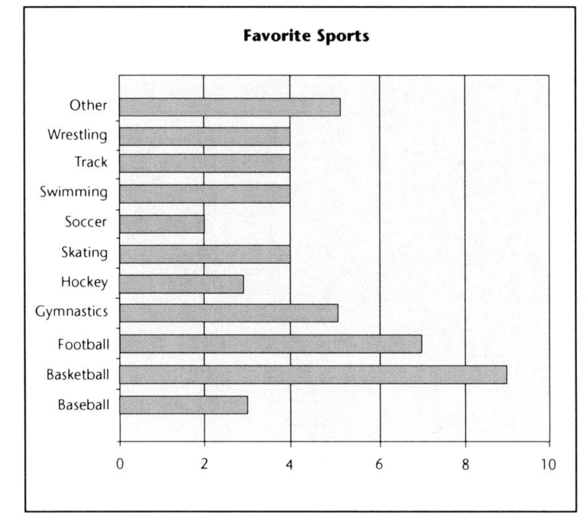

1. How many students were surveyed?
2. How many students chose gymnastics?

From the graph, we can compare the number of students that chose gymnastics as their favorite sport to the total number of students by using a fraction.

$\frac{5}{50}$ ← students who chose gymnastics / ← total number of students

Equivalent fractions are fractions that name the same number. You can find an equivalent fraction by dividing the numerator and denominator by the same number. A fraction is in **simplest form** when the only common factor of the numerator and the denominator is 1.

To write a fraction in simplest form, you can either:

- Method 1: Divide the numerator and denominator by common factors until the only common factor is 1.
- Method 2: Divide the numerator and denominator by the GCF.

Examples

A. The fraction of the students who chose gymnastics is $\frac{5}{50}$. To write the fraction in simplest form you need to divide the numerator and denominator by a common factor, 5.

$$\frac{5 \div 5 = 1}{50 \div 5 = 10} \qquad \frac{5}{50} = \frac{1}{10}$$

B. Write $\frac{18}{24}$ in simplest form.

Method 1
$\frac{18}{24} \div \frac{2}{2} = \frac{9}{12} \div \frac{3}{3} = \frac{3}{4}$

Method 2
The GCF of 18 and 24 is 6.
$\frac{18}{24} \div \frac{6}{6} = \frac{3}{4}$

Try THESE

Is each fraction in simplest form? Write *yes* or *no*.

1. $\frac{1}{5}$ 2. $\frac{2}{4}$ 3. $\frac{6}{30}$ 4. $\frac{3}{6}$ 5. $\frac{6}{10}$ 6. $\frac{2}{9}$

7. $\frac{7}{14}$ 8. $\frac{5}{7}$ 9. $\frac{1}{3}$ 10. $\frac{10}{25}$ 11. $\frac{3}{18}$ 12. $\frac{6}{11}$

Exercises

Write each fraction in simplest form.

1. $\frac{2}{4}$ 2. $\frac{3}{6}$ 3. $\frac{2}{8}$ 4. $\frac{2}{6}$ 5. $\frac{5}{10}$ 6. $\frac{3}{12}$

7. $\frac{4}{8}$ 8. $\frac{8}{10}$ 9. $\frac{2}{10}$ 10. $\frac{4}{10}$ 11. $\frac{15}{18}$ 12. $\frac{10}{12}$

13. $\frac{22}{28}$ 14. $\frac{2}{12}$ 15. $\frac{8}{16}$ 16. $\frac{6}{10}$ 17. $\frac{20}{40}$ 18. $\frac{4}{12}$

19. $\frac{9}{12}$ 20. $\frac{7}{14}$ 21. $\frac{18}{36}$ 22. $\frac{15}{45}$ 23. $\frac{18}{24}$ 24. $\frac{125}{1,000}$

Use the chart to complete each question. Write each fraction in simplest form.

1 year = 52 weeks
1 week = 7 days
1 day = 24 hours
1 hour = 60 minutes
1 minute = 60 seconds

25. What fraction of a week is 5 days?
26. What fraction of a minute is 30 seconds?
27. What fraction of a day is 12 hours?
28. What fraction of an hour is 45 minutes?
29. What fraction of a year is 39 weeks?

Problem SOLVING

30. A fraction is equivalent to $\frac{3}{4}$, and the sum of the numerator and denominator is 84. What is the fraction?

31. Lions sleep around $\frac{5}{6}$ of the day. About how many hours does a lion sleep?

Mind BUILDER

Paper Folding

When you fold a piece of paper in half as shown, you have two parts of the same size. Suppose you fold the piece of paper in half again. How many parts of the same size do you now have?

How many parts of the same size will you have if you fold a piece of paper in the following ways?

1. in half three times
2. in half four times
3. in half five times
4. in half seven times
5. How are the number of parts related to the number of folds?

Which One Does Not Belong?

In each group, three cubes are the same. One is different. Can you identify the cube that does not belong?

Extension

Make your own cube puzzle and have a friend solve it. First make your own cube. Trace the net, cut it out, draw numbers or pictures on each side and fold it into a cube. Then draw three sketches of your cube from different views and one sketch that does not match. Each sketch should show only three sides.

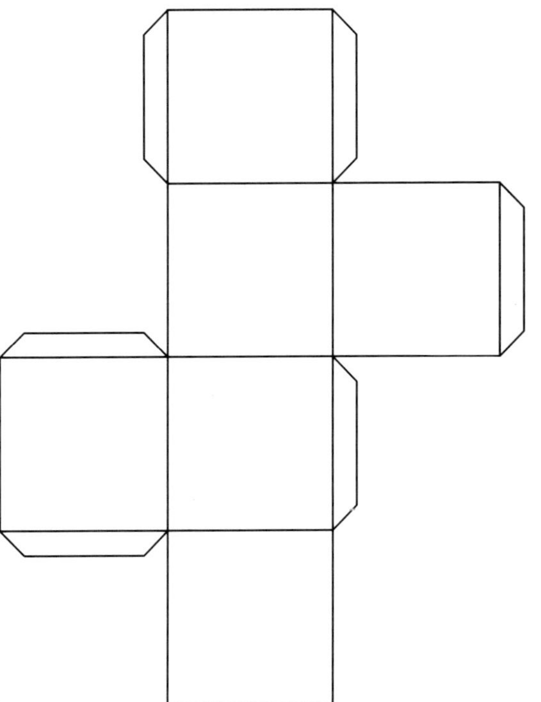

4.8 Mixed Numbers and Improper Fractions

Objective: to write mixed numbers as improper fractions; to write improper fractions as mixed numbers

Cassie practiced playing the piano for one hour plus three-fourths of an hour each week. Her brother Dale practiced the piano for seven-fourths of an hour each week.

The number of hours that Cassie plays is an example of a **mixed number**. A mixed number indicates the sum of a whole number and a fraction. It is written with a whole number part and a fraction part. Cassie plays $1\frac{3}{4}$ hours each week.

The number of hours that her brother plays is an example of an **improper fraction**. An improper fraction is a fraction that has a value equal to or greater than one. The numerator is equal to or larger than the denominator. Dale plays $\frac{7}{4}$ hours each week.

You can write improper fractions as mixed numbers by using division.

$$\frac{7}{4} \longrightarrow 7 \div 4 = 1\text{ R}3 \longrightarrow 1\frac{3}{4}$$

Divide 7 by 4. Use the remainder as the numerator of the fraction.

You can also change a mixed number to an improper fraction. There are two methods:

■ Method 1:

- Change the whole number to an equivalent fraction with the same denominator as the fraction.
- Then add the fractions.

$$5\frac{1}{2} = 5 + \frac{1}{2} = \frac{10}{2} + \frac{1}{2} = \frac{11}{2}$$

■ Method 2:

- Find the numerator of the improper fraction by multiplying the whole number and denominator.
- Then add the original numerator to the product.

$$5\frac{1}{2} \longrightarrow (5 \times 2) + 1 = 11 \longrightarrow \frac{11}{2}$$

Note: When asked for simplest form, use a mixed number. An improper fraction is not considered the simplest form of a fraction.

Try THESE

1. Write $\frac{9}{2}$ as a mixed number.

2. Write $4\frac{1}{4}$ as an improper fraction.

Exercises

Express each improper fraction as a mixed number in simplest form.

1. $\frac{5}{3}$
2. $\frac{5}{4}$
3. $\frac{8}{5}$
4. $\frac{9}{4}$
5. $\frac{15}{5}$
6. $\frac{17}{8}$
7. $\frac{36}{3}$
8. $\frac{55}{8}$
9. $\frac{47}{10}$
10. $\frac{7}{6}$
11. $\frac{11}{8}$
12. $\frac{7}{3}$
13. $\frac{22}{9}$
14. $\frac{19}{4}$
15. $\frac{15}{4}$
16. $\frac{16}{6}$
17. $\frac{20}{12}$
18. $\frac{36}{16}$

Express each mixed number as an improper fraction.

19. $1\frac{1}{2}$ $\quad \frac{(1 \times 2) + 1}{2}$
20. $2\frac{3}{4}$ $\quad \frac{(2 \times 4) + 3}{4}$
21. $1\frac{1}{6}$
22. $3\frac{1}{7}$
23. $2\frac{1}{2}$
24. $6\frac{1}{8}$
25. $1\frac{2}{3}$
26. $1\frac{7}{8}$
27. $2\frac{2}{5}$
28. $1\frac{5}{6}$
29. $2\frac{4}{7}$
30. $3\frac{5}{8}$
31. $5\frac{3}{5}$
32. $8\frac{1}{9}$
33. $3\frac{7}{8}$
34. $7\frac{6}{7}$
35. $9\frac{2}{3}$
36. $10\frac{2}{5}$
37. $11\frac{1}{4}$
38. $15\frac{5}{9}$
39. Express six and three-fifths as an improper fraction.
40. Write one hundred sevenths as a mixed number.
41. Express the improper fraction equivalent to eight and five-sixths.

Problem SOLVING

42. Sarah bought 18 eggs from the store. How many dozen eggs did she buy?

★ 43. What fraction am I? As an improper fraction, the product of my numerator and denominator is 42. As a mixed number, the sum of my whole number, numerator, and denominator equals 9. My value is between 4 and 5.

Constructed RESPONSE

44. Joshua cannot decide if $\frac{12}{12}$ is an improper fraction. What do you think? Explain your answer.

4.9 Least Common Multiple

Objective: to find the least common multiple of two or more numbers

Carla counts band uniforms for the music teacher. Since they are packaged with 5 uniforms in each box, she counts by fives: 5, 10, 15, 20, 25, 30, 35, 40, 45, 50, . . .

A **multiple** of a number is the product of the number and any whole number. When you learned to multiply, you learned multiples of numbers. What is the next multiple of 5?

Maria is counting hats that are part of the band uniform. However, they are packaged in groups of three: 3, 6, 9, 12, 15, 18, 21, 24, 27, 30, . . .

Common multiples are multiples shared by two different numbers. Notice that 3 and 5 share 15 and 30 as common multiples. What is the next common multiple of 3 and 5?

The smallest common multiple shared by two numbers is the **least common multiple (LCM)** of the numbers. The LCM of 3 and 5 is 15.

Examples

■ Method 1: Find the LCM by making a list.

A. Find the LCM of 9 and 12 by making a list.

Multiples of 9: **9, 18, 27, 36, 45, 54, 63, 72, 81, . . .**

Multiples of 12: **12, 24, 36, 48, 60, 72, 84, 96, . . .**

The LCM of 9 and 12 is 36.

■ Method 2: Use prime factorization.

B. Find the LCM of 16 and 28 by using prime factorization.

Step 1	Step 2	Step 3
Find the prime factors. $16 = 2 \times 2 \times 2 \times 2$ $28 = 2 \times 2 \times 7$	Circle the common factors and each remaining factor. $16 = 2 \times 2 \times 2 \times 2$ $28 = 2 \times 2 \times 7$	Multiply all circled factors using each common factor only once. $2 \times 2 \times 2 \times 2 \times 7 = 112$ The LCM of 16 and 28 is 112.

Try THESE

Find the LCM of each group of numbers.

1. Multiples of 15: 15, 30, 45, 60, 75, 90, 105, 120, 135, 150, …
   Multiples of 25: 25, 50, 75, 100, 125, 150, 175, 200, 225, …

2. $6 = 2 \times 3$
   $8 = 2 \times 2 \times 2$

Exercises

Find the LCM of each group of numbers. Use any method.

1. 2 and 10
2. 9 and 54
3. 3 and 4
4. 7 and 9
5. 8 and 36
6. 16 and 20
7. 5 and 12
8. 18 and 24
9. 21 and 28
10. 4, 8, and 12
11. 12, 15, and 30
12. 15, 25, and 75

Find the GCF and LCM for each set of numbers. Use any method.

13. 4 and 6
14. 8 and 20
15. 50 and 55
16. 5 and 9
17. 9 and 21
18. 9, 12, and 15

Problem SOLVING

19. Find a pair of numbers that have 30 as their LCM.
20. Find a pair of numbers that have 24 as their LCM and 2 as their GCF.

Constructed RESPONSE

21. Both Jim and Matt are finding the LCM of 9 and 12. Who is correct? Explain your reasoning.

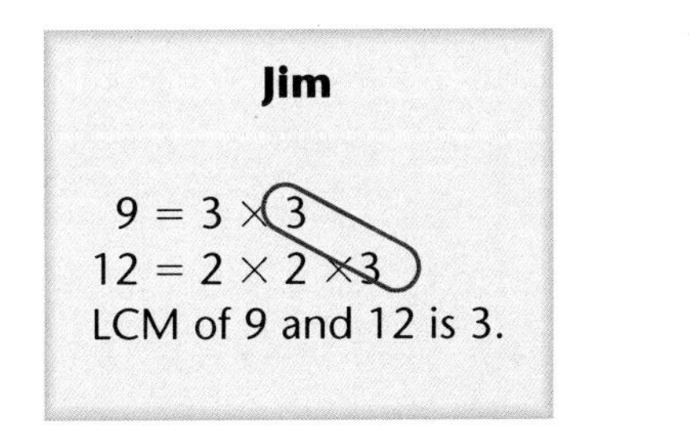

Matt

$9 = 3 \times 3$
$12 = 2 \times 2 \times 3$
LCM is $2 \times 2 \times 3 \times 3 = 36$.

4.10 Comparing and Ordering Fractions

Objective: to compare and order fractions

Player	On-Base Average
Will	1/2
Joey	3/8
Calvin	1/8
Matt	2/3
Pete	2/5

When comparing fractions that have the same denominator, you simply look at the numerator to tell which is larger. In baseball, an important statistic is on-base average. It compares how often a batter gets on base, either by getting a base hit, a walk, or hit by a pitch. Who has a better on-base average, Joey or Calvin?

The fraction $\frac{3}{8}$ is larger than $\frac{1}{8}$, because 3 is larger than 1. Joey has the higher on-base average. But what if you compare Joey and Pete? The fractions $\frac{3}{8}$ and $\frac{2}{5}$ have different denominators. You must change the fractions so that they have the same denominator before you can compare them.

In choosing a new denominator for the two fractions, find a **common denominator**. A common denominator is a common multiple of the denominators. To reduce the need for simplifying fractions later, it is best to use the **least common denominator (LCD)**, or the LCM of the denominators.

The common multiples of 5 and 8 are 40, 80, 120, 160, and so on. Use 40 since it is the smallest multiple, or LCM, of 5 and 8. Now you can change the fractions to equivalent fractions with a denominator of 40.

Step 1	Step 2	Step 3
Find a common multiple, preferably the LCM. **5:** 5, 10, 15, 20, 25, 30, 35, 40, 45, 50, ... **8:** 8, 16, 24, 32, 40, 48, 56, 64, 72, 80, ...	Rename each fraction using the common multiple as the denominator, preferably the LCD. $\frac{3}{8} \times \frac{5}{5} = \frac{15}{40}$ $\quad \frac{2}{5} \times \frac{8}{8} = \frac{16}{40}$	Compare the numerators. $\frac{15}{40} < \frac{16}{40}$

Since $\frac{15}{40} < \frac{16}{40}$, Pete has a higher on-base average than Joey.

Examples

A. Compare $\frac{3}{8}$ and $\frac{9}{16}$ using <, >, or = .

- Step 1: Find the LCD. (8: 8, 16, 24, 32; 16: 16, 32, 48, 64; the LCD is 16.)
- Step 2: Rewrite the fractions. ($\frac{3}{8} \times \frac{2}{2} = \frac{6}{16}$; $\frac{9}{16}$ already has a denominator of 16)
- Step 3: Compare numerators. ($\frac{6}{16} < \frac{9}{16}$) so $\frac{3}{8} < \frac{9}{16}$

B. Order the fractions $\frac{1}{2}$, $\frac{9}{14}$, $\frac{3}{4}$, and $\frac{5}{7}$ from least to greatest.

- Step 1: Find the LCD. (The LCM of 2, 14, 4, and 7 is 28.)
- Step 2: Rewrite the fractions.

$\frac{1}{2} = \frac{14}{28}$ $\frac{9}{14} = \frac{18}{28}$ $\frac{3}{4} = \frac{21}{28}$ $\frac{5}{7} = \frac{20}{28}$

- Step 3: Compare numerators. (14 < 18 < 20 < 21) so $\frac{1}{2} < \frac{9}{14} < \frac{5}{7} < \frac{3}{4}$

Try THESE

1. Compare $\frac{4}{7}$ and $\frac{9}{14}$ using <, >, or = .
2. Order the fractions $\frac{1}{2}$, $\frac{5}{6}$, $\frac{2}{3}$, and $\frac{3}{5}$ from least to greatest.

Exercises

Find a common denominator for each pair of fractions.

1. $\frac{1}{2}$, $\frac{3}{4}$
2. $\frac{3}{4}$, $\frac{5}{6}$
3. $\frac{2}{3}$, $\frac{3}{9}$
4. $\frac{2}{9}$, $\frac{5}{6}$

Compare using <, >, or = .

5. $\frac{11}{12} \blacksquare \frac{5}{6}$
6. $\frac{6}{9} \blacksquare \frac{2}{3}$
7. $\frac{1}{10} \blacksquare \frac{2}{5}$
8. $\frac{4}{14} \blacksquare \frac{2}{7}$
9. $\frac{4}{15} \blacksquare \frac{1}{3}$
10. $\frac{8}{18} \blacksquare \frac{4}{9}$
11. $\frac{1}{2} \blacksquare \frac{2}{3}$
12. $\frac{3}{5} \blacksquare \frac{2}{3}$
13. $\frac{3}{4} \blacksquare \frac{3}{5}$
14. $\frac{1}{8} \blacksquare \frac{1}{12}$
15. $\frac{5}{4} \blacksquare \frac{7}{6}$
16. $\frac{5}{6} \blacksquare \frac{9}{10}$
17. $\frac{5}{8} \blacksquare \frac{10}{16}$
18. $\frac{4}{7} \blacksquare \frac{3}{5}$
19. $\frac{4}{3} \blacksquare \frac{5}{4}$
20. $\frac{20}{12} \blacksquare \frac{3}{5}$

Order the fractions from least to greatest.

21. $\frac{5}{8}, \frac{3}{8}, \frac{4}{8}, \frac{7}{8}$
22. $\frac{1}{3}, \frac{5}{6}, \frac{1}{2}, \frac{2}{3}$
23. $\frac{1}{2}, \frac{5}{12}, \frac{3}{4}, \frac{2}{3}$
24. $\frac{5}{8}, \frac{1}{6}, \frac{7}{8}, \frac{3}{4}$

Problem SOLVING

25. Which is smaller, three-fifths or three-twentieths?

26. There are two ways to travel home. One way will take $\frac{2}{3}$ of an hour, and the other way will take $\frac{8}{15}$ of an hour. Which way is quicker?

27. Marissa is building a deck and needs the longest wood screw to finish the job. The screws are the following lengths: $3\frac{1}{2}$ in., $3\frac{9}{16}$ in., and $3\frac{17}{32}$ in. Which screw should she use?

★28. A fraction is in simplest form. Its numerator and denominator have a difference of 2. The sum of the numerator and denominator is 12. What is the fraction?

Constructed RESPONSE

29. Describe under what circumstances it is necessary to find the LCD before comparing fractions. In your explanation, name some fractions that are difficult to compare without finding the LCD.

Cumulative Review

Write in words.

1. 3,012
2. 120,082
3. 0.6
4. 9.003

Estimate.

5. $\begin{array}{r} 8{,}299 \\ +\ 6{,}612 \\ \hline \end{array}$

6. $\begin{array}{r} 237.18 \\ +\ 805.73 \\ \hline \end{array}$

7. $\begin{array}{r} 7{,}641.8 \\ -\ \ 547.3 \\ \hline \end{array}$

8. $\begin{array}{r} 668.7 \\ \times\ 29.5 \\ \hline \end{array}$

9. $84\overline{)3{,}192}$

Compute.

10. $\begin{array}{r} 36 \\ +\ 53 \\ \hline \end{array}$

11. $\begin{array}{r} 67.24 \\ +\ \ 8.75 \\ \hline \end{array}$

12. $\begin{array}{r} 36.406 \\ -\ \ 2.218 \\ \hline \end{array}$

13. $\begin{array}{r} \$20{,}055 \\ -\ \ \ 7{,}882 \\ \hline \end{array}$

14. $\begin{array}{r} 81.8 \\ \times\ 3.2 \\ \hline \end{array}$

15. $\begin{array}{r} 706 \\ \times\ 315 \\ \hline \end{array}$

16. $5.3\overline{)23.32}$

17. $48\overline{)1{,}152}$

18. $2.3 + 17 + 4.06$

19. $84 - 17.003$

20. $\$7.83 + \$4.66 + \$0.25$

21. $734.44 \div 60.2$

22. 137.3×7.3

23. $65{,}125 \div 521$

Use factor trees or division to write the prime factorization. Write using exponents where possible.

24. 30
25. 56
26. 81
27. 175

Solve.

28. Colleen buys 4 pairs of jeans. She gives the clerk $100 and gets back $16. What is the average cost of each pair of jeans? Which method of computation did you use?

29. Eduardo needs to make at least $150 from his part-time job to buy his own bicycle. If he works 10 hours a week, how many weeks will it take him if he earns $2 per hour?

30. Lettuce costs $0.89 a head and tomatoes cost $0.99 a pound. Find the cost of 2 heads of lettuce and 3 pounds of tomatoes.

31. There are 128 sixth graders at Janis Middle School. How many more students would bring the total to 6 classes of 25 each?

32. Shelley collects aluminum cans for recycling. Find the average number of pounds she collects each day.

Day	Pounds
Mon.	23
Tues.	35
Wed.	58
Thurs.	47

4.11 Writing Decimals as Fractions

Objective: to write decimals as fractions and mixed numbers in simplest form

Decimals and fractions can often be used to represent the same number. Decimals like 0.34, 0.15, 0.009, and 0.4 can be written as fractions with denominators of 10, 100, 1000, and so on.

To write a decimal as a fraction, you can follow these steps:

- Step 1: Identify the place value of the last (right) decimal place.
- Step 2: Write the decimal as a fraction using the last place value as the denominator.
- Step 3: Simplify the fraction if necessary.

Examples

A. Write 0.2 as a fraction.

1000	100	10	1	0.1	0.01	0.001	0.0001
thousands	hundreds	tens	ones	tenths	hundredths	thousandths	ten-thousandths
0	0	0	0 .	2	0	0	0

The place-value chart shows that the last decimal place is tenths. This means 0.2 is 2 tenths, or $\frac{2}{10}$.

Simplify. $\frac{2}{10} \div \frac{2}{2} = \frac{1}{5}$

In simplest form, 0.2 is written as $\frac{1}{5}$.

B. Write 9.25 as a fraction.

1000	100	10	1	0.1	0.01	0.001	0.0001
thousands	hundreds	tens	ones	tenths	hundredths	thousandths	ten-thousandths
0	0	0	9 .	2	5	0	0

$9.25 = 9\frac{25}{100}$

$\frac{25}{100} \div \frac{25}{25} = \frac{1}{4}$

In simplest form, 9.25 is written as $9\frac{1}{4}$.

Try THESE

Express each decimal as a fraction or mixed number in simplest form.

1. 0.3
2. 0.45
3. 1.5
4. 0.375

Exercises

Express each decimal as a fraction or mixed number in simplest form.

1. 0.7
2. 0.06
3. 0.2
4. 0.8
5. 0.9
6. 0.01
7. 0.63
8. 5.25
9. 7.2
10. 0.015
11. 2.625
12. 6.44
13. 7.08
14. 50.506
15. 65.342

Problem SOLVING

16. Stock market prices used to be written as fractions and mixed numbers. One share of BooksRUs stock is listed at 32.125 points. Write 32.125 as a mixed number.

17. A baseball player has a batting average of 0.320. Write 0.320 as a fraction.

18. If 0.35 of the people surveyed use a certain toothpaste, what fraction of the people used this toothpaste?

19. There are 100 students in the sixth grade. Seventy of them went to a basketball game. Ben said that 0.70 sixth graders went to the game. Ashley said $\frac{7}{10}$ went to the game. Who is correct? Explain your answer.

20. Derrick bought 25.5 feet of rope. He used 15.25 feet to tie his bike to the bed of his pickup truck. Write the remaining rope as a mixed number.

21. Five boys in a homeroom were making pinewood derby cars. The weights of the cars in ounces were as follows: 4.65, 4.75, 4.8, 4.95, and 5.0. Find the total weight of the cars, and then write the decimal as a mixed number.

Constructed RESPONSE

22. How does saying or writing the decimal in words help you to write the decimal as a fraction? For example, 0.23 is twenty-three hundredths.

4.12 Writing Fractions as Decimals

Objective: to write fractions and mixed numbers as decimals

Decimals and fractions can often be used to represent the same number. Have you ever watched a baseball game? A baseball player's batting average can be written as a decimal or a fraction. Cal Ripken Jr.'s best batting average was the year he had 113 hits in 332 at bats. His batting average that year was 0.340.

To express a fraction or mixed number as a decimal, divide the numerator by the denominator.

Examples

A. Write $\frac{11}{20}$ as a decimal.

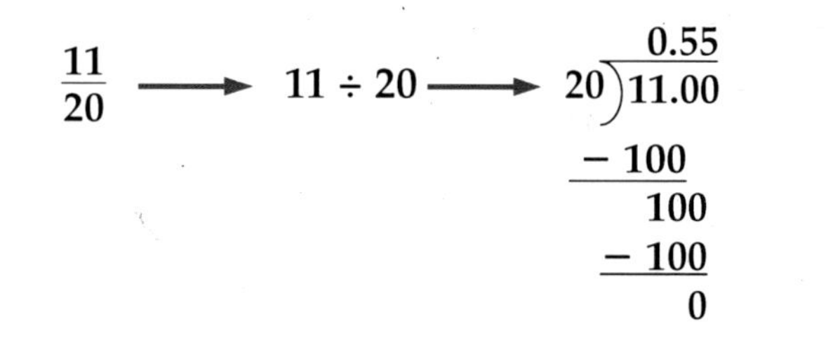

B. Write $1\frac{1}{2}$ as a decimal.

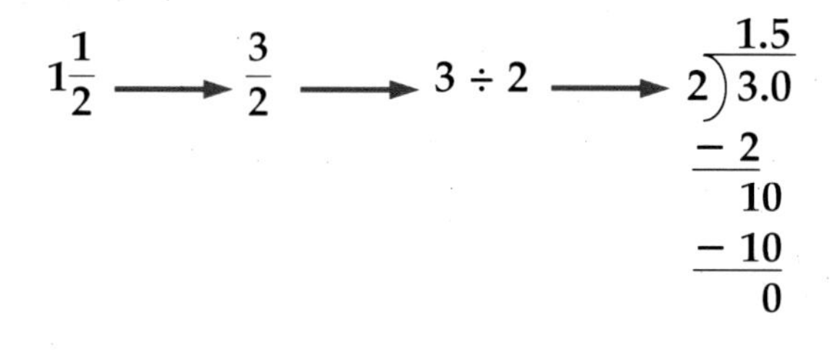

C. Write $\frac{1}{8}$ as a decimal.

$$\begin{array}{r} 0.125 \\ 8\overline{)1.000} \\ -8 \\ \hline 20 \\ -16 \\ \hline 40 \\ -40 \\ \hline 0 \end{array}$$

$\frac{1}{8} = 0.125$

D. Write $3\frac{3}{4}$ as a decimal.

$$\begin{array}{r} 3.75 \\ 4\overline{)15.00} \\ -12 \\ \hline 3\,0 \\ -2\,8 \\ \hline 20 \\ -20 \\ \hline 0 \end{array}$$

How do you rename $3\frac{3}{4}$ as $\frac{15}{4}$?

$3\frac{3}{4} = 3.75$

Try THESE

Express each fraction or mixed number as a decimal.

1. $\frac{1}{10}$ 2. $\frac{2}{5}$ 3. $4\frac{5}{8}$ 4. $2\frac{7}{10}$

Exercises

Express each fraction or mixed number as a decimal.

1. $\frac{9}{10}$ 2. $\frac{31}{100}$ 3. $\frac{1}{4}$ 4. $\frac{3}{8}$ 5. $\frac{3}{20}$

6. $\frac{3}{4}$ 7. $\frac{42}{100}$ 8. $\frac{1}{20}$ 9. $\frac{31}{50}$ 10. $5\frac{3}{5}$

11. $9\frac{3}{10}$ 12. $5\frac{1}{5}$ 13. $11\frac{5}{8}$ 14. $6\frac{7}{8}$ 15. $\frac{3}{2}$

Problem SOLVING

The table shows batting averages for two seasons. Use the table for questions 16–17.

16. Which players improved their batting averages in season 2?

17. Which player had the highest batting average in either season?

Player	Season 1	Season 2
Jen	0.380	$\frac{3}{8}$
Max	0.271	$\frac{1}{4}$
Sally	0.290	$\frac{7}{20}$
Mark	0.420	$\frac{2}{5}$

Constructed RESPONSE

18. The decimal equal to $\frac{1}{25}$ is 0.04. The decimal equal to $\frac{2}{25}$ is 0.08. Without dividing, find the decimal equal to $\frac{6}{25}$. Explain how you found your answer.

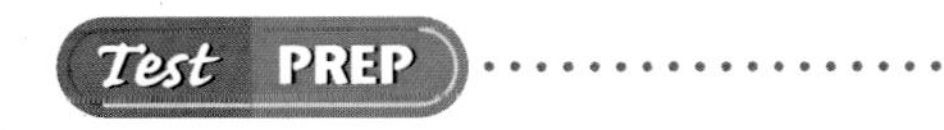

19. It rained $\frac{5}{16}$ of an inch overnight. Which decimal represents $\frac{5}{16}$?

a. $0.\overline{3}$ b. 0.325 c. 0.3125 d. $0.\overline{3125}$

4.13 Compare and Order Mixed Fractions and Decimals

Objective: to compare and order mixed fractions and decimals

Sometimes we need to compare measurements, but some are written as fractions and others are written as decimals. You have a 2-liter bottle, but you need to make $\frac{1}{2}$ gallon of sports drink. Is the 2-liter bottle large enough to hold a $\frac{1}{2}$ gallon?

You find out that 1 liter is about 0.26 gallon, so 2 liters is about 0.52 gallon. How can we compare the decimal 0.52 with the fraction $\frac{1}{2}$?

To compare mixed fractions and decimals, you can follow these steps:

- Step 1: Change the fractions to decimals.
- Step 1: Then compare the decimals.

So, since $\frac{1}{2} = 0.5$, and $0.5 < 0.52$, we know that $\frac{1}{2} < 0.52$. A 2-liter bottle is large enough to hold a $\frac{1}{2}$ gallon.

Example

A. Compare $\frac{1}{8}$ and 0.132 using <, >, or = .

- Step 1: Change $\frac{1}{8}$ to a decimal. $\quad \frac{1}{8} \longrightarrow 1 \div 8 = 0.125$
- Step 2: Compare the decimals.

0.125

0.132

$\frac{1}{8} < 0.132$

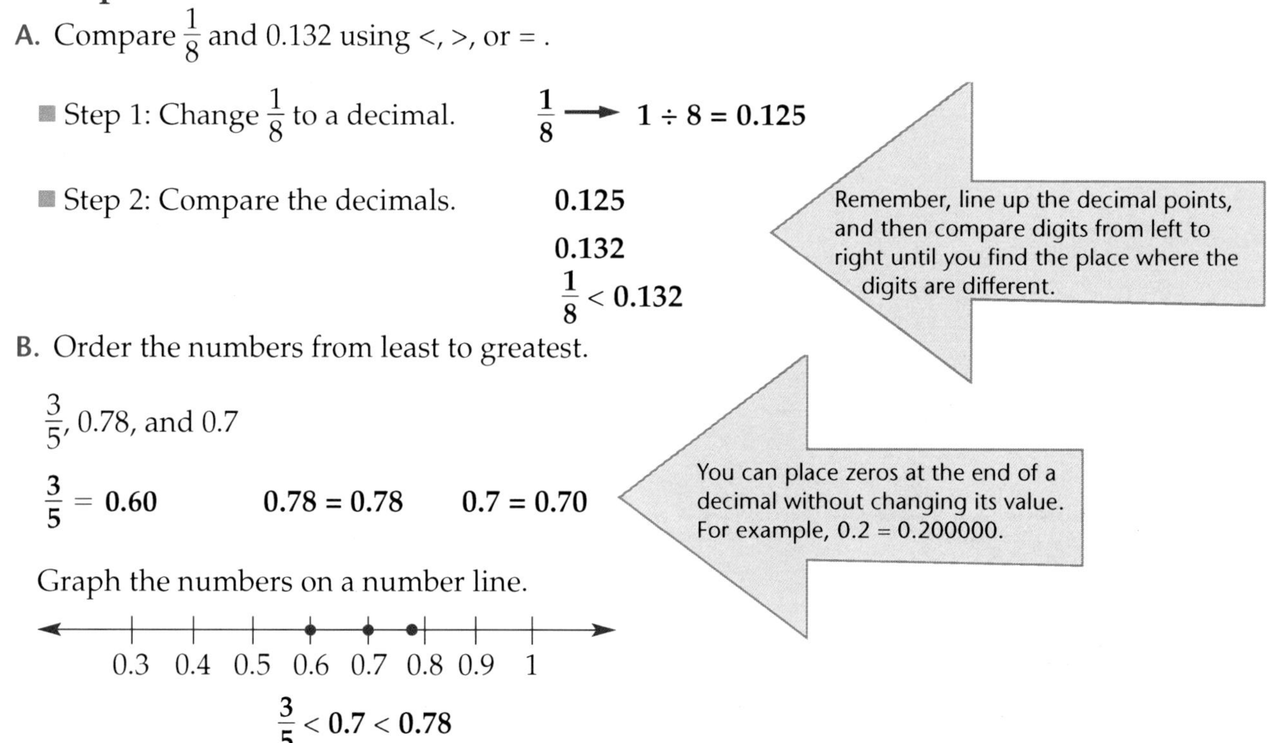

B. Order the numbers from least to greatest.

$\frac{3}{5}$, 0.78, and 0.7

$\frac{3}{5} = 0.60 \qquad 0.78 = 0.78 \qquad 0.7 = 0.70$

Graph the numbers on a number line.

0.3 0.4 0.5 0.6 0.7 0.8 0.9 1

$\frac{3}{5} < 0.7 < 0.78$

Try THESE

Express each fraction as a decimal.

1. $\frac{3}{20}$
2. $\frac{7}{8}$
3. $\frac{16}{25}$

Compare the fractions and decimals using <, >, or = .

4. $0.2 \blacksquare \frac{1}{4}$
5. $0.628 \blacksquare \frac{5}{8}$
6. $\frac{13}{5} \blacksquare 2.5$

Order the numbers from least to greatest.

7. $\frac{1}{2}$, 0.4, and $\frac{1}{4}$

Exercises

Compare the fractions using <, >, or = .

1. $\frac{2}{3} \blacksquare \frac{5}{6}$
2. $\frac{2}{5} \blacksquare \frac{4}{7}$
3. $\frac{7}{10} \blacksquare \frac{3}{5}$
4. $\frac{5}{14} \blacksquare \frac{3}{8}$

Compare the decimals using <, >, or = .

5. $3.8 \blacksquare 3.79$
6. $0.088 \blacksquare 0.1$
7. $1.82 \blacksquare 1.777$
8. $2.08 \blacksquare 2.008$

Compare the fractions and decimals using <, >, or = .

9. $\frac{6}{11} \blacksquare 0.6$
10. $0.902 \blacksquare \frac{9}{10}$
11. $\frac{1}{2} \blacksquare 0.495$
12. $0.35 \blacksquare \frac{7}{20}$
13. $2\frac{1}{5} \blacksquare 2.\overline{2}$
14. $6.02 \blacksquare 6\frac{1}{10}$
15. $2.25 \blacksquare 2\frac{1}{3}$
16. $5\frac{3}{7} \blacksquare 5.428$

Order the numbers from least to greatest.

17. $\frac{1}{2}$, 0.4, $\frac{1}{4}$
18. 0.25, 0.4, $\frac{1}{5}$
19. 0.025, 0.63, $\frac{1}{8}$, $\frac{7}{8}$
20. $\frac{5}{6}$, $\frac{4}{5}$, 0.83
21. $2\frac{5}{10}$, $2\frac{3}{5}$, 2.7
22. 0.5, 0.05, $\frac{1}{25}$, $\frac{2}{33}$

Problem SOLVING

23. Gold purity is measured in karats. 24-karat gold is considered pure. Which has a higher purity, 18-karat gold or 0.8 pure gold?

24. Samantha did a project on calculating the use of water in her home. She found that $\frac{1}{8}$ of the water was used in the kitchen, and 0.15 was used outside. Was more water used in the kitchen or outside?

25. The four most popular movies of the summer of 2005 were the following lengths: 2 hr 26 min, 117 min, $2\frac{1}{4}$ hr, and 2.05 hr. Write the times in order from least to greatest.

4.14 Problem-Solving Strategy: Choose a Strategy

Objective: to solve a problem by choosing a strategy

Samantha bakes cookies in a bakery. Oatmeal cookies take 15 minutes to bake and raisin cookies take 10 minutes to bake. If the timers for both batches of cookies go off at noon, what are the next two times they will go off together?

The timer for the oatmeal cookies goes off every 15 minutes. The timer for the raisin cookies goes off every 10 minutes.

Make an organized list of the times when both timers will go off.

	12:00	12:15	12:30	12:45	1:00
Oatmeal					
Raisin					
	12:00	12:10	12:20	12:30	12:40	12:50	1:00

The two timers will go off together at 12:30 and at 1:00.

The two timers go off together every 30 minutes. 30 is the LCM of 10 and 15.

Solve.

1. Lena needs 17 yards of fabric to make cheerleader uniforms. The fabric is sold in 3-yard pieces or 5-yard pieces. A 3-yard piece costs $8.97, and a 5-yard piece costs $12.95. What should Lena buy to get the lowest cost and the least waste?

2. Antonio has a white shirt, a blue shirt, a pair of brown pants, a pair of blue jeans, a tweed sports jacket, and a jean jacket. How many different outfits can he make if he always wears a shirt, pants, and a jacket?

Solve

1. Lenora works 35 hours each week. If she is paid $5.50 per hour, how much is she paid after 4 weeks and after 6 weeks?

★2. Complete the following sequences:

   a. O, T, T, F, F, S, S, __?__, __?__

   b. 0, 1, 10, 11, 100, 101, __?__, __?__

3. There are 30 students in Ms. Stovall's classroom. Do two of her students *have* to have the same first initial? Why?

4. Phil is a cashier in a store. His cash register has $5 in quarters, $3 in dimes, $4 in nickels, and $1 in pennies. How many coins are there in all?

5. Anna bought 42 yards of fencing to go around her yard. If the length of her yard is twice the width, what is the length and width of her yard?

6. Using three 9s, arrange them so they form 10. You may add, subtract, multiply, or divide.

7. A belt and a vest cost $30. The vest costs twice as much as the belt. How much does the belt cost?

8. The numbers 1, 3, 6, and 10 are called triangular numbers because of the following pattern.

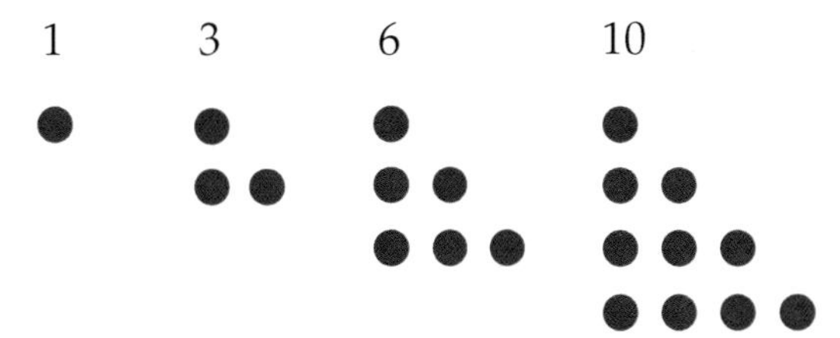

   a. What number can you add to 10 to get the next triangular number?

   b. What are the next three triangular numbers?

9. A baseball team has six pitchers, A through F, and three catchers, X, Y, and Z. From how many pitcher-catcher combinations can the coach choose?

Constructed RESPONSE

10. You are planning to save for a new video game system, which costs around $500. Your plan is to save $2 the first week, $4 the second week, $8 the third week, $16 the fourth week, and so on.

   a. How long will it take you to save enough money for the game system?

   b. How could you predict the amount that you will save each week using exponents?

Chapter 4 Review

Language and Concepts

Choose a word from the list at the right that best completes each sentence.

1. Fractions that name the same number are called __?__.
2. A __?__ number has more than two factors.
3. __?__ have a numerator that is greater than or equal to the denominator.
4. A fraction is in __?__ when the only common factor of the numerator and denominator is 1.
5. A __?__ has exactly two factors, 1 and the number itself.

common factor
common multiple
composite
divisible
equivalent fractions
even
improper fractions
prime number
simplest form

Skills and Problem Solving

Tell whether each number is divisible by 2, 3, 4, 5, 6, 9, or 10. Then classify the number as *even* or *odd*. (Section 4.1)

6. 14
7. 756
8. 420
9. 2,130

Find the prime factorization of each number. Use exponents whenever possible. (Section 4.3)

10. 16
11. 63
12. 70
13. 120

Find the greatest common factor (GCF) for each set of numbers. (Section 4.5)

14. 8, 10
15. 12, 16
16. 3, 4, 8
17. 4, 10, 15

Write each fraction in simplest form. If the fraction is already in simplest form, write *simplest form*. (Section 4.7)

18. $\frac{2}{10}$
19. $\frac{3}{12}$
20. $\frac{20}{80}$
21. $\frac{3}{11}$
22. $\frac{5}{10}$
23. $\frac{5}{25}$

Express each improper fraction as a mixed number in simplest form. (Section 4.8)

24. $\frac{5}{2}$
25. $\frac{18}{4}$
26. $\frac{29}{5}$
27. $\frac{48}{8}$
28. $\frac{56}{9}$
29. $\frac{44}{6}$

Find the least common multiple (LCM) of each group of numbers. Use any method. (Section 4.9)

30. 4 and 9 **31.** 7 and 14 **32.** 12 and 18

Compare using <, >, or = . (Section 4.10)

33. $\frac{2}{3} \bullet \frac{1}{3}$ **34.** $\frac{3}{10} \bullet \frac{9}{30}$ **35.** $\frac{9}{20} \bullet \frac{2}{5}$ **36.** $\frac{2}{6} \bullet \frac{6}{18}$

Express each fraction or mixed number as a decimal. (Section 4.12)

37. $\frac{8}{10}$ **38.** $\frac{7}{8}$ **39.** $\frac{9}{20}$ **40.** $2\frac{1}{4}$ **41.** $\frac{5}{2}$

Express each decimal as a fraction or mixed number in simplest form. (Section 4.11)

42. 0.7 **43.** 0.75 **44.** 0.14 **45.** 9.5

Compare using <, >, or = . (Section 4.13)

46. $0.25 \bullet \frac{2}{10}$ **47.** $\frac{3}{10} \bullet 0.31$ **48.** $\frac{1}{5} \bullet 0.2$ **49.** $\frac{3}{5} \bullet 0.35$

Use a Venn diagram to solve. (Section 4.4)

50. Sixth grade students were asked which magazine they read. A total of 29 students read *Jets*, a total of 31 read *Wheels*, and total of 34 read *Youth*. 9 read all three, 6 read only *Jets* and *Wheels*, 10 read only *Jets* and *Youth*, and 8 read only *Wheels* and *Youth*.

a. How many students were asked?

b. How many students read only *Jets*?

Make a list to answer the following problem. (Section 4.14)

51. Jeanne walks 5 miles every 3 days. Joan walks 3 miles every 2 days. If they begin walking the same day, who has walked farther

a. after 6 days?

b. after 2 weeks?

In Your Own Words

52. Explain how to find a common denominator of two fractions.

Chapter 4 Test

Tell whether each number is divisible by 2, 3, 4, 5, 6, 9, or 10. Then classify the number as *even* or *odd*.

1. 50 **2.** 51 **3.** 78 **4.** 232

Find the prime factorization of each number. Use exponents whenever possible.

5. 18 **6.** 30 **7.** 24 **8.** 56

Find the greatest common factor (GCF) and least common multiple (LCM) for each set of numbers.

9. 3, 12 **10.** 12, 15 **11.** 5, 7, 10 **12.** 8, 20, 25

Express each mixed number as a fraction. Express each improper fraction as a mixed number in simplest form.

13. $5\frac{3}{4}$ **14.** $\frac{9}{4}$ **15.** $2\frac{1}{5}$ **16.** $\frac{20}{3}$ **17.** $\frac{100}{8}$ **18.** $7\frac{5}{6}$

Compare using <, >, or = .

19. $\frac{2}{5} \bullet \frac{1}{5}$ **20.** $\frac{2}{5} \bullet \frac{5}{10}$ **21.** $\frac{3}{4} \bullet \frac{24}{32}$ **22.** $\frac{2}{7} \bullet \frac{4}{11}$

Express each fraction or mixed number as a decimal.

23. $\frac{3}{10}$ **24.** $3\frac{1}{4}$ **25.** $\frac{8}{25}$ **26.** $2\frac{1}{8}$ **27.** $\frac{7}{4}$

Compare using <, >, or = .

28. $0.2 \bullet \frac{1}{5}$ **29.** $\frac{7}{10} \bullet 0.07$ **30.** $\frac{5}{6} \bullet 0.8$ **31.** $\frac{2}{5} \bullet 0.25$

Solve.

32. At Rye High School, 32 freshmen take drafting, 28 take cooking, and 7 take both courses. If there are 200 freshmen at Rye High School, how many freshmen take neither drafting nor cooking?

33. Steven has $10 and saves $2 each week. His friend Jonathan saves $4 each week.

a. Make a list to show the amount of money Steven and Jonathan will have saved for 10 weeks.

b. After how many weeks will Jonathan and Steven have the same amount of money saved?

Change of Pace

Terminating and Repeating Decimals

The fraction $\frac{1}{4}$ can be written as the decimal 0.25. The decimal 0.25 is called a **terminating decimal** since the division ends or terminates when you get a remainder of 0.

$\frac{1}{4} \rightarrow 1 \div 4 = 0.25$

The fraction $\frac{2}{11}$ can be written as a decimal as shown at the right. Notice that the division does not terminate. Instead, the digits in the quotient repeat. This decimal is called a **repeating decimal.**

$$\frac{2}{11} \rightarrow \begin{array}{r} 0.1818 \\ 11\overline{)2.0000} \\ -1\,1 \\ \hline 90 \\ -88 \\ \hline 20 \\ -11 \\ \hline 90 \\ -88 \\ \hline 2 \end{array}$$

The remainders are never 0. They are either 9 or 2.

A repeating decimal is shown by a bar placed over the digits that repeat.

$\frac{2}{11} = 0.1818\ldots = 0.\overline{18}$

The digits 1 and 8 repeat.

Write the digits that repeat for each.

1. $0.\overline{8}$
2. $3.\overline{25}$
3. $4.08\overline{3}$
4. $0.\overline{142857}$

Write using bar notation.

5. 0.3333 . . .
6. 4.0555 . . .
7. 8.3636 . . .

Divide and write each fraction or mixed number as a terminating or repeating decimal.

8. $\frac{2}{5}$
9. $\frac{3}{8}$
10. $\frac{9}{10}$
11. $\frac{4}{9}$
12. $\frac{1}{3}$
13. $1\frac{11}{20}$
14. $8\frac{21}{25}$
15. $7\frac{3}{10}$
16. $4\frac{11}{15}$
17. $2\frac{8}{9}$
18. $1\frac{7}{8}$
19. $2\frac{8}{25}$
20. $2\frac{13}{25}$
21. $4\frac{21}{25}$
22. $1\frac{5}{7}$

Solve.

23. Write $\frac{1}{7}, \frac{2}{7}, \frac{3}{7}, \frac{4}{7}, \frac{5}{7}$, and $\frac{6}{7}$ as repeating decimals. What is the pattern of the digits that repeat?

Cumulative Test

1. Which number is *not* prime?
   a. 5
   b. 39
   c. 17
   d. 23

2. Which number is a prime number?
   a. 14
   b. 24
   c. 51
   d. none of the above

3. Which fraction is equivalent to $\frac{3}{4}$?
   a. $\frac{6}{8}$
   b. $\frac{15}{20}$
   c. both **a** and **b**
   d. neither **a** nor **b**

4. Express $\frac{9}{8}$ as a mixed number.
   a. $1\frac{1}{8}$
   b. $1\frac{8}{9}$
   c. $1\frac{17}{8}$
   d. none of the above

5. 216 is divisible by ____.
   a. 2
   b. 3
   c. both **a** and **b**
   d. neither **a** nor **b**

6. Which number is a common multiple for 3 and 7?
   a. 20
   b. 29
   c. 203
   d. 609

7. Adrian received a check for five thousand, five dollars. Write the amount in standard form.
   a. $5,005
   b. $5,050
   c. $5,500
   d. none of the above

8. The number named by the last four digits of Rosita's social security number is divisible by 5 and 3. What is the number?
   a. 2795
   b. 2950
   c. 2970
   d. 2975

9. Pedro's normal temperature is 98.6°F. While he had the flu, his temperature was 102.7°F. How much higher than normal was his temperature?
   a. 3.1°F
   b. 4.1°F
   c. 4.3°F
   d. 5.3°F

10. Bananas cost $0.39 a pound and strawberries cost $1.50 a carton. Find the cost of 5 pounds of bananas and 3 cartons of strawberries.
    a. $1.95
    b. $6.45
    c. $4.50
    d. $2.50

CHAPTER

Adding and Subtracting Fractions

Jessica Tate
Louisiana

5.1 Rounding Fractions and Mixed Numbers

Objective: to round fractions to 0, $\frac{1}{2}$, and 1

You and your friends decide to take a field trip and walk the trails at a bird sanctuary. One trail is $5\frac{1}{10}$ miles, and a second trail is $3\frac{3}{4}$ miles. *About* how far is each of the two trails?

Rounding fractions is similar to rounding decimals. Some fractions are small and close to 0, other fractions are larger and close to one, and some fractions are closer to $\frac{1}{2}$.

Tips

Round up	When the numerator is close to the denominator, round up to the next nearest whole number. Example: $\frac{5}{6}$ rounds up to 1.	5 is almost as large as 6.
Round to $\frac{1}{2}$	When the numerator is about half of the denominator, round the fraction to $\frac{1}{2}$. Example: $2\frac{3}{8}$ rounds up to $2\frac{1}{2}$.	3 is about half of 8.
Round to 0	When the numerator is much smaller than the denominator, round the number down to the whole number. Example: $\frac{1}{8}$ rounds to 0.	1 is much smaller than 8.

Examples

Round the following fractions to 0, $\frac{1}{2}$, or 1.

A. $\frac{1}{6}$

0 $\frac{1}{6}$ $\frac{2}{6}$ $\frac{3}{6}$ $\frac{4}{6}$ $\frac{5}{6}$ 1

$\frac{1}{3}$ $\frac{1}{2}$ $\frac{2}{3}$

$\frac{1}{6}$ is closer to zero.

The numerator of $\frac{1}{6}$ is much smaller than the denominator. $\frac{1}{6}$ rounds to 0.

B. $\frac{5}{12}$

The numerator of $\frac{5}{12}$ is close to half of the denominator. $\frac{5}{12}$ rounds to $\frac{1}{2}$.

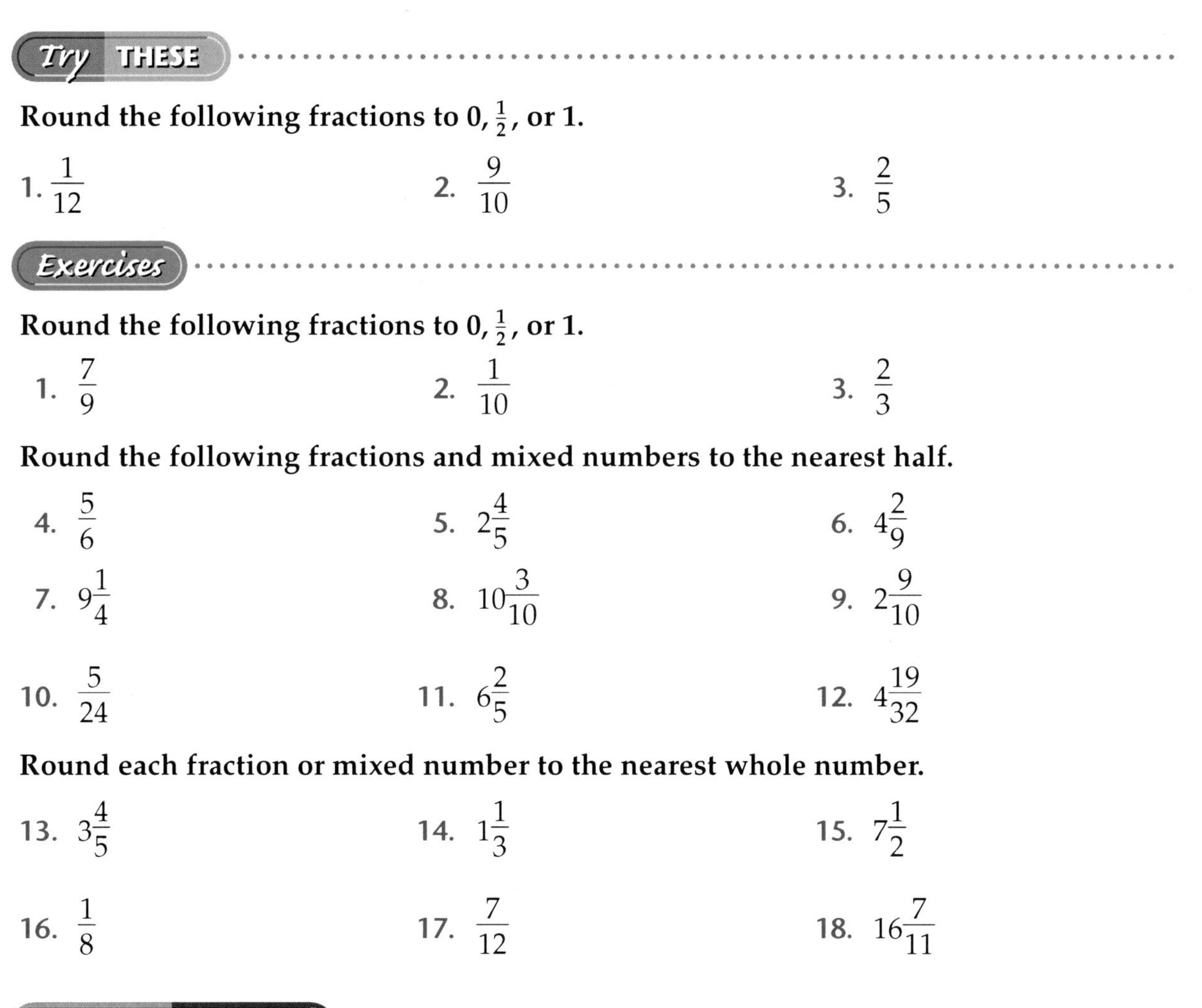

Try THESE

Round the following fractions to 0, $\frac{1}{2}$, or 1.

1. $\frac{1}{12}$ 2. $\frac{9}{10}$ 3. $\frac{2}{5}$

Exercises

Round the following fractions to 0, $\frac{1}{2}$, or 1.

1. $\frac{7}{9}$ 2. $\frac{1}{10}$ 3. $\frac{2}{3}$

Round the following fractions and mixed numbers to the nearest half.

4. $\frac{5}{6}$ 5. $2\frac{4}{5}$ 6. $4\frac{2}{9}$

7. $9\frac{1}{4}$ 8. $10\frac{3}{10}$ 9. $2\frac{9}{10}$

10. $\frac{5}{24}$ 11. $6\frac{2}{5}$ 12. $4\frac{19}{32}$

Round each fraction or mixed number to the nearest whole number.

13. $3\frac{4}{5}$ 14. $1\frac{1}{3}$ 15. $7\frac{1}{2}$

16. $\frac{1}{8}$ 17. $\frac{7}{12}$ 18. $16\frac{7}{11}$

Problem SOLVING

19. Which of the following numbers does not round to the same number as the others?
 $5\frac{1}{5}$, $4\frac{5}{6}$, $5\frac{6}{7}$, $4\frac{11}{12}$

20. A recipe for tacos calls for $1\frac{1}{4}$ pounds of ground beef. Should you buy a $1\frac{1}{2}$ pound package or a 1 pound package? Explain.

21. Name three mixed numbers that round to $4\frac{1}{2}$.

★22. The map shows a video game store and a home. How many ways are there to get from home to the video game store without backtracking?

★23. On a desk calendar, like the one shown, what are the numbers on the two cubes? Remember, all dates from 01–31 must be possible.

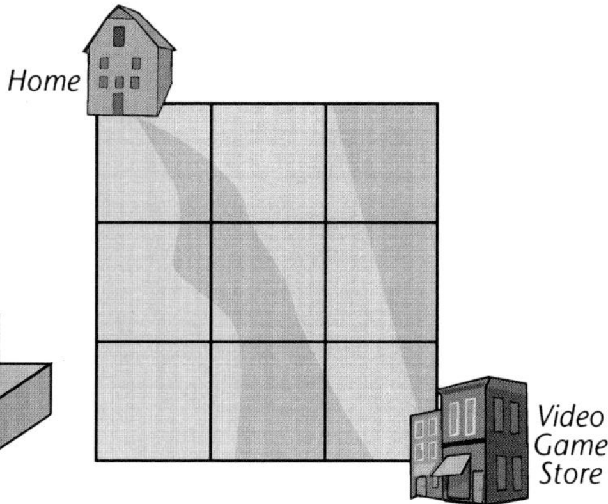

5.2 Estimating Fraction Sums and Differences

Objective: to estimate sums and differences of fractions and mixed numbers

You were thinking about the field trip to a bird sanctuary. One trail is $5\frac{1}{10}$ miles, and a second trail is $3\frac{3}{4}$ miles. *About* how far are the two trails *combined*? Now that you learned how to round fractions, you can estimate the total distance.

The mixed number $5\frac{1}{10}$ is close to 5, and the mixed number $3\frac{3}{4}$ is close to 4. You know this because $\frac{1}{10}$ rounds to 0, and $\frac{3}{4}$ rounds to 1. The estimated total distance is $5 + 4 = 9$ miles.

Examples

A. $\frac{8}{9} + \frac{2}{11}$

$\mathbf{\frac{8}{9} + \frac{2}{11}}$ $\frac{8}{9}$ rounds to 1, and $\frac{2}{11}$ rounds to zero.

$\mathbf{1 + 0 = 1}$

$\frac{8}{9} + \frac{2}{11}$ is about 1.

B. $10\frac{3}{10} + 9\frac{3}{4}$

$\mathbf{10\frac{3}{10} + 9\frac{3}{4}}$ $\frac{3}{10}$ rounds to $\frac{1}{2}$, and $\frac{3}{4}$ rounds to 1.

$\mathbf{10\frac{1}{2} + 10 = 20\frac{1}{2}}$

$10\frac{3}{10} + 9\frac{3}{4}$ is about $20\frac{1}{2}$.

Try THESE

Estimate each sum.

1. $\frac{3}{8} + \frac{5}{6}$

2. $2\frac{1}{8} + 1\frac{5}{8}$

3. $4\frac{5}{9} + \frac{1}{4}$

Exercises

Estimate each sum or difference.

1. $\frac{8}{9} + \frac{1}{6}$
2. $\frac{4}{9} - \frac{1}{12}$
3. $\frac{3}{7} + \frac{1}{12}$
4. $\frac{7}{8} - \frac{2}{5}$
5. $\frac{9}{10} + \frac{3}{5}$
6. $\frac{5}{8} - \frac{1}{3}$
7. $3\frac{4}{5} + 2\frac{1}{10}$
8. $2\frac{5}{6} - 2\frac{1}{2}$
9. $6\frac{3}{7} + 2\frac{2}{9}$
10. $12\frac{1}{6} - 11\frac{1}{5}$
11. $\frac{5}{11} + 2\frac{1}{7}$
12. $4\frac{1}{7} - 3\frac{9}{10}$

Estimate. Compare using < or >. Explain your reasoning.

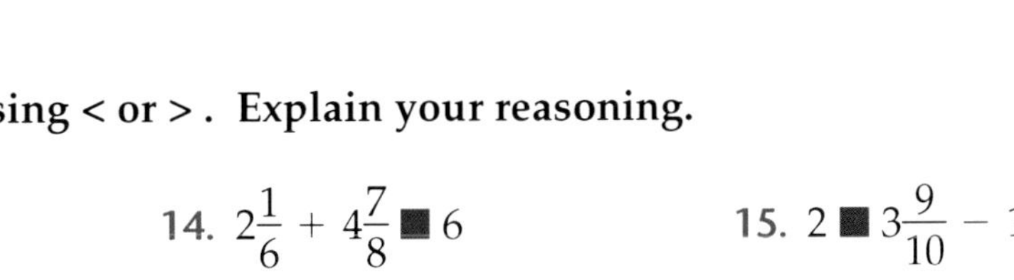

13. $\frac{3}{8} + \frac{9}{10}$ ■ 1
14. $2\frac{1}{6} + 4\frac{7}{8}$ ■ 6
15. 2 ■ $3\frac{9}{10} - 1\frac{1}{5}$

Problem SOLVING

16. Estimate the perimeter of the rectangle on the right.
17. Manuel is making a square frame. He wants the frame to be $6\frac{7}{8}$ inches on each side. About how much framing material should he buy?

$15\frac{7}{8}$ in.

$3\frac{1}{4}$ in.

To the right is a recipe for punch. Use the recipe to answer questions 18–20.

18. About how much more grape juice than orange juice is used in the punch?
19. About how much fruit juice is used in the punch?
20. About how many cups of punch will this recipe make?

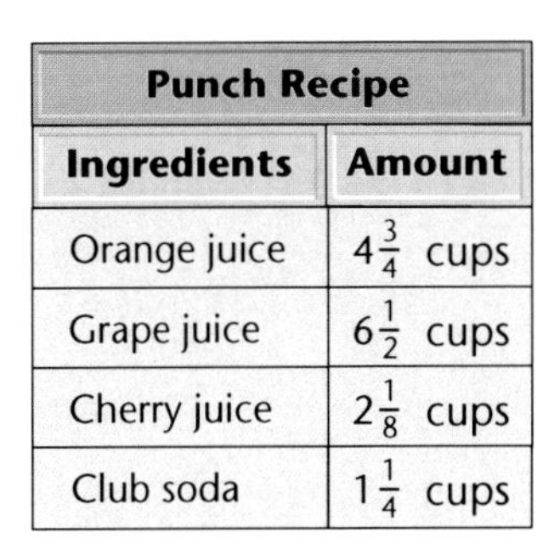

Punch Recipe	
Ingredients	**Amount**
Orange juice	$4\frac{3}{4}$ cups
Grape juice	$6\frac{1}{2}$ cups
Cherry juice	$2\frac{1}{8}$ cups
Club soda	$1\frac{1}{4}$ cups

Constructed RESPONSE

21. Why is it a good idea to estimate before adding or subtracting mixed numbers and fractions? Would it be more helpful to round to the nearest whole number or to the nearest $\frac{1}{2}$? Explain your reasoning.

Objective: to add and subtract fractions with like denominators

Eric baked an apple pie. He cut the pie into 8 equal slices. At lunch, 3 slices of pie were eaten. At dinner, 2 slices were eaten. What fraction of the pie was eaten altogether?

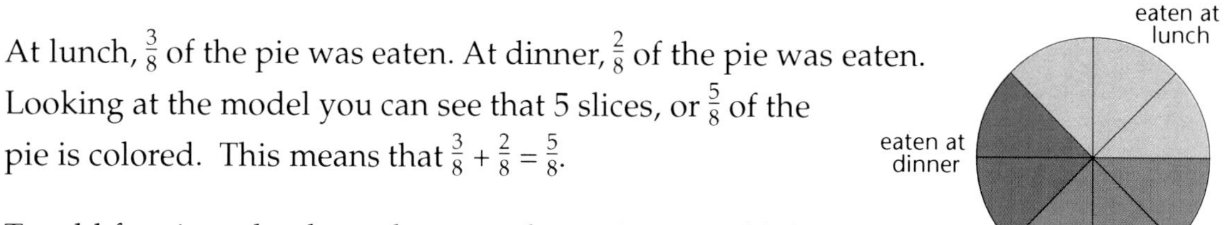

At lunch, $\frac{3}{8}$ of the pie was eaten. At dinner, $\frac{2}{8}$ of the pie was eaten. Looking at the model you can see that 5 slices, or $\frac{5}{8}$ of the pie is colored. This means that $\frac{3}{8} + \frac{2}{8} = \frac{5}{8}$.

To add fractions that have the same denominator, add the numerators and keep the same denominator. Notice that 2 + 3 = 5, and the denominators of the two fractions being added are the same as the denominator of their sum.

To subtract fractions that have the same denominator, subtract the numerators and keep the same denominator.

Examples

A. Find. $\frac{3}{6} - \frac{1}{6}$

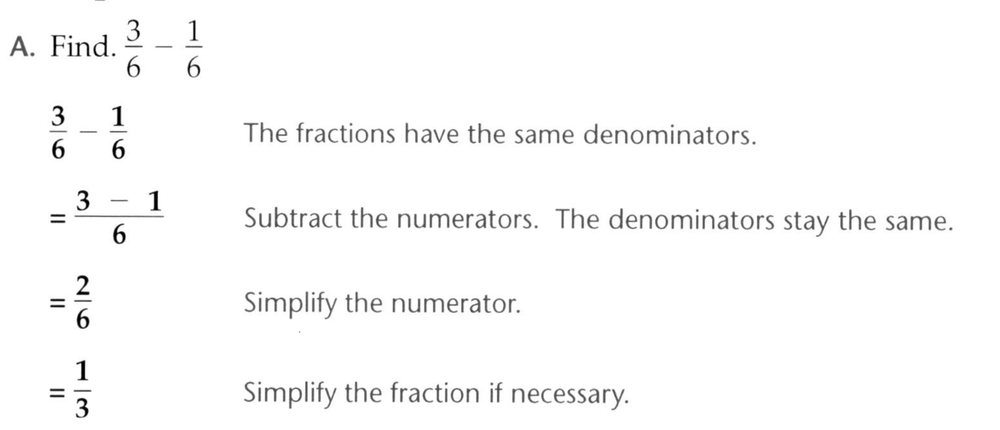

$\frac{3}{6} - \frac{1}{6}$ The fractions have the same denominators.

$= \frac{3 - 1}{6}$ Subtract the numerators. The denominators stay the same.

$= \frac{2}{6}$ Simplify the numerator.

$= \frac{1}{3}$ Simplify the fraction if necessary.

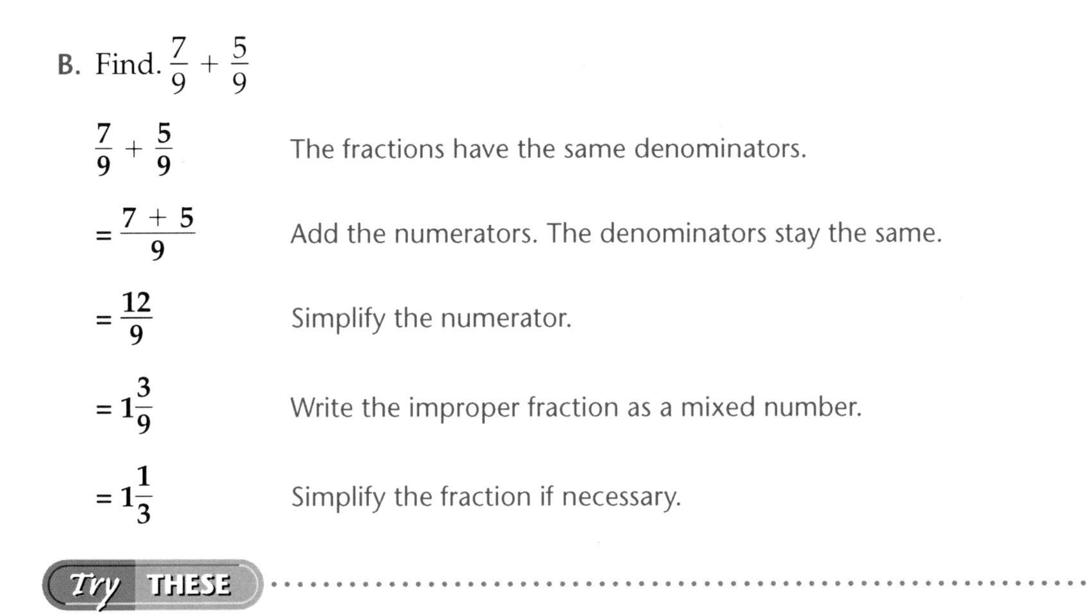

B. Find. $\frac{7}{9} + \frac{5}{9}$

$\frac{7}{9} + \frac{5}{9}$	The fractions have the same denominators.
$= \frac{7 + 5}{9}$	Add the numerators. The denominators stay the same.
$= \frac{12}{9}$	Simplify the numerator.
$= 1\frac{3}{9}$	Write the improper fraction as a mixed number.
$= 1\frac{1}{3}$	Simplify the fraction if necessary.

Try THESE

Write each sum or difference in simplest form.

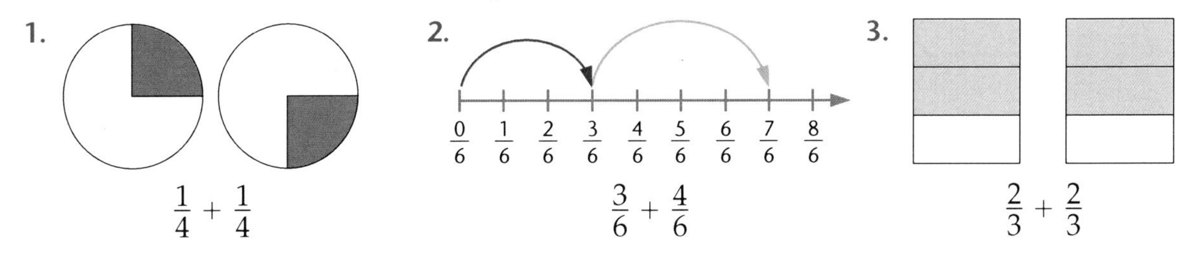

1. $\frac{1}{4} + \frac{1}{4}$

2. $\frac{3}{6} + \frac{4}{6}$

3. $\frac{2}{3} + \frac{2}{3}$

Exercises

Add or subtract. Write each sum or difference in simplest form.

1. $\frac{1}{4} + \frac{1}{4}$
2. $\frac{1}{3} + \frac{1}{3}$
3. $\frac{4}{9} + \frac{1}{9}$
4. $\frac{7}{8} + \frac{3}{8}$
5. $\frac{2}{5} - \frac{1}{5}$
6. $\frac{17}{25} - \frac{12}{25}$
7. $\frac{6}{5} - \frac{3}{5}$
8. $\frac{2}{3} - \frac{1}{3}$
9. $\frac{9}{10} + \frac{3}{10}$
10. $\frac{7}{12} + \frac{3}{12}$
11. $\frac{6}{7} + \frac{3}{7}$
12. $\frac{8}{12} + \frac{5}{12}$
13. $\frac{27}{100} - \frac{7}{100}$
14. $\frac{4}{15} - \frac{1}{15}$
15. $\frac{17}{24} - \frac{5}{24}$
16. $\frac{9}{16} - \frac{3}{16}$

Problem SOLVING

17. How much longer than $\frac{5}{8}$ inch is $\frac{7}{8}$ inch?
18. How much is $\frac{2}{3}$ cup plus $\frac{2}{3}$ cup?

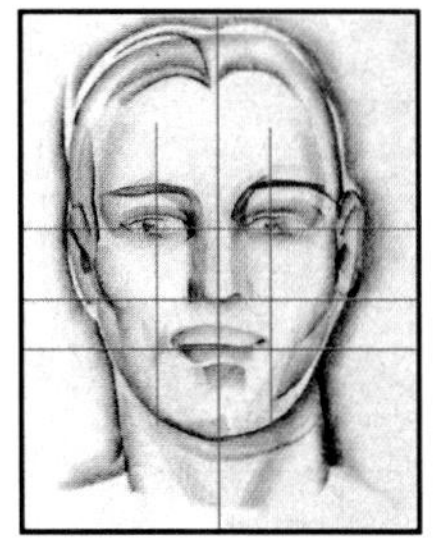

19. When drawing a portrait, the distance from the bottom of the nose to the chin is $\frac{1}{4}$ of the face, and the distance from the eyes to the bottom of the nose is also $\frac{1}{4}$ of the face. What fraction of the head is left from the top of the head to the eyes?
20. Plasma makes up $\frac{11}{20}$ of your blood, and blood cells make up the other $\frac{9}{20}$. How much more of your blood is plasma than blood cells?
21. Miguel needs $\frac{3}{4}$ cup of sugar to make pumpkin pie. His sister needs $\frac{3}{4}$ cup of sugar to make cookies. How much sugar do they need altogether?
22. The rainfall for Monday and Tuesday was $\frac{3}{8}$ inch. By the end of the day Wednesday the total rainfall for the week was $\frac{5}{8}$ inch. How much rain fell on Wednesday?

Constructed RESPONSE

23. Explain how to find the sum of $\frac{4}{7}$ and $\frac{6}{7}$.

Test PREP

24. A cake recipe calls for $\frac{1}{4}$ cup oil, $\frac{3}{4}$ cup water, and $\frac{3}{4}$ cup milk. How much liquid is used in the cake recipe?

    a. 2 c b. $2\frac{1}{4}$ c c. $1\frac{3}{4}$ c d. $3\frac{1}{4}$ c

25. Bill spent $\frac{5}{6}$ hour running and $\frac{2}{6}$ hour swimming. How much time did Bill spend exercising?

    a. $1\frac{1}{6}$ h b. $1\frac{5}{6}$ h c. $2\frac{5}{6}$ h d. 1 h

Mixed REVIEW

Express each fraction or mixed number as a decimal.

26. $\frac{4}{5}$ 27. $5\frac{1}{8}$ 28. $\frac{15}{4}$ 29. $\frac{21}{6}$

Cumulative Review

Round to the underlined place-value position.

1. 4,$\underline{8}$31
2. 13.2$\underline{0}$5
3. 158.8$\underline{9}$7
4. 0.6$\underline{0}$36

Compute. Use R to write each remainder.

5. 48 + 86
6. $1,207 + 958
7. 36.02 + 16.81
8. 4.063 + 11.215
9. 12,674 + 5,409
10. 13.8 − 9.3
11. 5,607 − 848
12. 22.043 − 4.130
13. $10,080 − 4,107
14. 0.8408 − 0.5612
15. 1.2 + 0.335 + 2
16. 14.17 − 0.24
17. $15.21 + $1.89
18. 523 × 7
19. $3,000 × 30
20. 4,215 × 1.2
21. 63.7 × 2.3
22. 42 × 14
23. 3)216
24. 24)316
25. 7.1)397.6
26. 62)7,936
27. 12.2)73.2

Estimate.

28. 9,408 + 8,553
29. 162.89 + 938.02
30. 6,831.8 − 2,440.7
31. 939.1 × 61.9
32. 42)2,720

Find the LCM for each group of numbers.

33. 3, 9
34. 6, 15
35. 7, 10
36. 5, 8, 10

Compare using <, >, or = .

37. $\frac{1}{3}$ ● $\frac{4}{12}$
38. $\frac{5}{6}$ ● $\frac{11}{18}$
39. $\frac{3}{5}$ ● $\frac{2}{8}$

Solve.

40. Frank wants to buy 4 pairs of socks that cost $1.89 each. The sales tax is $0.38. If Frank has $10, can he buy the socks?
41. Tara's new book has 224 pages. She wants to read the book in one week. How many pages does she have to read each day?
42. One serving of punch is 250 mL. Will ten servings fit in a 2 L bowl?
43. Brenda has $636.41 in her checking account. She writes checks for $32.01 and $12.68. What is her new balance?

Objective: to add and subtract fractions with unlike denominators

Asia, the world's largest continent, covers $\frac{3}{10}$ of the Earth's surface. Africa, the second largest continent, covers $\frac{1}{4}$ of the Earth's surface. To find the Earth's surface that is covered by both continents, you can add the two fractions $\frac{3}{10}$ and $\frac{1}{4}$. However, the fractions do not have the same denominator. They are fractions with *unlike* denominators.

To add or subtract unlike fractions, first rewrite the fractions as equivalent fractions with a common denominator.

Examples

A. What fraction of the Earth's surface is covered by Asia and Africa?

Add $\frac{3}{10}$ and $\frac{1}{4}$.

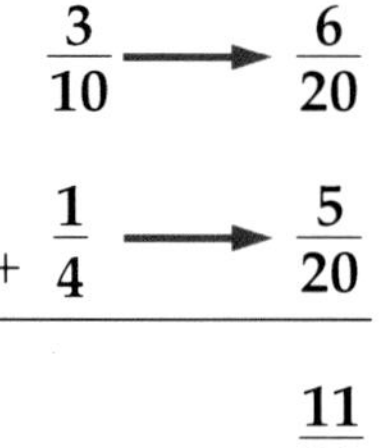

Find a common denominator for 10 and 4. You can use the LCM of 10 and 4, which is 20.

Write equivalent fractions with a common denominator of 20. 20 is the least common denominator, or LCD.

Add the numerators. Keep the common denominator.

You can use *any* common denominator to add and subtract unlike fractions. Choose a common denominator that is a multiple of the two unlike denominators. The least common denominator (LCD) is the least common multiple of the two denominators.

B. $\frac{2}{3} + \frac{1}{9}$

$\frac{2}{3} \rightarrow \frac{18}{27}$

$+\frac{1}{9} \rightarrow \frac{3}{27}$

$\frac{21}{27} = \frac{7}{9}$

One method of finding a common denominator is multiplying the two denominators.

C. $\frac{9}{10} - \frac{4}{5}$

$\frac{9}{10} \rightarrow \frac{9}{10}$

$-\frac{4}{5} \rightarrow \frac{8}{10}$

$\frac{1}{10}$

Another method of finding a common denominator is using the LCD.

Try THESE

Find the LCD. Then rename each pair of fractions.

1. $\frac{1}{5} - \frac{1}{10}$
2. $\frac{1}{3} + \frac{1}{6}$
3. $\frac{4}{7} - \frac{1}{3}$
4. $\frac{5}{6} + \frac{1}{8}$
5. $\frac{2}{4} - \frac{3}{20}$

Exercises

Add or subtract. Write each sum or difference in simplest form.

1. $\frac{1}{2} + \frac{1}{5}$
2. $\frac{1}{5} - \frac{1}{10}$
3. $\frac{1}{7} + \frac{1}{2}$
4. $\frac{2}{3} + \frac{1}{5}$
5. $\frac{1}{3} - \frac{1}{7}$
6. $\frac{1}{5} - \frac{1}{9}$
7. $\frac{2}{4} + \frac{3}{10}$
8. $\frac{13}{16} - \frac{3}{8}$
9. $\frac{3}{10} - \frac{2}{15}$
10. $\frac{5}{6} + \frac{1}{3}$
11. $\frac{7}{8} + \frac{1}{4}$
12. $\frac{3}{10} + \frac{1}{2} + \frac{2}{5}$
13. $\frac{5}{64} + \frac{5}{8} + \frac{3}{4}$
14. $\frac{3}{16} + \frac{7}{12} + \frac{5}{6}$
15. $\frac{2}{8} + \frac{1}{6}$
16. $\frac{2}{5} - \frac{1}{15}$
17. $\frac{9}{24} + \frac{5}{8}$
18. $\frac{13}{6} - \frac{14}{12}$
19. $\frac{3}{8} - \frac{3}{72}$

★20. $\left(\frac{9}{10} - \frac{9}{100}\right) + \frac{1}{2}$

★21. $\frac{4}{5} + \left(\frac{13}{20} - \frac{3}{20}\right)$

★22. $\frac{6}{5} - \left(\frac{1}{2} + \frac{1}{3}\right)$

Problem SOLVING

23. Suppose you used $\frac{1}{2}$ of a can of soup in a casserole. Then you ate $\frac{1}{3}$ of the can of soup for lunch. How much of the can of soup is left?

Solve. Use the circle graph.

24. What part is spent on lunch and entertainment?
25. Is more spent on savings or entertainment?
26. What part is spent on savings, school supplies, and lunch?
27. What part is *not* saved?

How Mary Spends Her Allowance

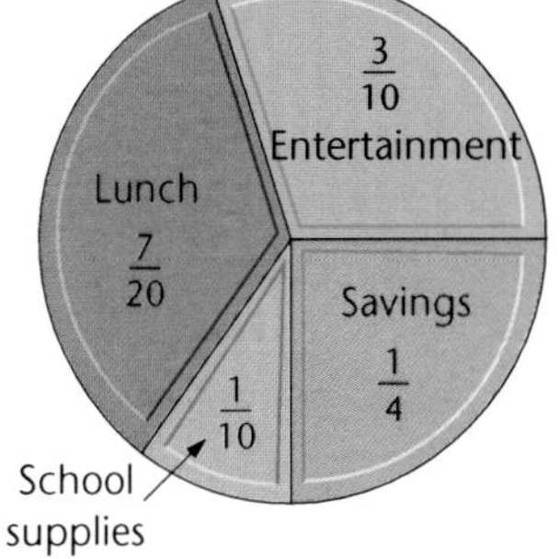

Constructed RESPONSE

28. To add $\frac{5}{6}$ and $\frac{7}{12}$, you can use the LCD, or another common denominator, such as 72. Find the sum of $\frac{5}{6}$ and $\frac{7}{12}$ both ways. Which do you prefer, and why?

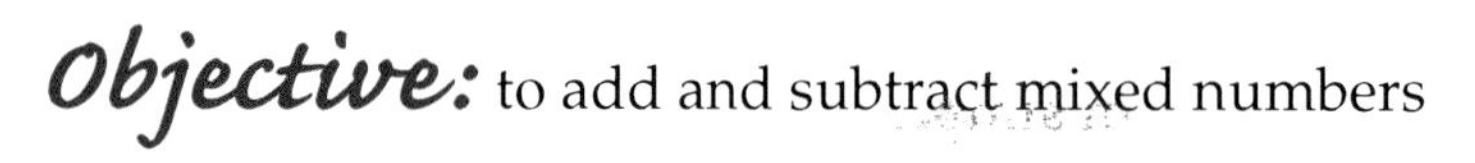

5.5 Adding and Subtracting Mixed Numbers

Objective: to add and subtract mixed numbers

Michelle is baking a cake and a batch of cookies for a bake sale. The cake requires $3\frac{3}{4}$ cups of flour, and the cookies require $2\frac{1}{8}$ cups of flour. How much flour does she need in all?

> To add or subtract mixed numbers, first add or subtract the fractions. Then add or subtract the whole numbers. Simplify if necessary.

Examples

A. How much flour does Michelle need in all?

Estimate. $3\frac{3}{4}$ rounds to 4, and $2\frac{1}{8}$ rounds to 2. $4 + 2 = 6$, so she needs about 6 cups of flour.

$$3\frac{3}{4} + 2\frac{1}{8}$$

Add the fractions.

$$\frac{3}{4} \longrightarrow \frac{6}{8}$$
$$+\ \frac{1}{8} \longrightarrow \frac{1}{8}$$
$$\frac{7}{8}$$

Add the whole numbers.

$$3\frac{6}{8}$$
$$+\ 2\frac{1}{8}$$
$$5\frac{7}{8}$$

Simplify if necessary.

B. Find the difference. Write the answer in simplest form.

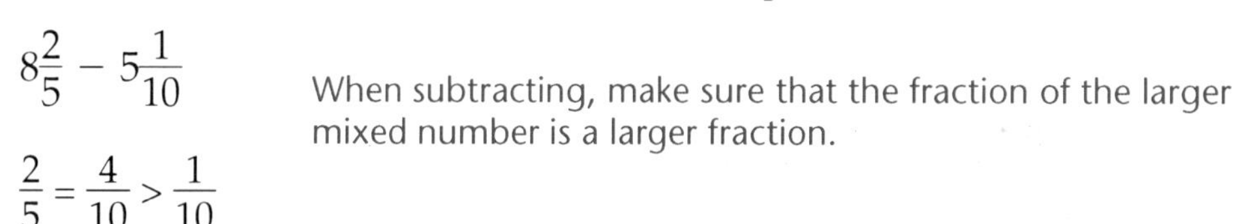

$$8\frac{2}{5} - 5\frac{1}{10}$$

When subtracting, make sure that the fraction of the larger mixed number is a larger fraction.

$$\frac{2}{5} = \frac{4}{10} > \frac{1}{10}$$

Subtract the fractions.

$$\frac{2}{5} \longrightarrow \frac{4}{10}$$
$$-\ \frac{1}{10} \longrightarrow \frac{1}{10}$$
$$\frac{3}{10}$$

Subtract the whole numbers.

$$8\frac{4}{10}$$
$$-\ 5\frac{1}{10}$$
$$3\frac{3}{10}$$

Simplify if necessary.

Try THESE

Add or subtract. Write each sum or difference in simplest form.

1. $3\frac{1}{8} + 1\frac{5}{6}$
2. $5\frac{2}{3} - 1\frac{1}{4}$
3. $2\frac{1}{2} + 4\frac{4}{5}$

Exercises

Add or subtract. Write each sum or difference in simplest form.

1. $5\frac{3}{8} + 4\frac{2}{8}$
2. $6\frac{1}{7} + 2\frac{2}{7}$
3. $10\frac{5}{10} + 4\frac{4}{10}$
4. $12\frac{1}{4} + 6\frac{2}{4}$
5. $6\frac{2}{5} + 13\frac{2}{5}$
6. $14\frac{3}{7} - 8\frac{2}{7}$
7. $12\frac{8}{12} - 9\frac{3}{12}$
8. $17\frac{7}{10} - 8\frac{2}{10}$
9. $11\frac{5}{8} - 7\frac{1}{8}$
10. $4\frac{3}{4} - 1\frac{1}{4}$
11. $3\frac{2}{3} + 2\frac{1}{6}$
12. $4\frac{3}{4} + 1\frac{1}{8}$
13. $6\frac{5}{8} + \frac{1}{2}$
14. $20\frac{2}{9} + 9\frac{1}{3}$
15. $11\frac{3}{10} + 5\frac{5}{6}$
16. $7\frac{3}{4} - 5\frac{3}{8}$
17. $12\frac{5}{6} - 9\frac{1}{3}$
18. $10\frac{5}{8} - 5\frac{1}{3}$
19. $17\frac{1}{2} + 11\frac{1}{5}$
20. $26\frac{2}{3} - 3\frac{1}{11}$

Problem SOLVING

21. Joe was making mixed nuts. He used $2\frac{1}{4}$ pounds of pecans and $3\frac{1}{5}$ pounds of cashews. How many pounds of mixed nuts did Joe make?
22. Susan and Samantha are making cookies using different recipes. Susan's recipe calls for $2\frac{1}{2}$ cups of sugar, but Samantha's recipe calls for $1\frac{3}{4}$ cups of sugar. How much more sugar is in Susan's recipe?
23. Tasha's bag weighs $30\frac{5}{8}$ pounds and Nicole's bag weighs $19\frac{1}{3}$ pounds. They want to put both bags on a cart, but the cart has a weight limit of 50 pounds. Can they both put their bags on the cart? Explain.
24. At a college track meet, the farthest men's shot put was a distance of $759\frac{1}{2}$ inches. The farthest women's shot put was $703\frac{1}{4}$ inches.
    a. What was the difference between the two throws?
    b. What was the difference between the two throws in feet? (12 inches = 1 foot)
25. If you cut an $8\frac{1}{2}$ by 11 inch sheet of paper into 11 1-inch strips and lay the strips end to end, how long would the paper be?

5.6 Renaming Mixed Numbers

Objective: to rename mixed numbers using improper fractions

Sometimes when you are subtracting mixed numbers, the larger mixed number has a smaller fraction. In order to subtract you will need to learn to rename the mixed number. Look at the following problem.

$$\begin{array}{r} 2\frac{1}{3} \\ -\ 1\frac{2}{3} \\ \hline \end{array}$$

Even though $2\frac{1}{3}$ is larger than $1\frac{2}{3}$, the fraction $\frac{1}{3}$ is smaller than $\frac{2}{3}$. You cannot subtract, yet.

By learning to rename, you will be able to complete problems like these.

Modeling Renaming

Use fraction bars to model $2\frac{1}{3}$.

You need to divide one whole into thirds. This will give you enough thirds to subtract.

So $2\frac{1}{3}$ is equal to $1\frac{4}{3}$. You could now subtract: $1\frac{4}{3} - 1\frac{2}{3} = \frac{2}{3}$.

Examples

A. Rename $2\frac{1}{6}$ using an improper fraction.

Use fraction bars to model $2\frac{1}{6}$.

$$2\frac{1}{6} = 1\frac{7}{6}$$

B. Rename $4\frac{3}{5}$ using an improper fraction. Use borrowing.

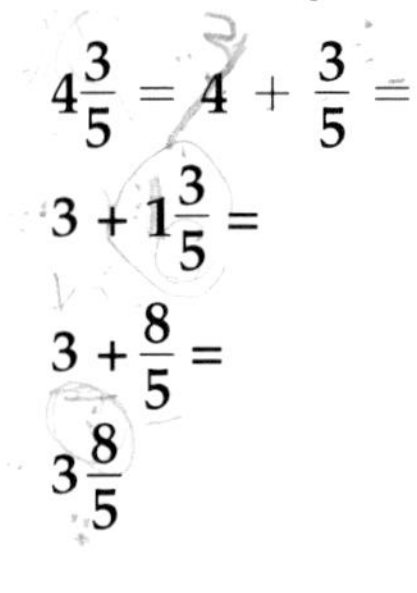

$4\frac{3}{5} = 4 + \frac{3}{5} =$

$3 + 1\frac{3}{5} =$ Borrow 1 from the 4 to make a mixed number.

$3 + \frac{8}{5} =$ Change the mixed number to an improper fraction.

$3\frac{8}{5}$ Simplify.

Try THESE

Rename.

1. $3\frac{3}{4} = 2\frac{■}{4}$
2. $10\frac{1}{2} = 9\frac{■}{2}$
3. $6\frac{5}{4} = 7\frac{■}{4}$

Exercises

Rename each as a mixed number in simplest form.

1. $\frac{10}{3}$
2. $\frac{21}{10}$
3. $\frac{36}{12}$
4. $\frac{16}{12}$
5. $\frac{14}{8}$
6. $\frac{9}{6}$
7. $\frac{15}{9}$
8. $\frac{21}{18}$
9. $\frac{28}{16}$
10. $\frac{22}{10}$
11. $1\frac{6}{5}$
12. $3\frac{11}{9}$
13. $5\frac{17}{12}$
14. $4\frac{15}{13}$
15. $2\frac{19}{17}$
16. $5\frac{9}{6}$
17. $7\frac{14}{10}$
18. $3\frac{18}{15}$
19. $6\frac{12}{8}$
20. $9\frac{16}{12}$
21. $2\frac{9}{9}$
22. $12\frac{14}{14}$
23. $1\frac{12}{6}$
24. $8\frac{24}{12}$
25. $11\frac{10}{4}$

Replace each $\blacksquare$ to rename each whole number or mixed number.

26. $1 = \frac{3}{\blacksquare}$
27. $2 = 1\frac{\blacksquare}{6}$
28. $3\frac{1}{5} = 2\frac{6}{\blacksquare}$
29. $8\frac{1}{6} = 7\frac{\blacksquare}{6}$
30. $20\frac{5}{6} = 19\frac{\blacksquare}{6}$
31. $17\frac{3}{5} = 16\frac{8}{\blacksquare}$
32. $12\frac{7}{9} - \blacksquare\frac{16}{9}$
33. $9\frac{2}{13} = \blacksquare\frac{15}{13}$
34. $10 = 9\frac{7}{\blacksquare}$
35. $5 = 4\frac{\blacksquare}{10}$
36. $9\frac{1}{12} = 8\frac{\blacksquare}{12}$
37. $10 = \blacksquare\frac{16}{16}$
38. $12\frac{5}{9} = 11\frac{\blacksquare}{9}$
39. $14\frac{4}{5} = 13\frac{9}{\blacksquare}$
40. $10\frac{14}{15} = 9\frac{\blacksquare}{15}$
41. $20\frac{13}{16} = 19\frac{\blacksquare}{16}$

Constructed RESPONSE

42. Do you need to rename before you subtract $3\frac{3}{8} - 1\frac{1}{8}$? Explain.
43. Do you need to rename before you subtract $4\frac{3}{5} - 1\frac{7}{10}$? Explain.

Mid-Chapter REVIEW

Compute. Write each answer in simplest form.

1. $\frac{1}{5} + \frac{2}{5}$
2. $\frac{1}{8} + \frac{3}{8}$
3. $\frac{7}{9} - \frac{2}{9}$
4. $\frac{7}{12} - \frac{5}{12}$
5. $\frac{1}{9} + \frac{2}{3}$
6. $\frac{3}{8} + \frac{3}{10}$
7. $\frac{5}{6} + \frac{4}{9}$
8. $\frac{9}{10} - \frac{5}{12}$
9. $\frac{4}{5} - \frac{2}{6}$

5.7 Subtracting Mixed Numbers with Renaming

Objective: to subtract mixed numbers involving renaming

Men and women, as well as boys and girls, enjoy playing the game of basketball. However, the basketballs are different sizes. A typical men's ball has a circumference of $29\frac{3}{4}$ inches, while a typical women's ball has a circumference of $28\frac{7}{8}$ inches. What is the difference in the circumference of the balls?

Examples

A. What is the difference in the circumference of a men's basketball and a women's basketball?

$$\begin{array}{r} 29\frac{3}{4} \longrightarrow \frac{6}{8} \\ -\ 28\frac{7}{8} \longrightarrow \frac{7}{8} \\ \hline \end{array}$$

When subtracting mixed numbers, it is a good idea, first, to rename the fractions in order to know if renaming is necessary.

$$29\frac{6}{8} = 28\frac{14}{8}$$

Rename if necessary.

$$\begin{array}{r} 28\frac{14}{8} \\ -\ 28\frac{7}{8} \\ \hline \frac{7}{8} \end{array}$$

Subtract the whole numbers.

Subtract the fractions.

Simplify if necessary.

The difference in circumference between the basketballs is $\frac{7}{8}$ inch.

B. Subtract $5\frac{1}{4} - 1\frac{2}{3}$. Write your answer in simplest form.

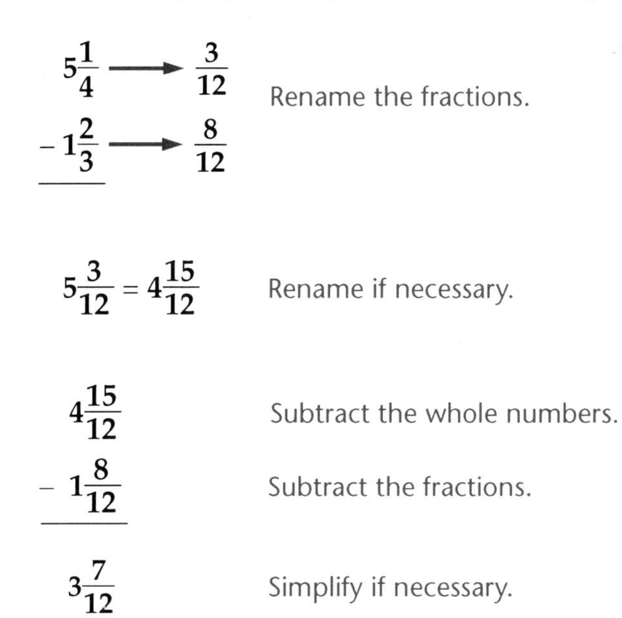

$$\begin{array}{r} 5\frac{1}{4} \longrightarrow \frac{3}{12} \\ -\,1\frac{2}{3} \longrightarrow \frac{8}{12} \end{array}$$

Rename the fractions.

$$5\frac{3}{12} = 4\frac{15}{12}$$

Rename if necessary.

$$\begin{array}{r} 4\frac{15}{12} \\ -\,1\frac{8}{12} \\ \hline 3\frac{7}{12} \end{array}$$

Subtract the whole numbers.

Subtract the fractions.

Simplify if necessary.

Try THESE

1. $7\frac{2}{5} - 2\frac{3}{5}$
2. $6\frac{1}{4} - 2\frac{1}{2}$
3. $7 - 3\frac{2}{5}$

Exercises

1. $3\frac{1}{2} - 1\frac{1}{3}$
2. $11\frac{2}{5} - 3\frac{1}{3}$
3. $4 - 2\frac{3}{4}$
4. $9\frac{1}{6} - \frac{2}{6}$
5. $8\frac{5}{6} - 6\frac{1}{3}$
6. $13\frac{2}{3} - 6\frac{1}{4}$
7. $14\frac{7}{12} - 8\frac{1}{6}$
8. $15\frac{1}{7} - 9\frac{2}{3}$
9. $12 - 4\frac{1}{8}$
10. $9 - 5\frac{5}{8}$
11. $10 - 9\frac{5}{16}$
12. $12 - 11\frac{11}{24}$
13. $7\frac{1}{4} - \frac{1}{2}$
14. $10\frac{5}{16} - \frac{5}{8}$
15. $6\frac{1}{6} - 3\frac{1}{4}$
16. $6\frac{1}{5} - 3\frac{4}{5}$
17. $3\frac{1}{3} - \frac{5}{6}$
18. $6\frac{2}{5} - 2\frac{9}{10}$
19. $6\frac{3}{8} - 2\frac{3}{5}$
20. $6\frac{1}{3} - 2\frac{5}{9}$
21. $12 - 5\frac{2}{7}$

Problem SOLVING

22. How much longer is $2\frac{1}{2}$ inches than $1\frac{1}{4}$ inches?
23. How much more is 7 miles than $5\frac{1}{8}$ miles?
24. A trip from Baltimore to New York City takes $3\frac{3}{10}$ hours. You have been traveling $2\frac{5}{6}$ hours. How much longer is your trip?
25. You and a partner are growing bean plants. The first week, your plant grew $1\frac{1}{2}$ cm, the second week your plant grew $2\frac{1}{4}$ cm, and the third week your plant grew $2\frac{5}{8}$ cm. After three weeks, your partner's plant was $6\frac{1}{4}$ cm. Whose plant is taller, and by how many centimeters?

Constructed RESPONSE

26. Your friend Lauren has a math tutor who showed her another way to subtract mixed numbers. You worked out the same problem. Explain why Lauren's method works. Then explain which method you feel works best for you and why.

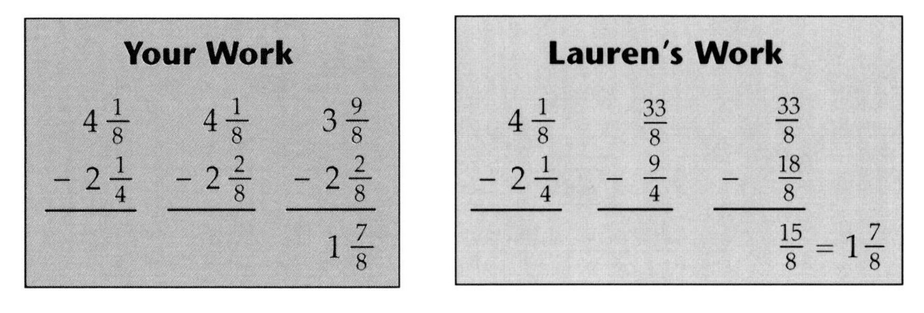

Test PREP

27. Julian studied for 3 hours for his math test and $1\frac{5}{8}$ hours for his science test. How many more hours did he study for his math test?

    a. $2\frac{5}{8}$ h b. $1\frac{3}{8}$ h c. 2 h d. $2\frac{1}{8}$ h

28. Helen bought $6\frac{4}{5}$ pounds of apples and $2\frac{8}{9}$ pounds of oranges. How many pounds of fruit did she buy?

    a. $8\frac{4}{5}$ lb b. $8\frac{12}{14}$ lb c. $8\frac{26}{45}$ lb d. $9\frac{31}{45}$ lb

Patsy's Conflict

Patsy has cheerleading practice on Monday and every fourth school day. She wants to be in the school play, but they have practice on Tuesdays and every sixth school day. Assuming the first school day is Wednesday, September 5, when would she have 2 meetings at the same time? Would she ever have 3 meetings at the same time? How many times would she have more than 1 meeting at the same time before the end of December?

Extension

If Patsy had to go to the dentist every eighth school day for a month, did she ever miss a meeting? If so, which meetings did she miss if her first visit was October 3?

5.8 Problem-Solving Strategy: Matrix Logic

Objective: to use matrix logic to solve problems

Stevie, Joey, Angie, and Camille all have a favorite sport. No two of them like the same sport. Here are some clues.

1. Angie does not like baseball or football.
2. Joey's favorite sport is not played with a round ball.
3. Camille does not care for basketball or soccer.
4. Stevie likes soccer.

Find the favorite sport of each person.
This problem can be solved using matrix logic. Begin by setting up a table.

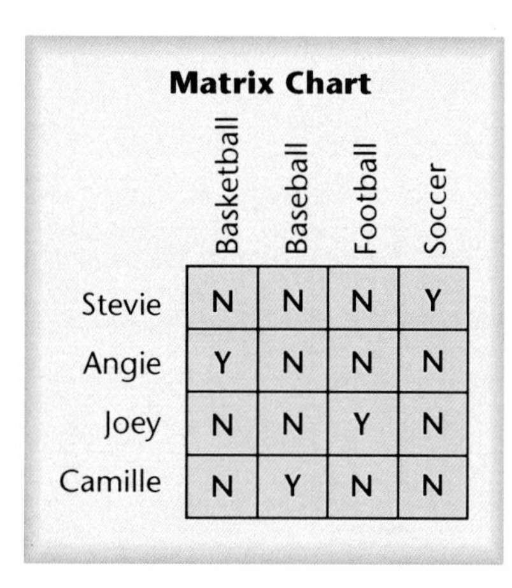

Matrix Chart

	Basketball	Baseball	Football	Soccer
Stevie	N	N	N	Y
Angie	Y	N	N	N
Joey	N	N	Y	N
Camille	N	Y	N	N

Write all the names and categories in the matrix. Find all of the clues that give you a definite yes or no. Write a **Y** for *yes* and an **N** for *no* in those boxes.

Remember, since no two like the same sport, once you write a **Y** in a box, you can write an **N** in all of the boxes in that row and column. Then complete the matrix.

Stevie likes soccer, Angie likes basketball, Joey likes football, and Camille likes baseball.
Set up a matrix for each set of clues below.
Solve the problems by completing the matrix.

1. Find the favorite musical group of each person.
   a. Sam's favorite group is The Heroes.
   b. Darryl does not like The Boys.
   c. The name of Mona's favorite does not begin with "The."
   d. Will does not listen to Bizarre or The Pops.
2. Find the favorite pet of each person.
   a. Kim is afraid of dogs and frogs.
   b. Carlton has a cat.
   c. Kevin's pet lives in the water.
   d. Ruth's pet is not a fish.

Exercises

Use matrix logic to solve the following problems.

1. Who likes which flower?
   a. Pat likes violets.
   b. Toby dislikes tulips.
   c. Ben grows roses, his favorite flower.
   d. Paul likes one flower.
   e. Someone does not like mums.
2. Who owns which vehicle?
   a. Matt drives in off-road sports.
   b. Mark does not like SUVs or trucks.
   c. Mike uses his vehicle to carry trees for landscaping.
   d. Murray has a sedan because a sports car does not have four doors.
   e. The vehicles are a sports car, an SUV, a truck, and a sedan.
3. Who likes which snack?
   a. Cindy does not like sweet treats.
   b. Carl loves fruit.
   c. Catherine does not like cantaloupe or cinnamon rolls.
   d. Chuck likes cake.
   e. Carrie likes icing.
   f. Someone likes cookies, and another likes cereal.

★ 4. There are five players on the basketball team named Tara, James, Jackie, Terry, and Corey. Their jersey numbers are 7, 11, 14, 24, and 42. Their positions are point guard, shooting guard, small forward, power forward, and center. Use the clues to match the players with their numbers and positions.
   a. The tallest player is the center.
   b. Tara is the smallest player.
   c. James has an odd numbered jersey.
   d. Terry has the smallest number.
   e. Jackie is not a forward.
   f. Corey is number 24.
   g. The point guard is number 11.
   h. The center is number 42.
   i. The power forward has an odd number.
   j. Corey is a small forward.

Mixed Review

Compare using <, >, or = .

5. $\frac{5}{9} \bullet \frac{7}{9}$ 6. $\frac{1}{2} \bullet \frac{1}{5}$ 7. $\frac{7}{12} \bullet \frac{2}{3}$ 8. $\frac{5}{6} \bullet \frac{7}{9}$ 9. $\frac{4}{7} \bullet \frac{1}{2}$

Chapter 5 Review

Language and Concepts

Write *true* or *false*. If false, replace the underlined word or words to make a true sentence.

1. The first common multiple, other than zero, of two numbers is called the <u>least common multiple</u>.
2. <u>Improper</u> fractions are less than one.
3. When subtracting fractions with <u>like</u> denominators, you must find the LCD and rename each fraction.
4. Using the <u>GCF</u> is one way to simplify fractions.
5. You can use <u>rounding</u> to estimate sums and differences of fractions.

Skills and Problem Solving

Compute mentally. Write each sum or difference in simplest form. Write only your answers. (Section 5.3)

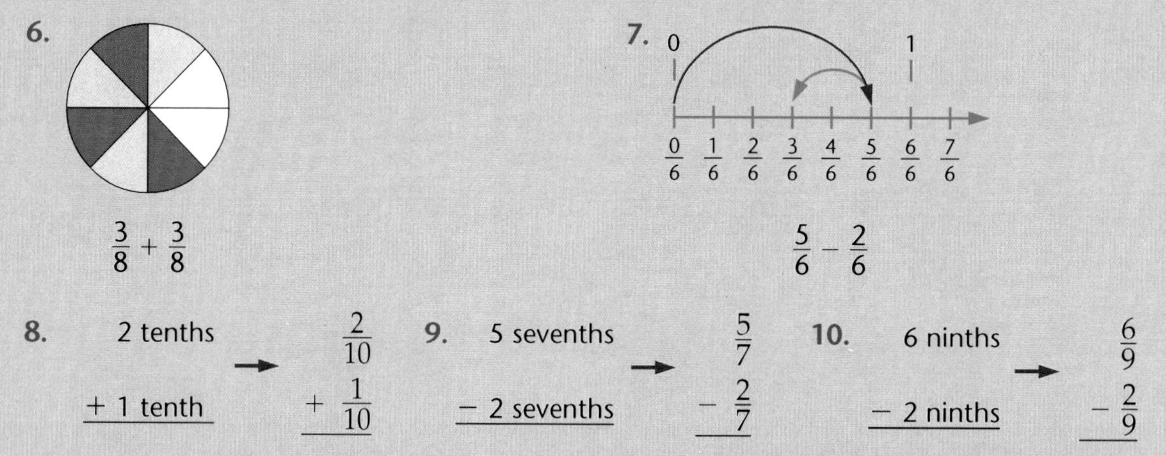

Replace the ■ to rename each whole number or mixed number. (Section 5.6)

11. $1 = \frac{■}{6}$

12. $4 = 3\frac{■}{7}$

13. $12\frac{1}{8} = 11\frac{■}{8}$

14. $10\frac{5}{18} = 9\frac{■}{18}$

Compute. Write each sum or difference in simplest form. (Sections 5.3–5.5, 5.7)

15. $\frac{2}{5} + \frac{1}{2}$

16. $\frac{1}{4} + \frac{2}{5}$

17. $\frac{6}{16} + \frac{1}{2}$

18. $\frac{5}{6} + \frac{1}{9}$

19. $\frac{7}{15} + \frac{6}{12}$

20. $\frac{3}{4} - \frac{1}{8}$

21. $\frac{7}{10} - \frac{3}{5}$

22. $\frac{3}{4} - \frac{2}{16}$

23. $\frac{5}{6} - \frac{1}{8}$

24. $\frac{5}{6} - \frac{3}{10}$

25. $2\frac{2}{5} + 4\frac{2}{5}$

26. $3 + 5\frac{4}{11}$

27. $7\frac{4}{15} + 3\frac{8}{15}$

28. $12\frac{9}{10} - 3\frac{4}{10}$

29. $9\frac{14}{21} - 2$

30. $6\frac{1}{2} + 2\frac{2}{5}$

31. $5\frac{2}{3} + 4\frac{1}{3}$

32. $2\frac{3}{4} + 10\frac{5}{12}$

33. $6\frac{3}{10} + 12\frac{7}{8}$

34. $10\frac{2}{3} - 7\frac{1}{4}$

35. $8 - 7\frac{3}{13}$

36. $10\frac{1}{6} - 2\frac{1}{4}$

37. $10\frac{1}{4} - 3\frac{7}{12}$

Solve. (Section 5.4)

38. Do the Kanes spend more on food or rent? What part of the budget is spent on rent, food, and clothing?

39. How much more is spent on food and rent than on clothing and savings?

40. What part of the Kanes' income is *not* saved?

41. You are driving on a toll road. The next toll will be 60¢. You cannot use half dollars or pennies. How many different combinations of coins can you use if at least one of the coins is a quarter?

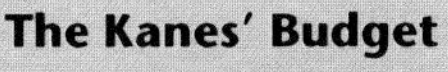

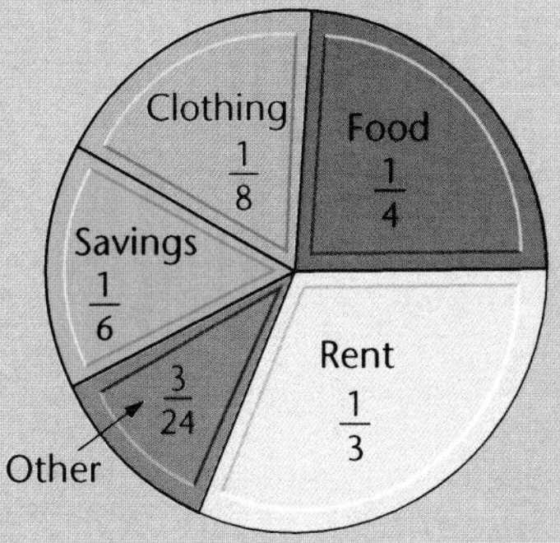

Use matrix logic to solve the following problems. (Section 5.8)

42. Ann, Tom, Joan, and Ken each have one of the following careers: teacher, plumber, pediatrician, and writer. Ann and Tom work with children. Joan works from home, while Ken travels to other people's homes during the day. Sometimes Ann's job requires her to get up in the middle of the night for emergencies. Match each person to his or her career.

Chapter 5 Test

Compute mentally. Write each answer in simplest form. Write only your answer.

1. $\frac{5}{12} + \frac{5}{12}$
2. $\frac{0}{8} + \frac{6}{8}$
3. $\frac{9}{7} - \frac{2}{7}$
4. $\frac{5}{4} - \frac{0}{4}$

Replace each ■ to rename each whole number or mixed number.

5. $3 = 2\frac{■}{7}$
6. $6 = 5\frac{■}{4}$
7. $11\frac{3}{8} = 10\frac{■}{8}$
8. $21\frac{4}{5} = 20\frac{■}{5}$

Compute. Write each sum or difference in simplest form.

9. $\frac{1}{2} + \frac{1}{6}$
10. $\frac{2}{5} + \frac{3}{10}$
11. $\frac{4}{7} + \frac{6}{14}$
12. $\frac{4}{6} + \frac{4}{9}$
13. $\frac{7}{8} + \frac{1}{3}$
14. $\frac{2}{3} - \frac{5}{9}$
15. $\frac{4}{5} - \frac{1}{10}$
16. $\frac{5}{6} - \frac{2}{4}$
17. $\frac{6}{5} - \frac{2}{10}$
18. $\frac{9}{10} - \frac{1}{4}$
19. $3\frac{1}{5} + 4\frac{3}{5}$
20. $9\frac{7}{8} + 4$
21. $7 + 3\frac{2}{11}$
22. $6\frac{7}{8} - 3\frac{3}{8}$
23. $48\frac{10}{15} - 9$
24. $4\frac{1}{5} + 5\frac{3}{10}$
25. $6\frac{5}{8} + 3\frac{3}{4}$
26. $8\frac{3}{8} + 5\frac{1}{6}$
27. $10\frac{8}{21} + 3\frac{2}{3}$
28. $12\frac{1}{8} - 7\frac{1}{3}$
29. $7\frac{3}{5} - 2\frac{3}{4}$
30. $9\frac{7}{8} - 4\frac{5}{6}$
31. $15\frac{3}{4} - 12\frac{6}{8}$

Solve.

32. Marlene bakes a cake for $\frac{5}{6}$ of an hour. Then she bakes it $\frac{1}{4}$ of an hour at a higher temperature. How long does it bake altogether?

33. Frank, Bob, Carol, and Lori are in rooms 3, 4, 10, and 11. Only Carol's and Frank's room numbers are prime numbers. Lori's room number is two digits. The sum of Carol's and Bob's room numbers is less than Frank's room number. Who is in which room?

34. Bruce completed the following problem incorrectly.

$$\begin{array}{r} 5 \\ -\ 2\frac{3}{4} \\ \hline 3\frac{3}{4} \end{array}$$

   a. Find his mistake, and complete the problem correctly.

   b. Explain why the answer you determined is correct. Use what you know about subtracting mixed numbers in your explanation. Use words, numbers, and/or symbols in your explanation.

Tessellations

A **tessellation** is a pattern formed by fitting shapes together without leaving any spaces or overlapping. These are used in designs for quilts and floor tiles. In nature, the honeycomb forms a tessellation.

There are only three regular polygons that form tessellations by themselves. These **regular tessellations** are formed by the square, the equilateral triangle, and the regular hexagon.

Another type of tessellation is formed by combinations of regular polygons. These are called **semi-regular tessellations**. The one shown at the right is formed by the square and the regular octagon.

For each of the tessellations shown below, write two equivalent fractions to represent the shaded part.

1. 

2.

3. 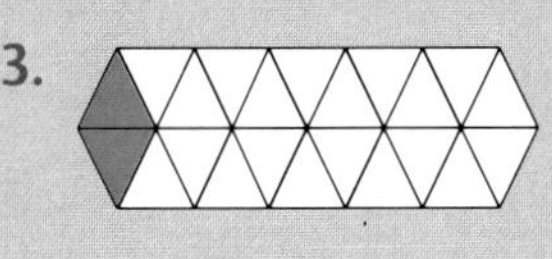

4. 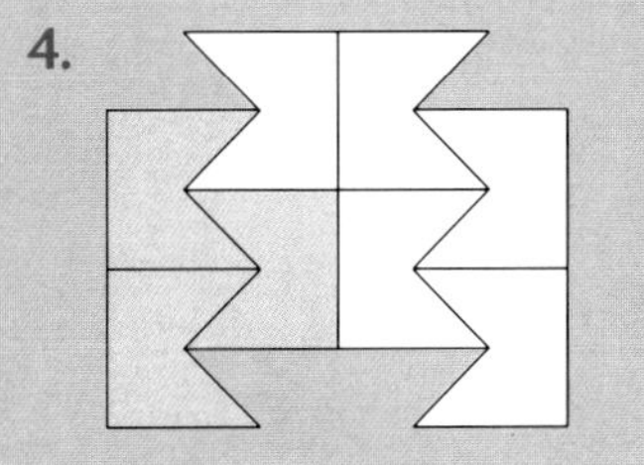

5. 

6. 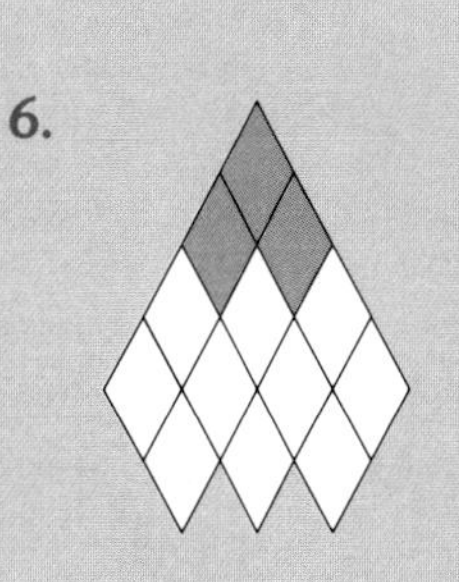

Draw a tessellation using the shapes described.

7. triangles
8. squares
9. triangles and hexagons
10. triangles and squares

Cumulative Test

1. $\frac{1}{2} + \frac{3}{4}$

a. $\frac{2}{3}$ b. $\frac{4}{6}$
c. $\frac{6}{8}$ d. $1\frac{1}{4}$

2. $5\frac{1}{4} - 3\frac{1}{5}$

a. $1\frac{4}{5}$ b. $2\frac{1}{20}$
c. $2\frac{1}{9}$ d. $8\frac{9}{20}$

3. Find the LCM of 8 and 12.

a. 4
b. 16
c. 24
d. 32

4. If a number is divisible by 2 and 3, it is also divisible by ____.

a. 5
b. 6
c. 12
d. 24

5. Which numbers are prime factors of 171?

a. 3
b. 3 and 9
c. 3 and 19
d. 3 and 57

6. Replace the ■ in $3\frac{■}{9}$ to rename $4\frac{5}{9}$.

a. 14
b. 15
c. 16
d. none of the above

7. Maria, Toni, Lucy, and Armando rented a boat together for \$30 per hour. What was the cost per person?

a. \$3.75
b. \$7.50
c. \$15
d. \$30

8. Tom needs $1\frac{1}{2}$ cups of sugar to make cookies. He has $\frac{2}{3}$ cup. How much more sugar does he need?

a. $\frac{1}{2}$ cup b. $\frac{1}{3}$ cup
c. $\frac{5}{6}$ cup d. none of the above

9. Mr. Ruiz bought 3 jackets for \$40 each. He had 4 at home. What was the total cost of his new jackets?

What is the extra fact in this problem?

a. He bought 3 jackets.
b. The jackets cost \$40 each.
c. He had 4 at home.
d. all of the above

10. How many hours is Lowell's open per week?

LOWELL'S

Hours:
10 A.M.–10 P.M. Mon.–Sat.
12 Noon–6 P.M. Sun.

a. 56 hours
b. 66 hours
c. 68 hours
d. 78 hours

CHAPTER

Multiplying and Dividing Fractions

Elisabeth Dye
California

6.1 Mental Math: Fraction Products

Objective: to compute fraction products mentally

Mark is planning his summer garden. He wants to plant 6 rows of corn. He knows that a row takes $\frac{2}{3}$ package of seeds. How many packages of seeds does Mark need to plant 6 rows? To find out, you can add or multiply.

Add: $\frac{2}{3} + \frac{2}{3} + \frac{2}{3} + \frac{2}{3} + \frac{2}{3} + \frac{2}{3}$

Multiply: $\frac{2}{3} \times 6$ **means** $\frac{2}{3}$ **of 6**

$\frac{1}{3}$ **of 6 is 2** → $\frac{2}{3}$ **of 6 is 4**

Mark needs 4 packages of seeds.

More Examples

A. Multiply $\frac{5}{8} \times 24$.

$\frac{5}{8} \times 24$ **means** $\frac{5}{8}$ **of 24**

$\frac{1}{8}$ **of 24 is 3** → $\frac{5}{8}$ **of 24 is 15**

B. Multiply $8 \times \frac{1}{4}$.

$8 \times \frac{1}{4}$ **means 8 fourths** ($\frac{8}{4}$)

$\frac{8}{4}$ **fourths is 2**

C. Multiply $\frac{4}{9} \times 2$.

$\frac{4}{9} \times 2$ **means** $\frac{4}{9}$ **of 2**

$\frac{1}{9}$ **of 2 is** $\frac{2}{9}$ → $\frac{4}{9}$ **of 2 is** $\frac{8}{9}$

Try THESE

Compute mentally. Write each product in simplest form. Write only your answers.

1. $\frac{2}{3} \times 3$
2. $6 \times \frac{1}{2}$
3. $\frac{2}{7} \times 3$
4. $\frac{3}{4} \times 4$
5. $8 \times \frac{1}{8}$
6. $4 \times \frac{2}{9}$
7. $\frac{1}{6} \times 12$
8. $\frac{7}{8} \times 16$
9. $10 \times \frac{4}{5}$
10. $\frac{5}{6} \times 18$

Exercises

Compute mentally. Write each product in simplest form. Write only your answers.

1. $8 \times \frac{1}{2}$
2. $\frac{1}{3} \times 9$
3. $\frac{5}{6} \times 12$
4. $10 \times \frac{2}{5}$
5. $\frac{3}{4} \times 8$
6. $\frac{2}{9} \times 3$
7. $8 \times \frac{1}{9}$
8. $42 \times \frac{6}{7}$
9. $25 \times \frac{1}{5}$
10. $\frac{3}{4} \times 36$
11. $\frac{3}{10} \times 2$
12. $20 \times \frac{3}{5}$
13. $\frac{11}{12} \times 12$
14. $\frac{7}{9} \times 18$
15. $\frac{2}{3} \times 1$
16. $0 \times \frac{1}{4}$

★17. Find the product of $6 \times 4 \times \frac{2}{3}$.

Problem SOLVING

Solve by computing mentally.

18. There are 5 dozen plants in clay pots. Seven-tenths of the plants are white. How many dozen plants are white?

19. Marie has a quart of vegetable oil. She uses $\frac{1}{4}$ quart to fry fish and $\frac{1}{8}$ quart to make salad dressing. How much oil is left?

20. Four-fifths of the school want to add a new color to the uniform. If there are 100 students, about how many want to add a new color to the uniform?

★21. Choose a number. Multiply that number by 2. Then add 5 and multiply by 5. Subtract 25 from the result, and divide by 10. Try this with three different numbers. What can you say about the result each time?

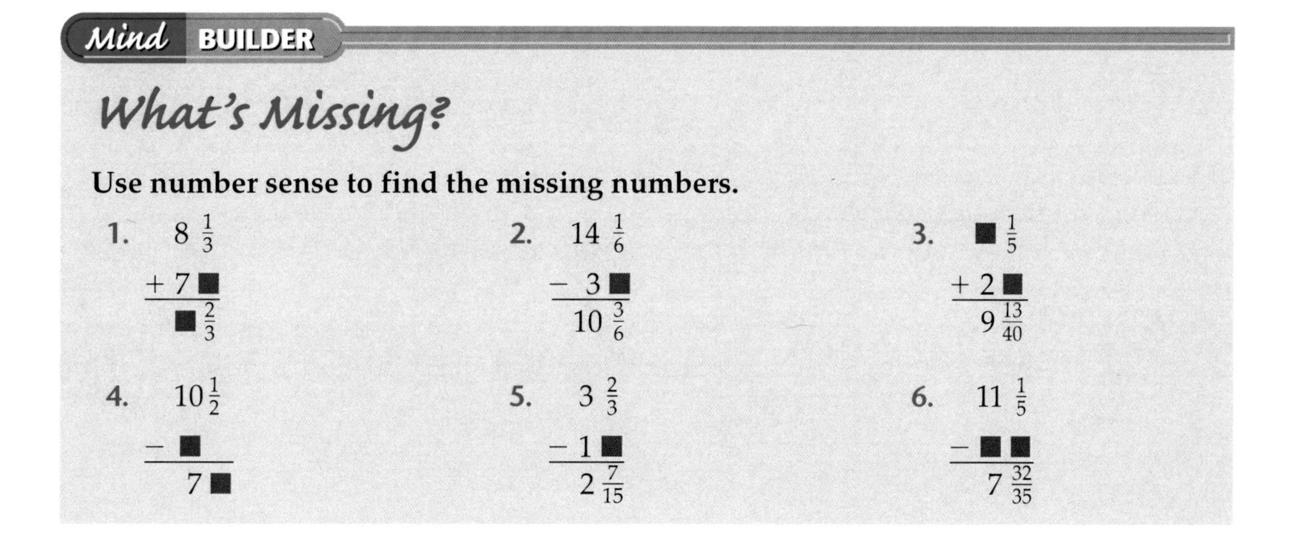

Mind BUILDER

What's Missing?

Use number sense to find the missing numbers.

1. $8\frac{1}{3} + 7\blacksquare = \blacksquare\frac{2}{3}$
2. $14\frac{1}{6} - 3\blacksquare = 10\frac{3}{6}$
3. $\blacksquare\frac{1}{5} + 2\blacksquare = 9\frac{13}{40}$
4. $10\frac{1}{2} - \blacksquare = 7\blacksquare$
5. $3\frac{2}{3} - 1\blacksquare = 2\frac{7}{15}$
6. $11\frac{1}{5} - \blacksquare\blacksquare = 7\frac{32}{35}$

6.2 Multiplying Fractions and Whole Numbers

Objective: to multiply fractions and whole numbers

In the previous lesson, you learned to multiply fractions and whole numbers. $\frac{1}{4} \times 12 = 3$ because $12 \div 4 = 3$. You will examine multiplying fractions and whole numbers a little further.

Exploration Exercise

Activity 1

Materials: compass, paper, pencil, scissors

1. Use the compass to draw a circle.
2. Cut out the circle.
3. Fold the circle in half.
4. Fold the circle in half again.
5. Open the paper.

Analysis

1. Into how many parts did you divide the circle?
2. What does $3 \times \frac{1}{4}$ mean? How would you shade the circle to show the product? Now shade the circle to show your answer.
3. Is $\frac{1}{4} \times 3$ the same as $3 \times \frac{1}{4}$? How would you show this using three circles? Draw three circles divided into fourths, and shade the circles to show your answer.

Activity 2

Materials: paper, pencil, colored pencils, scissors

1. Fold the paper into four equal parts by folding in half horizontally, then in half vertically.
2. Cut out the four parts.
3. Fold each part into thirds.
4. Color in 2 thirds on each paper.

Analysis

1. How does this activity model $4 \times \frac{2}{3}$?
2. What is $4 \times \frac{2}{3}$?
3. Write a rule for multiplying whole numbers and fractions.

Explanation

From the activities you completed, you developed your own method for multiplying fractions and whole numbers. Below is another way to multiply fractions and whole numbers.

$8 \times \frac{1}{4} = \frac{8}{4}$

- Step 1: Multiply the whole number and the numerator.
- Step 2: Write the product over the denominator.

$\frac{8}{4} = \frac{2}{1} = 2$

- Step 3: Simplify the fraction.

Examples

Multiply.

A. $4 \times \frac{1}{3} = \frac{4}{3}$

$\frac{4}{3} = 1\frac{1}{3}$

B. $7 \times \frac{2}{5} = \frac{14}{5}$

$\frac{14}{5} = 2\frac{4}{5}$

Try THESE

Use your rule for multiplying fractions and whole numbers to complete the following.

1. $8 \times \frac{1}{4}$
2. $5 \times \frac{2}{7}$
3. $12 \times \frac{2}{3}$

Exercises

Multiply. Write each product in simplest form.

1. $4 \times \frac{2}{3}$
2. $6 \times \frac{1}{2}$
3. $10 \times \frac{2}{3}$
4. $8 \times \frac{1}{5}$
5. $25 \times \frac{2}{5}$
6. $16 \times \frac{7}{8}$
7. $12 \times \frac{11}{12}$
8. $14 \times \frac{2}{7}$
9. $12 \times \frac{5}{6}$
10. $0 \times \frac{4}{5}$
11. $20 \times \frac{2}{3}$
12. $42 \times \frac{5}{7}$

Problem SOLVING

13. Jenny has 5 one-liter containers and each is two-thirds full of sports drink. How much sports drink is in the 5 containers?

Constructed RESPONSE

14. Judith noticed that when she multiplied $5 \times \frac{4}{5}$ her answer was 4. When she multiplied $12 \times \frac{3}{4}$ her answer was 9. She developed the following rule: "If the denominator of the fraction is a factor of the whole number, divide the whole number by the denominator, and then multiply the quotient by the numerator to get your answer." Is Judith correct? Explain.

6.3 Multiplying Fractions

Objective: to multiply fractions

Brian is baking cookies, but wants to cut the recipe in half. The recipe calls for $\frac{3}{4}$ cup of sugar. What is $\frac{1}{2}$ of $\frac{3}{4}$?

When you multiply fractions, you do not need to change the denominators to make them the same. You simply multiply. First, multiply the numerators, and then multiply the denominators.

Examples

A. Multiply $\frac{1}{2} \times \frac{3}{4}$.

Step 1	Step 2	Step 3
Multiply the numerators. $\frac{1 \times 3}{2 \times 4} = \frac{3}{}$	Multiply the denominators. $\frac{1 \times 3}{2 \times 4} = \frac{3}{8}$	Simplify. $\frac{3}{8}$ **is in simplest form.**

Brian should use $\frac{3}{8}$ cup of sugar to cut the recipe in half.

B. $\frac{5}{6} \times \frac{3}{20} = \frac{5 \times 3}{6 \times 20} = \frac{15}{120} = \frac{3}{24} = \frac{1}{8}$

C. You can simplify a numerator and denominator before you multiply.

$$\frac{15}{16} \times \frac{4}{5} = \frac{\overset{3}{\cancel{15}} \times \overset{1}{\cancel{4}}}{\underset{4}{\cancel{16}} \times \underset{1}{\cancel{5}}} = \frac{3 \times 1}{4 \times 1} = \frac{3}{4}$$

The GCF of 5 and 15 is 5.

The GCF of 4 and 16 is 4.

Divide both the numerator and denominator by 5 and then by 4.

Multiply. Write each product in simplest form.

1. $\frac{2}{7} \times \frac{2}{7}$

2. $\frac{1}{9} \times \frac{3}{4}$

3. $\frac{5}{6} \times \frac{11}{15}$

Exercises

Multiply. Write each product in simplest form.

1. $\frac{1}{2} \times \frac{1}{3}$
2. $\frac{1}{4} \times \frac{1}{2}$
3. $\frac{1}{5} \times \frac{2}{3}$
4. $\frac{3}{4} \times \frac{1}{5}$
5. $\frac{2}{5} \times \frac{3}{7}$
6. $\frac{3}{8} \times \frac{3}{5}$
7. $\frac{7}{8} \times \frac{2}{5}$
8. $\frac{5}{8} \times \frac{1}{5}$
9. $\frac{5}{9} \times \frac{1}{10}$
10. $\frac{7}{8} \times \frac{1}{6}$
11. $\frac{5}{9} \times \frac{3}{8}$
12. $\frac{7}{10} \times \frac{5}{4}$
13. $\frac{9}{10} \times \frac{5}{7}$
14. $\frac{5}{8} \times \frac{0}{3}$
15. $20 \times \frac{4}{5}$
16. $\frac{3}{4} \times 12$
17. $0 \times \frac{7}{8}$
18. $\frac{6}{7} \times \frac{7}{6}$
19. $\frac{4}{3} \times \frac{3}{4}$

Problem SOLVING

20. You pour a glass of water to $\frac{1}{2}$ full. After sitting on the counter for a few days, $\frac{1}{3}$ of the water you poured evaporated. How full is the glass now?

21. Mike plans to run $\frac{7}{8}$ mile. He begins running, but $\frac{2}{3}$ of the way he pulls his hamstring muscle. How far did he run?

22. Three-fourths of the bushes are azaleas. One-third of the azaleas are pink. What part of the bushes are pink azaleas?

23. A rectangle has a length of $\frac{3}{4}$ inch and a width of $\frac{1}{8}$ inch. Find the area of the rectangle. *Remember:* Area = length × width and your answer should be inches squared.

Constructed RESPONSE

24. Joey looks at the problem $\frac{1}{3} \times \frac{3}{8}$. His answer was $\frac{9}{8}$. What did he do wrong?

25. Anna completes the problem $\frac{1}{4} \times \frac{4}{5}$. Her answer is $\frac{1}{5}$, so she concludes that if the numerator of one fraction matches the denominator of the other, cross them out. The remaining numerator and denominator are the answer.

    Complete the following problems using the method that you learned, and then use Anna's method. Does Anna's method work? Explain.

    a. $\frac{1}{3} \times \frac{3}{8}$
    b. $\frac{3}{4} \times \frac{4}{5}$
    c. $\frac{8}{9} \times \frac{9}{11}$

MIXED REVIEW

Express the prime factorization of each number.

26. 15
27. 7
28. 4
29. 28
30. 35

6.4 Multiplying Mixed Numbers

Objective: to multiply fractions and mixed numbers

The dimensions of a typical U.S. flag are $1\frac{2}{3}$ feet by $3\frac{1}{6}$ feet. If you wanted to make a replica of the U.S. flag, what would the area of your flag be?

In this case, you need to multiply two mixed numbers.

Step 1	Step 2	Step 3
Rename mixed numbers as fractions. $1\frac{2}{3} \times 3\frac{1}{6} = \frac{5}{3} \times \frac{19}{6}$	Multiply. $\frac{5 \times 19}{3 \times 6} = \frac{95}{18}$	Rename as a mixed number. $\frac{95}{18} = 5\frac{5}{18}$

The area of your U.S. flag would be $5\frac{5}{18}$ in^2.

Examples

A. $3\frac{3}{5} \times 3\frac{1}{3}$ Estimate: $4 \times 3 = 12$

■ Method 1: Multiply first, and then simplify.

$$\frac{18}{5} \times \frac{10}{3} = \frac{180}{15} = \frac{12}{1} = 12$$

■ Method 2: Divide by common factors, and then multiply.

$$\frac{\overset{6}{\cancel{18}}}{\underset{1}{\cancel{5}}} \times \frac{\overset{2}{\cancel{10}}}{\underset{1}{\cancel{3}}} = \frac{12}{1} = 12$$

Compare with the estimate.

B. $4\frac{2}{3} \times 5$

$$\frac{14}{3} \times \frac{5}{1} = \frac{70}{3} = 23\frac{1}{3}$$

Multiply. Express each product as a mixed number in simplest form.

1. $2\frac{1}{5} \times 5$
2. $1\frac{1}{2} \times 1\frac{1}{2}$
3. $3\frac{1}{4} \times 2\frac{2}{3}$

Exercises

Multiply. Express each product as a mixed number in simplest form.

1. $1\frac{3}{8} \times 1$
2. $3 \times 1\frac{3}{4}$
3. $1\frac{1}{4} \times 3\frac{1}{2}$
4. $2\frac{3}{4} \times 1\frac{1}{5}$
5. $6 \times 7\frac{1}{2}$
6. $3\frac{1}{2} \times 1\frac{1}{2}$
7. $0 \times 2\frac{1}{3}$
8. $5\frac{1}{3} \times 1\frac{3}{5}$
9. $1\frac{3}{4} \times 11$
10. $1\frac{1}{3} \times \frac{5}{8}$
11. $3\frac{1}{4} \times 4$
12. $3\frac{3}{5} \times 3\frac{1}{3}$
13. $2 \times 3\frac{3}{4}$
14. $4\frac{1}{2} \times 2\frac{2}{5}$
15. $24 \times 2\frac{3}{8}$
16. $2\frac{4}{5} \times 1\frac{2}{3}$
17. $\frac{7}{3} \times 9$
18. $6\frac{4}{5} \times 0 \times 2\frac{9}{10}$
19. $1\frac{1}{2} \times 7 \times 2\frac{1}{3}$
20. $1\frac{1}{2} \times \frac{3}{4} \times 2\frac{1}{9}$
21. What is the common factor of the numerator and denominator of a fraction in simplest form?

Problem SOLVING

22. Peter filled 5 cups each with $1\frac{1}{2}$ pounds of candy. How much candy did all 5 cups contain together?
23. Kelsey discovered that an Alaskan brown bear is $1\frac{1}{8}$ times as long as a grizzly bear. If the average grizzly bear is 8 feet long, how long is an Alaskan brown bear?
24. The length of a high school basketball court is $1\frac{17}{25}$ times its width. If the width of the court is $4\frac{1}{6}$ yards, what is the length?

Constructed RESPONSE

25. Carlene was multiplying whole numbers and mixed numbers and discovered a short cut. Her work is shown. What property is Carlene using to solve the problem? Can this property be used to multiply two mixed numbers? Explain.

Carlene's Work

$$5\frac{1}{2} \times 6$$
$$= (5 \times 6) + (\frac{1}{2} \times 6)$$
$$= (30) + (3)$$
$$= 33$$

Test PREP

26. $6\frac{1}{4} \times 2\frac{4}{5} =$ ____.

a. $12\frac{4}{20}$ b. $12\frac{1}{5}$ c. $5\frac{1}{2}$ d. $17\frac{1}{2}$

Choosing the Right Tiles

Mr. Tessell owns a tile store. One day, the Hamiltons came into Mr. Tessell's store to buy new floor tile. He showed them a collection of tiles, and told them that some of the tiles would not cover their whole floor. Can you help the Hamiltons choose a tile, if they only want to use one type of tile? (*Hint:* Which shapes tessellate?)

Extension

Suppose you want to tile a floor with two types of tiles. You may use the tiles shown above or two polygons of your own choosing. Make a drawing of the tiles and the pattern you have chosen.

Cumulative Review

Write in words.

1. 630
2. 0.074
3. 52,022
4. 9.0016

Compare using <, >, or = .

5. 1,085 ● 1,805
6. 1.345 ● 1.350
7. 23,011 ● 23,010
8. 21.380 ● 20.380
9. $\frac{4}{5}$ ● $\frac{11}{15}$
10. $\frac{3}{4}$ ● $\frac{12}{16}$

Round to the underlined place-value position.

11. $0.0\underline{4}63$
12. $5\underline{4}.0218$
13. $1{,}465.\underline{3}81$
14. $52{,}000.\underline{4}49$

Estimate.

15. $\begin{array}{r} 249.8 \\ +\ 681.9 \\ \hline \end{array}$
16. $\begin{array}{r} 6{,}319 \\ -\ 4{,}801 \\ \hline \end{array}$
17. $\begin{array}{r} 137.5 \\ \times\ \ 2.8 \\ \hline \end{array}$
18. $\begin{array}{r} 6{,}542 \\ \times\ 207 \\ \hline \end{array}$
19. $9.9\overline{)1{,}199}$

Compute.

20. $\begin{array}{r} 37.019 \\ +\ 9.28 \\ \hline \end{array}$
21. $\begin{array}{r} 4{,}566 \\ +\ 8{,}209 \\ \hline \end{array}$
22. $\begin{array}{r} 12{,}836 \\ -\ 8{,}245 \\ \hline \end{array}$
23. $\begin{array}{r} 65.15 \\ -\ 31.06 \\ \hline \end{array}$
24. $\begin{array}{r} 847.08 \\ -\ 361.13 \\ \hline \end{array}$
25. $\begin{array}{r} 925 \\ \times\ 731 \\ \hline \end{array}$
26. $\begin{array}{r} 88.1 \\ \times\ \ 4.4 \\ \hline \end{array}$
27. $62\overline{)8{,}312}$
28. $4.3\overline{)206.4}$
29. $0.16\overline{)238.5}$

Find the LCM for each group of numbers.

30. 3, 7
31. 8, 10
32. 12, 16
33. 3, 5, 6

Add or subtract. Write each answer in simplest form.

34. $\frac{8}{11} + \frac{4}{11}$
35. $\frac{7}{8} - \frac{2}{3}$
36. $7\frac{1}{2} - 4\frac{1}{8}$
37. $5\frac{4}{7} + 3\frac{1}{14}$

Solve.

38. Jonathan spent \$25.42 on a new pair of tennis shoes and \$83.95 on a new tennis racquet. Find the total of Jonathan's purchases.
39. Patsy worked $4\frac{1}{2}$ hours on Friday, $3\frac{1}{4}$ hours on Saturday, and $3\frac{1}{4}$ hours on Sunday. She earns \$3.50 an hour. How much did Patsy earn altogether?
40. The drama club sold \$1,245 worth of play tickets. If 498 tickets were sold in all, how much did each ticket cost? Write the method of computation you used.
41. Ethan needs $4\frac{1}{2}$ cups of flour to make bread. He only has $2\frac{1}{8}$ cups. How much more flour does Ethan need?

6.5 Dividing Fractions

Objective: to divide fractions

Margaret is selling candy bars for a fundraiser. She has 10 candy bars, and wants to sell 2 candy bars to each person. To how many people can she sell candy bars? Think, how many 2s are in 10? You would solve this problem by dividing: $10 \div 2 = 5$.

But what if she wanted to sell $\frac{1}{2}$ a candy bar to each person? You still need to divide, but now you are dealing with a fraction. Think, how many $\frac{1}{2}$s are in 10?

1		1		1		1		1		1		1		1		1		1	
$\frac{1}{2}$	$\frac{1}{2}$	$\frac{1}{2}$	$\frac{1}{2}$	$\frac{1}{2}$	$\frac{1}{2}$	$\frac{1}{2}$	$\frac{1}{2}$	$\frac{1}{2}$	$\frac{1}{2}$	$\frac{1}{2}$	$\frac{1}{2}$	$\frac{1}{2}$	$\frac{1}{2}$	$\frac{1}{2}$	$\frac{1}{2}$	$\frac{1}{2}$	$\frac{1}{2}$	$\frac{1}{2}$	$\frac{1}{2}$

There are twenty $\frac{1}{2}$s in 10. But how can you divide any number by a fraction?

To divide a number by a fraction, multiply by the **reciprocal** of the fraction. Two numbers whose product is 1 are called reciprocals. You find the reciprocal of a number by writing the number as a fraction, and inverting the numerator and denominator.

Examples

A. $10 \div \frac{1}{2}$

$$\mathbf{10 \div \frac{1}{2} = 10 \times \frac{2}{1} = \frac{10}{1} \times \frac{2}{1} = \frac{20}{1} = 20}$$

$\frac{1}{2}$ and $\frac{2}{1}$ are reciprocals.

B. $\frac{3}{4} \div \frac{2}{3}$

$$\mathbf{\frac{3}{4} \div \frac{2}{3} = \frac{3}{4} \times \frac{3}{2} = \frac{9}{8} = 1\frac{1}{8}}$$

$\frac{2}{3}$ and $\frac{3}{2}$ are reciprocals.

C. $\frac{9}{4} \div \frac{3}{8}$

$\frac{3}{8}$ and $\frac{8}{3}$ are reciprocals.

$$\mathbf{\frac{9}{4} \div \frac{3}{8} = \frac{\overset{3}{\cancel{9}}}{\underset{1}{\cancel{4}}} \times \frac{\overset{2}{\cancel{8}}}{\underset{1}{\cancel{3}}} = \frac{6}{1} = 6}$$

Simplify before you multiply.

Try THESE

Find the reciprocal of each number.

1. $\frac{1}{3}$
2. $\frac{1}{4}$
3. $\frac{3}{5}$
4. 5
5. 9
6. $\frac{9}{3}$

Complete each sentence by writing the reciprocal of the divisor.

7. $8 \div \frac{1}{4} = 8 \times \frac{\blacksquare}{\blacksquare}$
8. $\frac{3}{4} \div \frac{1}{3} = \frac{3}{4} \times \frac{\blacksquare}{\blacksquare}$
9. $\frac{5}{6} \div 7 = \frac{5}{6} \times \frac{\blacksquare}{\blacksquare}$

Exercises

Divide. Write each quotient in simplest form.

1. $12 \div \frac{1}{2}$
2. $6 \div \frac{2}{3}$
3. $5 \div \frac{5}{8}$
4. $8 \div \frac{8}{9}$
5. $4 \div \frac{5}{6}$
6. $\frac{1}{4} \div \frac{3}{4}$
7. $\frac{7}{8} \div \frac{5}{8}$
8. $\frac{8}{3} \div \frac{2}{3}$
9. $\frac{6}{7} \div \frac{3}{7}$
10. $\frac{3}{4} \div \frac{2}{5}$
11. $\frac{7}{8} \div \frac{4}{5}$
12. $\frac{5}{7} \div \frac{1}{2}$
13. $\frac{2}{7} \div \frac{6}{9}$
14. $\frac{1}{4} \div \frac{4}{8}$
15. $\frac{5}{9} \div \frac{10}{11}$

Problem SOLVING

16. Flyers are folded into thirds before being stapled and mailed. If a page is 11 inches long, how wide is each section?

17. A meat company cuts slices of pepperoni with a thickness of $\frac{3}{8}$ inch. If the stick of pepperoni is 12 inches long, how many slices can be made from one stick?

18. A gourmet chocolate bar weighs $\frac{3}{4}$ pounds. There are 16 pieces on the chocolate bar. How much does each piece weigh?

19. Pizza crust recipes call for $\frac{1}{3}$ tablespoon of salt. The pizza is usually divided into 8 slices. How many tablespoons of salt are in one slice of pizza crust?

Constructed RESPONSE

20. Suppose it takes you $\frac{3}{4}$ of a minute to answer a multiple-choice question on a test. How many questions can you answer in 15 minutes? Explain.

Test PREP

21. Divide $\frac{1}{6}$ by $\frac{1}{3}$.

a. 2 b. 9 c. $\frac{1}{2}$ d. 18

22. Divide $\frac{2}{5}$ by $\frac{5}{8}$.

a. $\frac{16}{25}$ b. $\frac{2}{8}$ c. $\frac{10}{40}$ d. $\frac{1}{4}$

6.6 Dividing Mixed Numbers

Objective: to divide mixed numbers

Susan almost completed her summer reading. She read $14\frac{1}{4}$ books this summer in $6\frac{1}{3}$ weeks. How many books did she read each week?

You can find the answer by dividing $14\frac{1}{4}$ by $6\frac{1}{3}$.

Step 1	Step 2	Step 3
Change the mixed numbers to improper fractions. $14\frac{1}{4} \div 6\frac{1}{3} = \frac{57}{4} \div \frac{19}{3}$	Multiply by the reciprocal. $\frac{57}{4} \times \frac{3}{19}$	Simplify. $\frac{\overset{3}{\cancel{57}}}{4} \times \frac{3}{\underset{1}{\cancel{19}}} = \frac{9}{4} = 2\frac{1}{4}$

Estimating $14 \div 6$ is close to 2; the answer makes sense. The steps for dividing fractions and the steps for dividing mixed numbers are the same, but remember to change the mixed numbers to improper fractions first.

Examples

A. $2\frac{1}{4} \div 1\frac{1}{3}$

Estimate. $2 \div 1 = 2$.

$2\frac{1}{4} \div 1\frac{1}{3} = \frac{9}{4} \div \frac{4}{3}$

Change the mixed numbers to improper fractions.

$\frac{9}{4} \times \frac{3}{4} = \frac{27}{16} = 1\frac{11}{16}$

Simplify.

B. $7 \div 6\frac{1}{7}$

Estimate. $7 \div 6$ is close to 1.

$7 \div 6\frac{1}{7} = \frac{7}{1} \div \frac{43}{7}$

$\frac{7}{1} \times \frac{7}{43} = \frac{49}{43} = 1\frac{6}{43}$

Try THESE

Find the reciprocal of each number.

1. $3\frac{2}{5}$
2. $10\frac{1}{2}$
3. $3\frac{2}{3}$
4. $8\frac{2}{9}$
5. $12\frac{1}{9}$
6. $22\frac{5}{7}$

Exercises

Divide. Express each quotient as a mixed number in simplest form.

1. $7\frac{1}{2} \div 3\frac{3}{4}$
2. $9 \div 2\frac{1}{4}$
3. $10\frac{1}{2} \div 2\frac{5}{8}$
4. $1\frac{7}{8} \div \frac{3}{8}$
5. $2\frac{1}{3} \div 1\frac{1}{6}$
6. $7\frac{1}{8} \div \frac{3}{8}$
7. $1\frac{1}{2} \div 1\frac{1}{4}$
8. $1\frac{1}{2} \div 1\frac{2}{7}$
9. $6\frac{3}{4} \div 6$
10. $6 \div 1\frac{1}{4}$
11. $2\frac{1}{7} \div 1\frac{3}{7}$
12. $7\frac{1}{2} \div 4$
13. $1\frac{3}{4} \div 2\frac{4}{5}$
14. $2\frac{7}{9} \div 4\frac{1}{6}$
15. $0 \div 5\frac{2}{5}$
16. $4\frac{3}{5} \div 4\frac{3}{5}$
17. What is the reciprocal of $4\frac{3}{4}$?
18. Find the quotient for $8\frac{3}{16} \div 1$.

Problem SOLVING

19. Sean runs $3\frac{1}{2}$ miles in 28 minutes. How long does it take him to run 1 mile?
20. $8\frac{1}{4}$ pounds of flour are used to make $4\frac{1}{8}$ papier-mâché paste recipes. How much flour is used to make one recipe?
21. Cameron buys $\frac{3}{4}$ pound of almonds and divides it evenly among 5 friends. Find the amount of almonds each friend will get.
22. Ten people share $3\frac{1}{2}$ buckets of chicken.
    a. What fraction of a bucket does each person receive?
    b. If each bucket contains 10 pieces of chicken, how many pieces of chicken does each person receive?

Mid-Chapter REVIEW

Multiply or divide. Write each product or quotient in simplest form.

1. $12 \times \frac{5}{6}$
2. $\frac{3}{4} \times 15$
3. $\frac{3}{4} \times \frac{1}{9}$
4. $\frac{5}{8} \times \frac{2}{3}$
5. $4 \times 1\frac{2}{5}$
6. $7 \times 3\frac{4}{9}$
7. $4\frac{1}{2} \times 5\frac{1}{3}$
8. $\frac{3}{8} \times 2\frac{1}{7}$
9. $5 \div \frac{1}{8}$
10. $9 \div \frac{1}{4}$
11. $\frac{5}{6} \div \frac{1}{6}$
12. $\frac{4}{5} \div \frac{2}{5}$
13. $\frac{4}{9} \div \frac{2}{7}$
14. $\frac{5}{6} \div \frac{2}{5}$
15. $3\frac{1}{5} \div 6$
16. $2\frac{2}{3} \div 1\frac{1}{3}$

6.7 Problem-Solving Strategy: Work Backward

Objective: to solve problems by working backward

Chris went to the mall to spend his birthday money. He purchased a cinnamon roll with a drink for \$5. Then he bought a CD for \$18. Finally, Chris bought a video game for \$20. When he returned home he had \$7 left. How much money did Chris have before he went to the mall?

Some problems involve a sequence of information that leads to a final result. Then you are asked to find what happened in the beginning. To solve a problem like this, start from the end of the problem and **work backward** to the beginning.

Example

How much money did Chris start with in the problem above?

You know how much money Chris had when he returned home. You also know how much money he spent at the mall on various items. You need to find how much money he had before he went to the mall.

To find the amount of money that Chris had at the beginning, you will work backward. Start at the last clue and reverse them.

3. SOLVE

\$7 + \$20 = \$27	Chris had \$7 left. He spent \$20 on a game.
\$27 + \$18 = \$45	Chris spent \$18 on a CD.
\$45 + \$5 = \$50	Chris spent \$5 on a snack. He started with \$50.

4. CHECK

Look at the clues and start from the beginning.

\$50 – \$5 = \$45	Chris started with \$50. He spent \$5 on a snack.
\$45 – \$18 = \$27	Chris spent \$18 on a CD.
\$27 – \$20 = \$7	He spent \$20 on a game. Chris had \$7 left.

Try THESE

1. I am thinking of a number. If I divide the number by 5, and then add 5 to my number, the result is 11. What is my number?
2. My number is multiplied by 2, and then 4 is subtracted. The result is 6. What is my number?

Solve

Work backward to solve each problem.

1. I am thinking of a number. If I add 5 to my number, and then add 10, the result is 22. What is my number?
2. I am thinking of a number. If I multiply my number by 11, the result is 3.3. What is my number?
3. I am thinking of a number. If I divide my number by $\frac{2}{3}$, the result is 2. What is my number?
4. Sue gives $\frac{1}{2}$ of her plants to Lois, who gives $\frac{1}{4}$ of her plants to Kent. Kent receives 3 plants. How many plants does Sue have at the beginning?
5. Charles picks cucumbers. Each day he picks twice as many as the day before. On the fifth day, he picks 64 cucumbers. How many does he pick the first day?
6. The Smith family has four children. Madison is 5 years older than Michael. Tabitha is half as old as Michael. Tiffany is 2 years older than Tabitha. Tiffany is 8. How old are the other children?

★ 7. Gabriella brought cupcakes to class and asked for help passing them out. Nick took two-thirds of the cupcakes. He gave Henry half of those cupcakes. Henry kept three cupcakes, and gave six to Alexander. How many cupcakes did Gabriella bring to class?

Solve. Use any strategy.

8. A landscaper wants to arrange 12 square bricks to make a patio with the least perimeter possible. How many bricks will be in a row?
9. A rectangular area rug has an area of 98 square feet. The length of the rug is twice as long as the width. What are the dimensions of the rug?

Constructed RESPONSE

10. The sum of Kayla's age and her younger sister's age is 40. The difference between their ages is 6. How old is Kayla? Explain how you found your answer.
11. Refer to the table at the right. If the pattern continues, what will be the output when the input is 5? Explain the pattern.

Input	Output
0	7
1	9
2	11
3	
4	

6.8 Customary Units of Length

Objective: to convert measures of length

The Washington Monument is 555 ft $5\frac{1}{8}$ in. tall. How many inches tall is the structure? How many yards tall is the structure?

You are probably very familiar with the customary system. The United States is one of the few countries in the world that uses this system of measurement. Customary measurements usually involve fractions, as compared to the metric system which uses decimals.

The most commonly used units of customary length are **inches**, **feet**, **yards**, and **miles**.

Customary Units of Length			
Name of Unit	**Abbreviation**	**Equivalents**	**Measurable Objects**
inch	in.	12 in. = 1 ft 36 in. = 1 yd	length of soda bottle cap
foot	ft	1 ft = 12 in. 3 ft = 1 yd	length of adult male foot
yard	yd	1 yd = 3 ft 1,760 yd = 1 mi	width of typical door
mile	mi	1 mi = 1,760 yd 1 mi = 5,280 ft	length of 14 football fields length of 16 small city blocks

To change from a larger unit to a smaller one, multiply. To change from a smaller unit to a larger one, divide.

Examples

A. About how many yards tall is the Washington Monument?

555 feet = ________________ yards — Since feet are smaller than yards, divide.

$555 \div 3 = 185$ — There are 3 feet in 1 yard.

555 feet = 185 yards

B. A bowling alley is 20 yards in length from the foul line to the head pin. How many inches long is the bowling alley?

20 yards = ______________ inches — Since yards are larger than inches, multiply.

$20 \times 36 = 720$ — There are 36 inches in 1 yard.

20 yards = 720 inches

Try THESE

Choose an appropriate unit for each length. Explain.

1. length of a football
2. length of a paper clip
3. length of a kitchen
4. length of an airplane
5. length of a street
6. height of a tree
7. height of a small child
8. width of a car license plate

Exercises

Find each missing value.

1. 6 feet = ____ inches
2. 26,400 feet = ____ miles
3. 84 inches = ____ feet
4. 768 inches = ____ feet
5. 2 miles = ____ feet
6. $1\frac{1}{2}$ feet = ____ inches
7. $7\frac{3}{8}$ feet = ____ inches
8. $3\frac{3}{4}$ feet = ____ yards
9. $\frac{2}{5}$ miles = ____ yards
10. 300 inches = ____ yards
11. $1\frac{1}{4}$ miles = ____ inches

★12. 95,040 inches = ____ miles

Problem SOLVING

13. A fast garden snail can travel 12 inches in 2 minutes. If there are 60 minutes in an hour, at this pace how many feet will the snail travel in an hour?
14. On a baseball diamond, the distance from first base to third base is approximately 127 feet. How many inches is the distance from first base to third base?
15. The length of a chopstick is $\frac{3}{4}$ feet. How many inches long is a chopstick?
16. The official length of a marathon is $26\frac{1}{5}$ miles. How long is the marathon in feet?

Constructed RESPONSE

17. On a basketball court, the painted area from the baseline to the free throw line is a rectangle 15 feet long and 12 feet wide. Find the area in square yards. Explain how you found your answer.

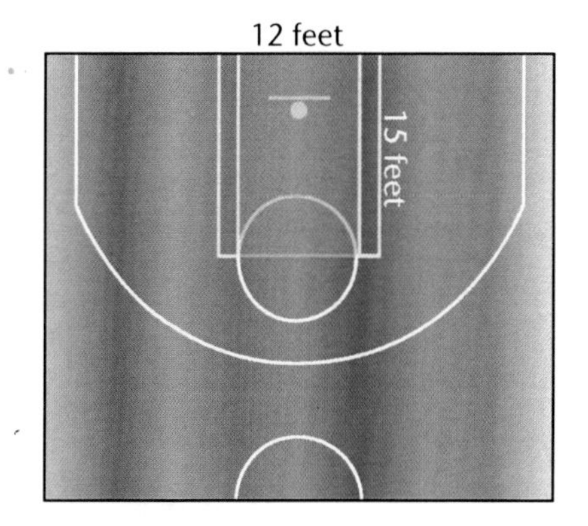

6.9 Weight and Capacity in Customary Units

Objective: to convert units of weight and capacity

According to the Department of Sanitation in New York City, New York, they collect 13,000 tons of garbage every day. That is a lot of trash! How many pounds of trash do they collect every day?

The most commonly used customary units for weight are ounces, pounds, and tons.

Customary Units of Weight			
Name of Unit	**Abbreviation**	**Equivalents**	**Measurable Objects**
ounce	oz	16 oz = 1 lb	weight of a slice of bread
pound	lb	1 lb = 16 oz	weight of a loaf of bread
ton	T	1 T = 2,000 lb	weight of a pickup truck

Examples

A. 13,000 tons = ________ pounds — Tons are larger than pounds, so multiply.

$13{,}000 \times 2000 = 26{,}000{,}000$ — There are 2,000 pounds in one ton.

13,000 tons = 26,000,000 pounds

To measure liquids, we use fluid ounces, cups, pints, quarts, and gallons.

Customary Units of Capacity			
Name of Unit	**Abbreviation**	**Equivalents**	**Measurable Objects**
fluid ounce	fl oz	1 fl oz = 2 tbsp	capacity of a cap of mouthwash
cup	c	1 c = 8 fl oz	capacity of a cup of milk
pint	pt	1 pt = 2 c	capacity of a bottle of water
quart	qt	1 qt = 2 pt	capacity of a jar of spaghetti sauce
gallon	gal	1 gal = 4 qt	capacity of a large can of paint

B. 32 quarts = ________ gallons — Quarts are smaller than gallons, so divide.

$32 \div 4 = 8$ — There are 4 quarts in one gallon.

32 quarts = 8 gallons

C. 56 ounces = ________ pounds

$56 \div 16 = 3\frac{3}{4}$

56 ounces = $3\frac{3}{4}$ pounds

D. 6 pints = ________ cups

$6 \times 2 = 12$

6 pints = 12 cups

Try THESE

Choose an appropriate unit for each weight. Explain.

1. bag of oranges
2. large car
3. package of gum

Choose an appropriate unit for each capacity. Explain.

4. tablespoon
5. gasoline in a car
6. tube of toothpaste

Exercises

Find each missing value.

1. 14 quarts = _____ gallons
2. 80,000 pounds = _____ tons
3. 8 cups = _____ pints
4. 14 quarts = _____ pints
5. 72 ounces = _____ pounds
6. 12 gallons = _____ quarts
7. 4 tons – _____ pounds
8. 5,000 pounds – _____ tons
9. 24 fluid ounces = _____ cups
10. 5 cups = _____ fluid ounces
11. 9 pints = _____ cups
12. 40 ounces = _____ pounds
13. 17 pints = _____ quarts
14. 3 pounds = _____ ounces
15. 48 fluid ounces = _____ quarts
16. 2 tons = _____ ounces
17. 10 quarts = _____ cups
18. 5,000 ounces = _____ tons
19. 2 gallons = _____ cups
20. 4 lb 4 oz = _____ ounces

Problem SOLVING

21. From a gallon of milk, Cheryl takes 1 pint to put on her cereal. How much milk does she have remaining?
22. A can of fruit weighs 40 ounces. How many pounds does the can weigh?
23. A can of soda has 12 fluid ounces. How many cups are in the can of soda?
24. A turkey weighs 12 lb 8 oz. How many ounces does the turkey weigh?
25. It is suggested that you drink $3\frac{1}{2}$ gallons of water every week. How many cups should you drink every day?
26. A large aquarium tank, such as those used for dolphins, can hold 15 million gallons of water. How many cups of water would this tank hold?

Constructed RESPONSE

27. An Olympic-sized swimming pool can hold 648,000 gallons of water. How many fluid ounces of water can the Olympic-sized swimming pool hold? Explain how you calculated your answer.

6.10 Adding and Subtracting Measures

Objective: to add and subtract units of measure

Patrick caught two fish, one weighing 2 lb 9 oz, and one 4 lb 8 oz. What was the weight of both fish?

To find the total weight add.

4 lb 8 oz
+ 2 lb 9 oz
6 lb 17 oz → 7 lb 1 oz

Simplify your answer since 17 oz is greater than 1 lb.

More Examples

A. **4 ft 10 in.**
+ 2 ft 6 in.
6 ft 16 in. = 7 ft 4 in.

B. **~~12~~ 11 yd ~~1~~ 4 ft**
− 2 yd 2 ft
9 yd 2 ft

Rename 12 yd 1 ft so you can subtract.

Use this summary of measures to help you complete the exercises.

Time
1 min = 60 s
1 h = 60 min
1d = 24 h
1wk = 7 d

Length
1 ft = 12 in.
1 yd = 3 ft or 36 in.
1 mi = 5,280 ft or 1,760 yd

Capacity
1c = 8 fl oz
1 pt = 2 c
1 qt = 2 pt
1 gal = 4 qt

Weight
1 lb = 16 oz
1 T = 2,000 lb

Try THESE

Add.

1. 3 d 6 h
+ 8 d 4 h

2. 16 ft 10 in.
+ 2 ft 8 in.

3. 8 pt 1 c
+ 4 pt 1 c

Subtract.

4. 3 yd 2 ft
− 1 yd 1 ft

5. 1 h 25 min
− 14 min

6. 2 T 1,000 lb
− 1 T 1,500 lb

Exercises

Add.

1. $\begin{array}{r} 14\text{ T } 1{,}000\text{ lb} \\ +\ 2\text{ T } 1{,}000\text{ lb} \\ \hline \end{array}$

2. $\begin{array}{r} 1\text{ mi } 100\text{ yd} \\ +\ 7\text{ mi } 1{,}700\text{ yd} \\ \hline \end{array}$

3. $\begin{array}{r} 12\text{ lb } 12\text{ oz} \\ +\ 17\text{ lb } 8\text{ oz} \\ \hline \end{array}$

4. $\begin{array}{r} 9\text{ d } 17\text{ h} \\ +\ 4\text{ d } 12\text{ h} \\ \hline \end{array}$

5. $\begin{array}{r} 7\text{ pt } 1\text{ c} \\ +\ 4\text{ pt } 1\text{ c} \\ \hline \end{array}$

6. $\begin{array}{r} 72\text{ ft } 7\text{ in.} \\ +\ 8\text{ ft } 5\text{ in.} \\ \hline \end{array}$

Subtract.

7. $\begin{array}{r} 18\text{ yd } 1\text{ ft} \\ -\ 17\text{ yd } 2\text{ ft} \\ \hline \end{array}$

8. $\begin{array}{r} 3\text{ h } 42\text{ min} \\ -\ 1\text{ h } 12\text{ min} \\ \hline \end{array}$

9. $\begin{array}{r} 7\text{ c } 3\text{ fl oz} \\ -\ 2\text{ c } 4\text{ fl oz} \\ \hline \end{array}$

10. $\begin{array}{r} 4\text{ lb } 4\text{ oz} \\ -\ 2\text{ lb } 7\text{ oz} \\ \hline \end{array}$

11. $\begin{array}{r} 7\text{ d } 12\text{ h} \\ -\ 5\text{ d } 18\text{ h} \\ \hline \end{array}$

12. $\begin{array}{r} 7\text{ T } 200\text{ lb} \\ -\ 2\text{ T } 700\text{ lb} \\ \hline \end{array}$

★13. $\begin{array}{r} 6\text{ hr } 12\text{ min } 25\text{ sec} \\ +\ 2\text{ hr } 55\text{ min } 45\text{ sec} \\ \hline \end{array}$

★14. $\begin{array}{r} 5\text{ gal } 2\text{ qt } 7\text{ c} \\ -\ 3\text{ gal } 1\text{ qt } 9\text{ c} \\ \hline \end{array}$

Problem SOLVING

15. Ada buys 3 pounds 8 ounces of ham and 4 pounds 2 ounces of cheese for a party. How much ham and cheese did she buy?

16. Mary Alice has a spool of red ribbon with 18 feet 10 inches on it and a spool of green ribbon with 12 feet 8 inches. How much ribbon does she have?

Constructed RESPONSE

17. Jack went to see a movie that was 2 h 32 min in length. His sister Jackie went shopping at a mall. She needed 1 h 45 min to shop. The mall is 20 minutes from the movie theater. Does Jackie have enough time to drop off Jack, go to the mall and shop, and then pick up her brother before the movie ends? Explain your answer.

Chapter 6 Review

Language and Concepts

Choose the correct word to complete each sentence.

1. It is (helpful, not helpful) to simplify before you multiply fractions so that the product is simplified.
2. When multiplying mixed numbers, first (simplify, rename) any mixed numbers as a fraction.
3. Simplify fractions by dividing the numerator and denominator by common (multiples, factors).
4. To (divide, multiply) by a fraction, multiply by its reciprocal.
5. The product of a number and its reciprocal is (one, zero).
6. To work backward start with the end result and then (undo, eliminate) each step.

Skills and Problem Solving

Compute mentally. Write each product in simplest form. Write only your answers. (Section 6.1)

7. $\frac{1}{5} \times 15$ **8.** $\frac{2}{3} \times 12$ **9.** $\frac{4}{7} \times 21$ **10.** $\frac{5}{9} \times 36$

Name the reciprocal of each number. (Section 6.5)

11. $\frac{1}{5}$ **12.** 7 **13.** $\frac{3}{8}$ **14.** $\frac{11}{7}$ **15.** 8

Compute. Write each answer is simplest form. (Sections 6.2–6.6)

16. $1\frac{1}{7} \times 5$ **17.** $4 \times \frac{1}{8}$ **18.** $\frac{2}{5} \times 3$ **19.** $12 \times \frac{2}{3}$

20. $\frac{1}{2} \times \frac{1}{3}$ **21.** $\frac{2}{3} \times \frac{1}{3}$ **22.** $\frac{1}{4} \times \frac{4}{5}$ **23.** $\frac{3}{4} \times \frac{5}{6}$

24. $2 \times 2\frac{1}{2}$ **25.** $3\frac{1}{4} \times 2$ **26.** $5 \times 1\frac{2}{3}$ **27.** $7\frac{1}{2} \times 4$

28. $1\frac{1}{2} \times 2\frac{1}{2}$ **29.** $2\frac{1}{3} \times 1\frac{1}{7}$ **30.** $3\frac{1}{4} \times 2\frac{2}{5}$ **31.** $1\frac{1}{8} \times 2\frac{2}{3}$

32. $5 \div \frac{1}{4}$ **33.** $3 \div \frac{6}{7}$ **34.** $\frac{3}{4} \div \frac{1}{2}$ **35.** $\frac{6}{5} \div \frac{4}{5}$

36. $\frac{3}{8} \div 3$ **37.** $\frac{2}{3} \div \frac{4}{3}$ **38.** $2\frac{1}{2} \div 5$ **39.** $3\frac{1}{3} \div 2$

40. $3\frac{1}{2} \div \frac{1}{4}$ **41.** $4\frac{1}{3} \div \frac{1}{6}$ **42.** $5\frac{1}{5} \div 1\frac{1}{4}$ **43.** $1\frac{7}{8} \div 3\frac{1}{4}$

Find each missing value. (Sections 6.8–6.9)

44. 3 feet = ____ inches
45. 13,200 feet = ____ miles
46. 144 inches = ____ feet
47. 2 miles = ____ feet
48. 28 quarts = _____ gallons
49. 20,000 pounds = _____ tons
50. 16 cups = _____ pints
51. 4 quarts = _____ pints
52. 64 ounces = _____ pounds
53. 9 gallons = _____ quarts
54. 3 tons = _____ pounds
55. 2 pints = _____ fluid ounces

Add or subtract. (Section 6.10)

56.
```
   9 lb  12 oz
 + 4 lb   9 oz
 ------------
```

57.
```
   12 T    500 lb
 −  5 T  1,000 lb
 ----------------
```

58.
```
   4 mi  1,000 yd
 + 3 mi  1,500 yd
 ----------------
```

Solve. (Section 6.7)

59. I am thinking of a number. If I subtract 5 from my number, and then multiply by 5, the result is 25. What is my number?

60. Michelle is packing gift baskets. Each basket contains either 3 candy bars and 1 bag of popcorn or 2 candy bars and 3 bags of popcorn. She uses 15 candy bars. What is a possible number of gift baskets that she could make? Explain how your found your answer.

61. Half of the students in the sixth grade are boys. One-fourth of the boys, or nineteen, have blonde hair. How many students are in the sixth grade?

Chapter 6 Test

Compute mentally. Write each answer in simplest form. Write only your answers.

1. $\frac{2}{3} \times 18$
2. $14 \times \frac{2}{7}$
3. $\frac{8}{5} \times 0$
4. $40 \times \frac{5}{8}$

Multiply or divide. Write each answer in simplest form.

5. $\frac{3}{4} \times 16$
6. $\frac{1}{6} \times \frac{5}{6}$
7. $\frac{5}{12} \times \frac{3}{4}$
8. $4\frac{3}{4} \times 8$
9. $5\frac{1}{5} \times 4$
10. $6 \times 3\frac{2}{3}$
11. $1\frac{2}{7} \times 2\frac{1}{3}$
12. $6\frac{2}{3} \times 2\frac{1}{10}$
13. $1\frac{1}{4} \times 2\frac{1}{2} \times 8$
14. $7 \div \frac{7}{9}$
15. $\frac{2}{5} \div \frac{4}{5}$
16. $\frac{3}{4} \div 9$
17. $\frac{7}{12} \div \frac{2}{9}$
18. $10 \div 1\frac{3}{7}$
19. $5\frac{1}{3} \div \frac{1}{6}$
20. $6\frac{2}{3} \div 1\frac{4}{7}$

Find each missing value.

21. 16 quarts = _____ gallons
22. 20,000 pounds = _____ tons
23. 18,480 feet = ____ miles
24. 200 inches = ____ feet

Add or subtract.

25.
```
   5 lb  8 oz
 − 2 lb  9 oz
```

26.
```
   5 T  1,500 lb
 + 3 T  1,000 lb
```

27.
```
   6 mi 2,000 yd
 + 2 mi   500 yd
```

Solve.

28. I am thinking of a number. If I add 3 to my number, and then multiply by 4, the result is 60. What is my number?
29. Michael is baking cakes and cookies. A cake uses 5 eggs and a batch of cookies uses 2 eggs. Michael uses 20 eggs. What are three possibilities for the number of cakes and batches of cookies that he made?
30. Maxwell and his sister Maxine are traveling across the country. Maxwell will drive $280\frac{1}{5}$ miles daily over a period of 5 days. Maxine will drive $220\frac{1}{2}$ miles daily over a period of $6\frac{1}{4}$ days.
    a. How far will Maxwell and Maxine each drive separately?
    b. What is the difference in their overall driving distances?
    c. Explain why the answer you determined is correct. Use what you know about mixed numbers in your explanation. Use words, numbers, and/or symbols in your explanation.

Change of Pace

Möbius Strip

A band of paper that has only *one side* and *one edge* is an example of a **Möbius strip.**

The Möbius strip was discovered in 1858 by a German astronomer and mathematician named August Ferdinand Möbius.

To make a Möbius strip, cut a long narrow strip from a piece of paper. The strip should be at least $1\frac{1}{2}$ inches by 11 inches. Label one end **A** and the other **B**.

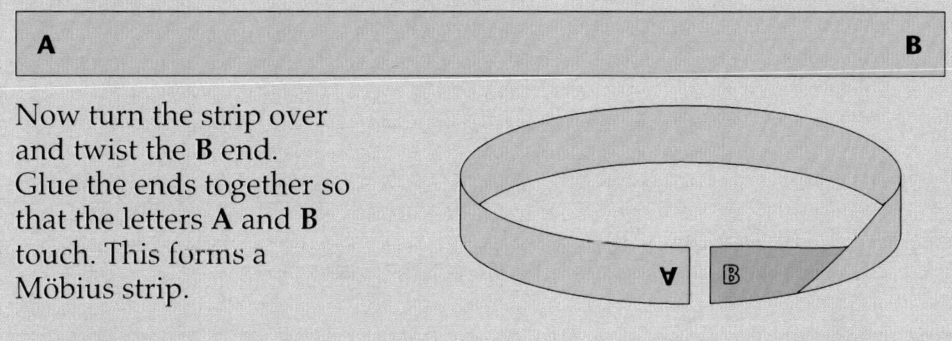

Now turn the strip over and twist the **B** end. Glue the ends together so that the letters **A** and **B** touch. This forms a Möbius strip.

Complete.

1. Make a Möbius strip. Mark a point in the center of the strip. Draw a line from that point until you return to that point. How does this show that the strip has only one side?

2. Cut the strip along the line you have drawn. How many strips do you have? How many twists are there? Are there now one or two sides?

3. Cut the strip down the middle again. How many strips do you have?

4. Make another Möbius strip. This time, start cutting it one-third of the way between the edges and cut until you return to the starting point. What is the result this time?

Research.

5. Because of the Möbius strip, fan belts on cars, conveyer belts, and even belts on sweepers have one twist in them. See if you can find out why.

Cumulative Test

1. $4 \times 1\frac{3}{4} =$ ____

a. $5\frac{3}{4}$ b. 6

c. 7 d. $8\frac{1}{2}$

2. Which is the reciprocal of $1\frac{1}{2}$?

a. $\frac{2}{3}$ b. $\frac{3}{2}$

c. $1\frac{2}{1}$ d. $2\frac{1}{1}$

3. $1\frac{1}{4} \div \frac{7}{8} =$ ____

a. $\frac{7}{10}$ b. $\frac{10}{7}$

c. $\frac{5}{3}$ d. $\frac{3}{5}$

4. Which is the even number?

a. 3 b. 6

c. 9 d. 11

5. Which is the standard form for MCMLX?

a. 1,960

b. 2,060

c. 2,140

d. none of the above

6. Which is the least common multiple of 4 and 10?

a. 2 b. 4

c. 20 d. 40

7. Geoffrey had nine dollar bills, a dime, and a nickel. Julie had eight dollar bills, three quarters, and four dimes. Which of these is true?

a. Geoffrey and Julie have the same amount of money.

b. Geoffrey has less money than Julie.

c. Geoffrey has more money than Julie.

d. Julie has more money than Geoffrey.

8. Which expression can be written as $6 \times (8 + 3)$?

a. $6 + (8 \times 3)$

b. $6 \times 8 + 3$

c. $(6 + 8) \times (6 + 3)$

d. $(6 \times 8) + (6 \times 3)$

9. The sum of two numbers is 20. The product of the sum and one of the numbers is 260. What are the numbers?

a. 5, 15

b. 6, 14

c. 7, 13

d. none of the above

10. Which inequality is true?

a. $0.025 < 0.031$

b. $0.39 < 0.053$

c. $2.019 > 2.038$

d. none of the above

CHAPTER 7

Ratios, Proportions, and Percents

William Martinez
Puerto Rico

7.1 Ratios

Objective: to name and write ratios; to write ratios as fractions

A volleyball team has 6 players. A baseball team has 9 players. The **ratio** of the number of players on a volleyball team to the number of players on a baseball team is 6 to 9.

A ratio is the comparison of two numbers. You can write ratios in the following ways.

6 to 9 **6 out of 9** **6:9** $\frac{6}{9}$

Example

The order of the numbers in a ratio shows the order in which the numbers are being compared.

players on a volleyball team → $\frac{6}{9}$ ← players on a baseball team. Read $\frac{6}{9}$ as six to nine.

players on a baseball team → $\frac{9}{6}$ ← players on a volleyball team. Read $\frac{9}{6}$ as nine to six.

players on a volleyball team → $\frac{6}{15}$ ← all players on both teams

all players on both teams → $\frac{15}{9}$ ← players on a baseball team

You have learned to use ratio tables to find equivalent ratios or an unknown part of the equivalent ratio by looking for patterns.

Volleyball players	6	12	18	24	?	36
Baseball players	9	18	27	36	45	?

How many volleyball players will there be if there are 45 baseball players?

How many baseball players will there be if there are 36 volleyball players?

Try THESE

Name the ratio that compares the number of objects in each phrase.

1. your ears to your eyes
2. your fingers to your feet
3. doors to your house
4. a car to its tires
5. centimeters to a meter
6. vowels to consonants in your name
7. boys to girls in your family
8. shoes to feet
9. phones in your house to number of people
10. inches in a foot to letters in the alphabet

Exercises

Write a phrase that would describe each ratio.

1. 2 to 6
2. 4 out of 5
3. 8:1
4. 14 to 7

Write each ratio in three other ways.

5. 3:7
6. 2:12
7. 4:5
8. 10:11

For each given ratio, write two equivalent ratios.

9. 2:3
10. 7:9
11. 4:9
12. 20:40

Use the chart to write each ratio as a fraction.

13. basketball to volleyball
14. soccer to softball
15. all uniforms to soccer
16. tennis to volleyball
17. softball to all uniforms

Sports Uniforms	
Type	**Number**
Basketball	12
Soccer	8
Volleyball	16
Softball	24
Tennis	36

Complete.

18.

Feet	2	4	6	?	?
Bodies	1	2	?	?	?

19.

Fruit	12	24	?	?	?
Apples	3	6	?	?	?

★20. 5 is to ■ as 30 is to 42

★21. ■ is to 15 as 4 is to 3

Problem SOLVING

22. Four baseballs cost \$7.50. What is the cost of 24 baseballs? $\frac{4}{7.50} = \frac{24}{■}$

23. The ratio of baseball gloves to players on the baseball team is 8 to 7. Is this ratio equivalent to 64 to 49?

24. The school sold 25 mixer tickets for \$37.50. What is the price of 40 tickets?

Constructed RESPONSE

25. Using the information in problem 24, determine how much money was raised by selling 180 tickets. Explain your work.

7.2 Rates

Objective: to find and use unit rates

Mark drove from Baltimore to Washington, DC. He traveled 40 miles and used 2 gallons of gas. A **rate** is a ratio that compares two quantities measured in different units. The rate $\frac{40 \text{ miles}}{2 \text{ gallons}}$ compares miles to gallons. The rate for one unit of a given quantity is called the **unit rate**. A unit rate has a denominator of 1.

$$\frac{\textbf{40 miles}}{\textbf{2 gallons}} \longrightarrow \mathbf{40 \div 2 = 20}$$

The unit rate is 20 miles per gallon.

Examples

Suppose a box of cereal contains 420 grams and 14 servings.

A. Write a rate that compares the grams and the servings for the box of cereal.

Since there are 420 grams and 14 servings, you can write a rate comparing the two quantities:

$$\frac{\textbf{420 grams}}{\textbf{14 servings}}$$

B. Find the unit rate for 420 grams and 14 servings.

$$\frac{\textbf{420 grams}}{\textbf{14 servings}} \longrightarrow \mathbf{420 \div 14 = 30}$$

The unit rate is $\frac{30 \text{ grams}}{1 \text{ serving}}$ or 30 grams per serving.

C. One serving of cereal contains 30 grams. How many grams of cereal would equal 5 servings?

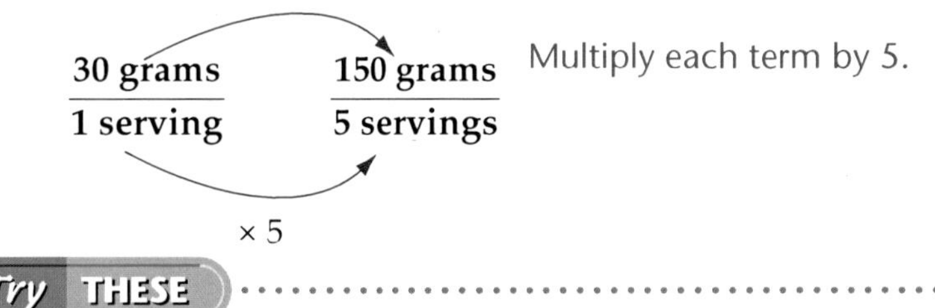

***Try* THESE**

Find the unit rate for each situation.

1. 36 inches in 3 feet
2. 300 calories in 2 servings
3. 330 miles on 15 gallons
4. $3.84 for 32 ounces

Exercises

Write each situation as a ratio. Then find the unit rate.

1. 180 heartbeats in 3 minutes
2. 200 students on 4 buses
3. 200 miles in 5 hours
4. $20 for 2 T-shirts

Find the unit rate for each situation.

5. $32 for 16 ounces
6. 3 pounds for $3.09
7. $84 for 12 hours
8. 27 miles in 6 hours
9. 50 players on 10 teams
10. 100 pages in 12 hours

Find each unit price. Round to the nearest cent. Then determine the better buy.

11. donuts: 6 for $2.95
    12 for $5.75
12. oranges: 5 pounds for $2.49
    8 pounds for $3.89
13. juice: 48 ounces for $2.99
    64 ounces for $4.49
14. pencils: 12 for $1.49
    100 for $9.99

Tell which unit rate is greater.

15. Olivia earns $138 in 12 hours. Jackie earns $281.25 in 25 hours.
16. Zoe types 175 words in 5 minutes. Ana types 1,800 words in one hour.
17. Shannon runs 2.5 miles in 18 minutes. Tom runs 9.25 miles in 111 minutes.
18. Lenny scored 279 points in 15 weeks. Larry scored 150 points in 8 weeks.

Constructed RESPONSE

19. Answer the following questions about unit rate.
    a. What is a unit rate? How do you find a unit rate?
    b. How are ratios and rates alike? How are they different?

Test PREP

20. You shovel snow in winter. After shoveling 8 driveways, you have earned $60. How much do you earn for each driveway?

    a. $7 b. $7.25 c. $7.50 d. $7.75

21. In the school, there are 160 students and 20 teachers. What is the unit rate?

    a. $\frac{\text{150 students}}{\text{20 teachers}}$ b. $\frac{\text{75 students}}{\text{10 teachers}}$ c. $\frac{\text{8 students}}{\text{1 teacher}}$ d. $\frac{\text{7.5 students}}{\text{1 teacher}}$

7.3 Proportions

Objective: to identify and solve proportions

Last season, Brad made 4 out of every 7 field goals that he attempted. If Brad continues at the same rate this season, how many field goals will he make out of 21 tries?

The ratio of the number of field goals made to the total number of tries is 4 to 7. If f represents the number of field goals Brad will make this season, you can write the ratio $\frac{f}{21}$ to represent a ratio that is equivalent to $\frac{4}{7}$.

An equation that states that two ratios are equivalent is called a **proportion**. We stated that the ratio $\frac{4}{7}$ is equivalent to $\frac{f}{21}$. The proportion would be written $\frac{4}{7} = \frac{f}{21}$.

Examples

A. Solve the proportion $\frac{4}{7} = \frac{f}{21}$.

Method 1: Use equivalent fractions.	**Method 2: Use cross products.**
$\frac{4}{7} = \frac{f}{21}$ $\frac{4 \times 3}{7 \times 3} = \frac{f}{21}$ $f = 12$	You can use cross products to solve a proportion. In a proportion, the cross products are equal. To find the cross products, multiply diagonally. $\frac{4}{7} = \frac{f}{21}$ $4 \times 21 = 7 \times f$ $84 = 7 \times f$ $f = 84 \div 7$ inverse operation $f = 12$

B. Solve the proportion $\frac{5}{6} = \frac{m}{24}$.

$\frac{5}{6} = \frac{m}{24}$

$\frac{5 \times 4}{6 \times 4} = \frac{20}{24}$

$m = 20$

$\frac{5}{6} = \frac{m}{24}$

$5 \times 24 = 6 \times m$

$120 = 6 \times m$

$m = 120 \div 6$

$m = 20$

Try THESE

Use cross products to see if the ratios form a proportion. Replace each ● with = or ≠ (*not equal to*).

1. $\frac{5}{9} \bullet \frac{3}{5}$
2. $\frac{4}{9} \bullet \frac{4}{10}$
3. $\frac{2}{3} \bullet \frac{6}{9}$
4. $\frac{15}{18} \bullet \frac{5}{6}$
5. $\frac{2}{1} \bullet \frac{4.2}{2.1}$
6. $\frac{5}{0.2} \bullet \frac{45}{3}$
7. $\frac{10}{7} \bullet \frac{3.5}{2.45}$
8. $\frac{0.65}{0.9} \bullet \frac{1.3}{1.8}$

Exercises

Solve each proportion by finding equivalent fractions.

1. $\frac{4}{q} = \frac{48}{60}$
2. $\frac{8}{9} = \frac{m}{90}$
3. $\frac{5}{13} = \frac{45}{w}$
4. $\frac{6}{7} = \frac{n}{42}$

Solve each proportion by using cross products.

5. $\frac{t}{3} = \frac{52}{78}$
6. $\frac{9}{b} = \frac{54}{96}$
7. $\frac{39}{247} = \frac{3}{y}$
8. $\frac{18}{30} = \frac{v}{15}$

Solve. Use any method.

9. $\frac{49}{56} = \frac{7}{u}$
10. $\frac{c}{22} = \frac{81}{198}$
11. $\frac{a}{6} = \frac{29}{174}$
12. $\frac{4.5}{9} = \frac{10}{t}$
13. $\frac{2.3}{9.2} = \frac{d}{1.2}$
14. $\frac{1.05}{10.5} = \frac{z}{15}$
15. ★ $\frac{5\frac{1}{2}}{16\frac{1}{2}} = \frac{1}{f}$
16. ★ $\frac{2\frac{3}{4}}{1\frac{3}{8}} = \frac{18}{k}$

Problem SOLVING

17. How many oranges are needed to serve punch to 100 people?
$\frac{6}{25} = \frac{r}{100}$

18. How many cups of sugar are needed to serve punch to 125 people?

19. Your friend is having a poster made from a photograph that is 4 inches wide and 6 inches long. If the poster will be 22 inches wide, how long will the poster be?

Fruit Punch

(Serves 25 people)

$1\frac{1}{2}$ cups water	juice of 6 lemons
$1\frac{1}{2}$ cups sugar	juice of 6 oranges
1 quart grape juice	1 pint tea
2 quarts chilled water	2 cups crushed pineapple

Constructed RESPONSE

20. A school has 180 students and 12 teachers. The school is adding 30 students next year and needs to keep the same student to teacher ratio. How many teachers need to be hired for next school year? Explain how you found your answer.

7.4 Similar Figures

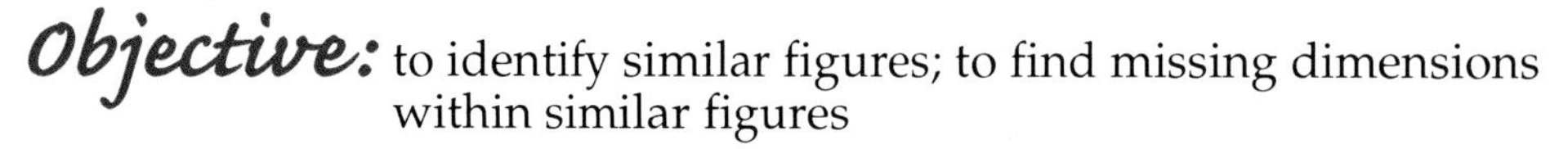

Objective: to identify similar figures; to find missing dimensions within similar figures

The two baseball diamonds shown are the same shape but not the same size. **Similar figures** have the same shape, but may differ in size. The lengths of the **corresponding sides**, or matching sides, of similar figures are proportional.

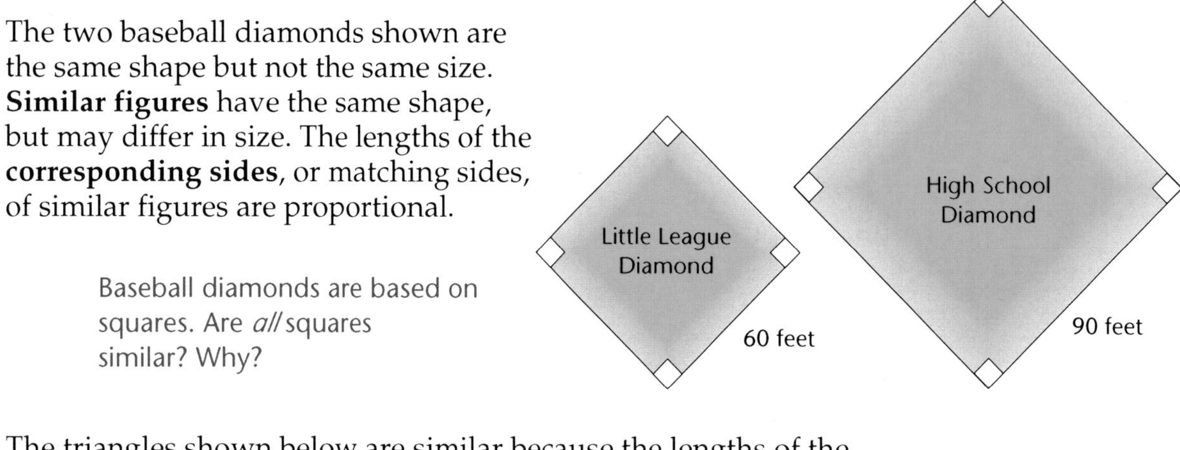

Baseball diamonds are based on squares. Are *all* squares similar? Why?

The triangles shown below are similar because the lengths of the *corresponding* or matching sides can be written as equivalent ratios.

$$\frac{4\text{ cm}}{8\text{ cm}} = \frac{3\text{ cm}}{6\text{ cm}} = \frac{5\text{ cm}}{c}$$

> The lengths of the corresponding sides of similar figures are proportional.

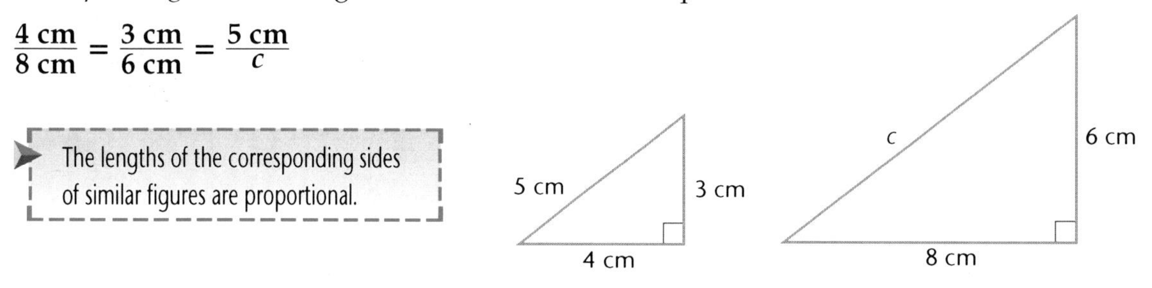

You can find the missing length by writing and solving a proportion. You can use $\frac{4}{8} = \frac{5}{c}$ or $\frac{3}{6} = \frac{5}{c}$.

$\frac{4}{8} = \frac{5}{c}$	$\frac{3}{6} = \frac{5}{c}$
$c \times 4 = 8 \times 5$	$c \times 3 = 6 \times 5$
$c \times 4 = 40$	$c \times 3 = 30$
$c = 10$ $c = 40 \div 4$	$c = 10$ $c = 30 \div 3$

Both proportions show that the missing length is 10 cm.

Try THESE

Are the figures similar? Write *yes* or *no*.

1. 2. 3.

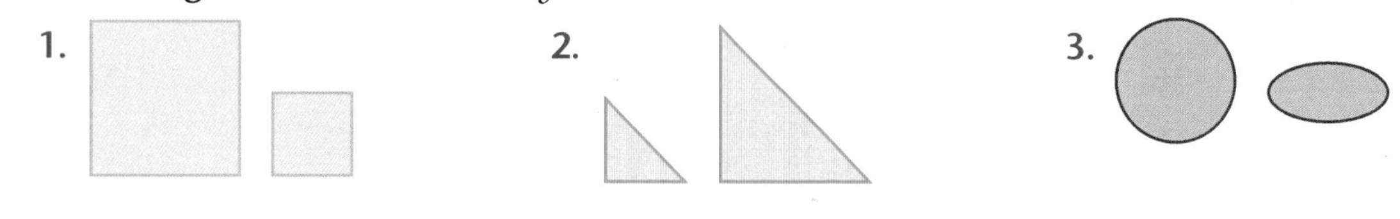

Exercises

For each pair of similar figures, use proportions to find the missing length.

1. 2.

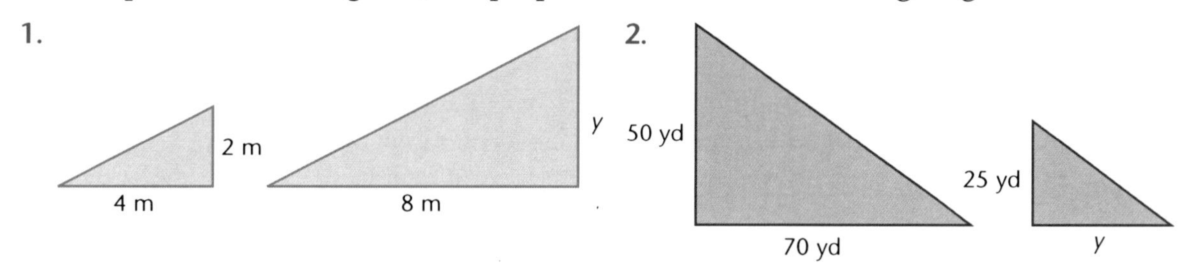

Solve. Use the similar triangles and proportions.

3. How long is the ladder?

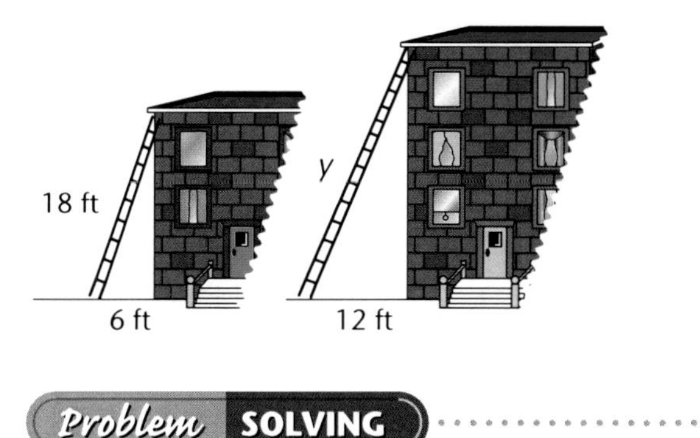

★ 4. How high is the building?

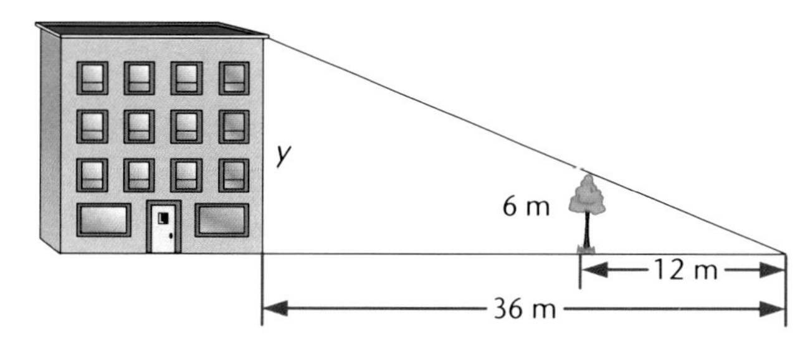

Problem SOLVING

5. The shadow cast by a man who is 6 feet tall is 10 feet. If the shadow of a flagpole is 20 feet, how tall is the flagpole?

6. Bernard has 2 envelopes. One is 6 cm by 22 cm. The other is 14 cm by 60 cm. Are the envelopes similar?

7. Jackie had a photo that was 3 in. wide and 5 in. long. She had the photo enlarged, and the enlargement was 9 in. wide. How long was the enlargement?

8. James is completing the following problem, and found the answer for the missing side. James says the length of x is 126 ft. Is he correct? Explain using words, symbols, and your knowledge of similar figures and proportions.

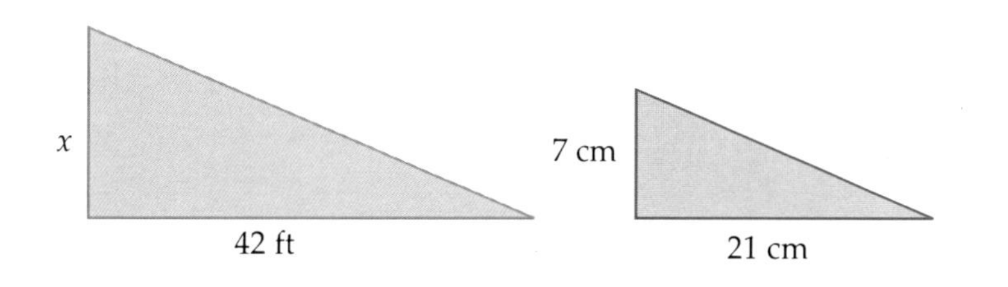

7.5 Scale Drawings

Objective: to use scale drawings to find missing measurements

The figure below is a **scale drawing** of a new swimming pool at Trevino Park. A scale drawing is an enlarged or reduced drawing of an object that is similar to the actual object. Scale drawings are used to picture things that are too large or too small to be shown in actual size.

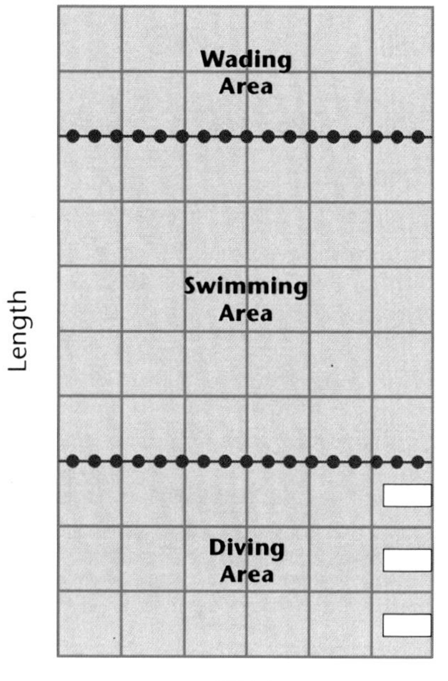

Each blue square in the scale drawing is 1 centimeter on a side. The following ratio indicates that 1 centimeter on the scale drawing represents 4 meters on the actual pool.

$\frac{1}{4}$ scale drawing distance in centimeters / actual distance in meters

> Can you think of some uses for scale drawings? A great example of a scale drawing is a map. Can you think of any others?

You can use proportions to find actual distances on scale drawings. Use the proportion $\frac{1}{4} = \frac{6}{w}$ to find the actual width (w) of the pool.

scale drawing → $\frac{1}{4} = \frac{6}{w}$ ← scale drawing
actual → ← actual

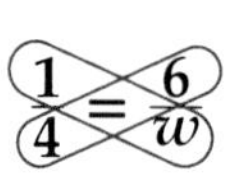

$$w \times 1 = 4 \times 6$$

$$w = 24$$

The actual width of the pool is 24 meters.

Try THESE

Complete. Use the scale drawing.

1. What is the actual length (ℓ) of the entire pool?
$\frac{1}{4} = \frac{10}{\ell}$

2. What is the actual width (w) of the diving area?
$\frac{1}{4} = \frac{6}{w}$

3. What is the actual length of the wading area?

4. What is the actual length of the swimming area?

5. Find the actual area of the entire pool in square meters. (Area = length × width)

Exercises

On the map of Texas, 1 centimeter represents 150 kilometers. Use a ruler and a proportion to find each distance.

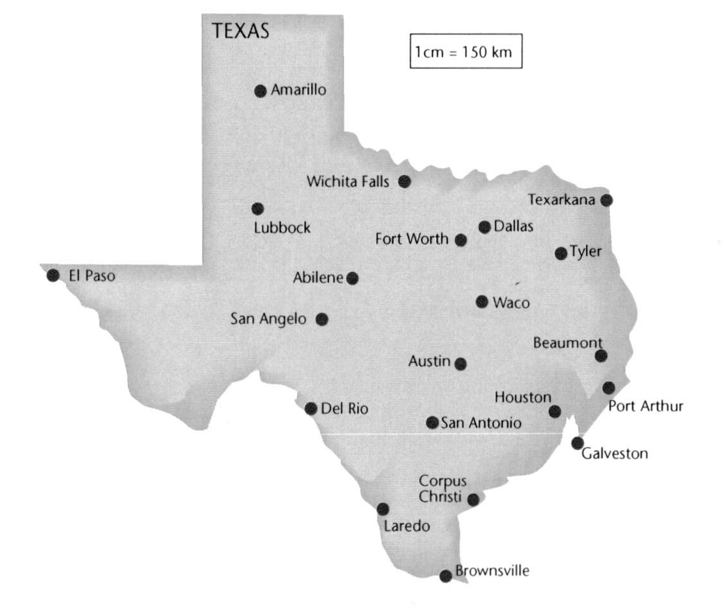

1. San Antonio to Galveston

   map → $\frac{1}{150} = \frac{2}{d}$ ← map
   actual → ← actual

2. Houston to Beaumont

   map → $\frac{1}{150} = \frac{1}{d}$ ← map
   actual → ← actual

3. Waco to Tyler
4. El Paso to Lubbock
5. Dallas to San Antonio

Solve. Use any strategy.

6. An Indy car must have a length near 192 inches. Jeanne is making a scale model of an Indy car using the scale 1 inch:6 inches. How long will her model car be?

7. Jeanne also plans to make a model of the Indianapolis Motor Speedway, an oval that is 2.5 miles around. Using the same scale, 1 inch:6 inches, what would be the distance for a lap on her model track?

Constructed RESPONSE

8. When you are using a map to find actual distances, are your calculations exact answers or approximate distances? Explain your answer.

Mid-Chapter REVIEW

Write each ratio as a fraction.

1. 8 to 9
2. 5 out of 5

Replace each ■ so that the ratios are equivalent.

3. $\frac{1}{5} = \frac{■}{10}$
4. $\frac{9}{6} = \frac{3}{■}$

5. The triangles at the right are similar. Find the missing length.

14 m, 50 m, b — 7 m, 25 m, 24 m

Solve.

6. A truck travels 85 km in 1 hour. At this rate, how far will it travel in 4 hours?

7. On a map, 1 inch represents 8 miles. The map distance is 6 inches. Find the actual distance.

Always a Winner

Ryan knows that his gym teacher always chooses a team captain in the same way. She asks the students to line up and count off until everyone has a number. Then she starts with the first student in line and has the first student remain standing, the second student sit down, the third student remain standing, the fourth student sit down, and so on. She continues until only one student is standing. If there are 18 students in gym class, where should Ryan stand to be chosen captain?

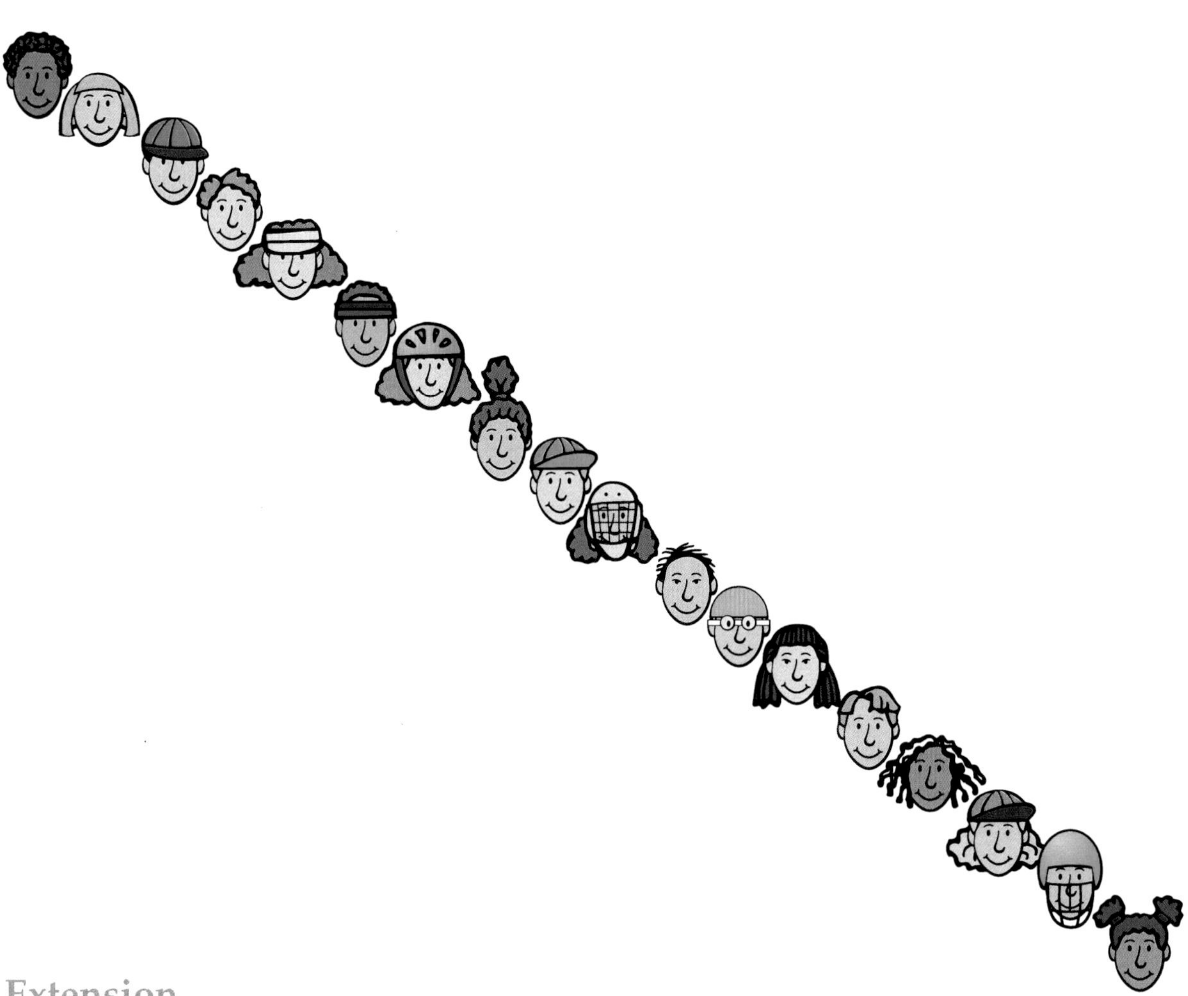

Extension

Where should Ryan stand in line if there are 8 students in gym class? 10 students? 13 students? 16 students? 17 students? 24 students?

Cumulative Review

Estimate.

1. $\begin{array}{r} 468 \\ +\ 801 \\ \hline \end{array}$
2. $\begin{array}{r} 9{,}320 \\ -\ 4{,}105 \\ \hline \end{array}$
3. $\begin{array}{r} 91 \\ \times\ 6 \\ \hline \end{array}$
4. $\begin{array}{r} 83 \\ \times\ 39 \\ \hline \end{array}$
5. $4\overline{)1{,}217}$

Compute.

6. $\begin{array}{r} 956 \\ +\ 623 \\ \hline \end{array}$
7. $\begin{array}{r} 848 \\ 65 \\ +\ 107 \\ \hline \end{array}$
8. $\begin{array}{r} \$807 \\ -\ 494 \\ \hline \end{array}$
9. $\begin{array}{r} 9{,}000 \\ -\ 1{,}232 \\ \hline \end{array}$
10. $\begin{array}{r} 13.5 \\ 0.6 \\ +\ 9.2 \\ \hline \end{array}$
11. $\begin{array}{r} 68 \\ \times\ 7 \\ \hline \end{array}$
12. $\begin{array}{r} 209 \\ \times\ 4 \\ \hline \end{array}$
13. $\begin{array}{r} 61 \\ \times\ 28 \\ \hline \end{array}$
14. $\begin{array}{r} 0.45 \\ \times\ 31 \\ \hline \end{array}$
15. $\begin{array}{r} 10.8 \\ \times\ 6.5 \\ \hline \end{array}$
16. $17.26 - 9.08$
17. $27.2 + 9.42 + 0.15 + 6.78$
18. $5.4 \times 1{,}000$
19. $890 \div 100$
20. $8\overline{)53.6}$
21. $0.9\overline{)51.3}$

Compute. Write each answer in simplest form.

22. $\frac{2}{3} + \frac{5}{6}$
23. $1\frac{3}{4} + \frac{5}{8}$
24. $1\frac{2}{3} - \frac{1}{2}$
25. $1\frac{5}{8} - \frac{15}{16}$
26. $\frac{1}{3} \times \frac{4}{5}$
27. $2\frac{1}{3} \times 3$
28. $\frac{7}{8} \div \frac{3}{4}$
29. $2\frac{1}{2} \div 4$

Copy and complete.

30. 2 mg = ■ g
31. 5 m = ■ cm
32. 18 L = ■ mL
33. 2 km = ■ m

Solve.

34. A recipe calls for 12 oz of fresh cranberries to make 2 cups of cranberry sauce. How many cranberries are needed to make 6 cups of cranberry sauce?

35. Ramona buys 2 cans of tennis balls. How much change does she receive from $10?

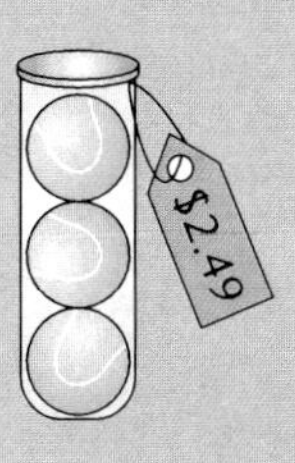

7.6 Problem-Solving Strategy: Use a Formula

Objective: to solve problems by writing an equation

Sam Wheatly enjoys collecting antiques. Sometimes he has to drive many miles to get to an antique show. Sam was told that it would take 6 hours to get from Amarillo to Pecos if he drove 55 miles per hour. How many miles is it from Amarillo to Pecos?

The relationship between distance, rate, and time can be written as a **formula**. A formula is a general statement showing how certain quantities are related. Knowing a formula will help you remember what steps to take in solving a problem. Writing an **equation** is a way to organize the information needed to solve a problem.

Examples

A. How many miles is it from Amarillo to Pecos?

distance = rate × time

$d = r \times t$	distance = mileage, rate = speed, time = hours
$d = 55 \times 6$	Replace *r* with 55 mph and *t* with 6 hours.
$d = 330$	It is 330 miles from Amarillo to Pecos.

B. Ms. Romano drove her car 88.8 miles. If the car used 6 gallons of gasoline, how many miles per gallon (mpg) does her car get?

Use the mpg formula, $s = m \div g$.

miles per gallon = miles ÷ gallons

$s = m \div g$	
$s = 88.8 \div 6$	Replace *m* with 88.8 miles and *g* with 6 gallons.
$s = 14.8$	Ms. Romano's car gets 14.8 miles per gallon.

C. Your math teacher gave you two formulas for converting temperature between Celsius and Fahrenheit: $C = \frac{5}{9}(F - 32)$ and $F = \frac{9}{5}C + 32$. The problem in your homework asks you to change 100°C to Fahrenheit. What is 100°C in Fahrenheit?

Use the formula $F = \frac{9}{5}C + 32$ to change from Celsius to Fahrenheit.

$F = \frac{9}{5}C + 32$	
$F = \frac{9}{5}(100) + 32$	Replace the *C* with 100.
$F = 180 + 32$	
$F = 212$	100°C = 212°F

Try THESE

Match each situation with the appropriate equation.

1. A box of chocolates costs \$5. There are 10 chocolates in the box. What is the price for each chocolate?
2. Sugar costs \$5 per bag. Find the price for 10 bags.
3. You and four friends shared a \$10 pizza. If everyone paid the same amount, how much did each of you pay?
4. Mary has 10 red jelly beans and 5 white jelly beans. How many jelly beans does she have altogether?

a. $P = 5 + 10$

b. $P = 5 \div 10$

c. $P = 10 \div 5$

d. $P = 5 \times 10$

Solve

Solve. Use the appropriate formula.

1. It took Carla 3 hours at 55 miles per hour to reach Salty State Park. How many miles did Carla drive?
2. Bob drove his van 52 miles on 5 gallons of gasoline. How many miles per gallon does the van get?
3. Suppose you drove 200 miles in 4 hours. What was your speed? Use the formula $r = d \div t$.

   (rate = distance ÷ time)
4. Howard drove 222.6 miles. His car gets 26.5 miles per gallon. How many gallons did his car use? Use the formula $g = m \div s$.

   (gallons = miles ÷ mpg)
5. What should your blood pressure be? Use $B.P. = 110 + \frac{A}{2}$.

   ($B.P.$ = blood pressure and A = age)
6. Normal room temperature is 72°F. What would this temperature be if measured in Celsius? Round to the nearest degree.

Use the pictograph.

7. Write an expression for the pictograph that represents the total number of cards collected.
8. If Barb collects 84 cards, how many pictures would be shown on the pictograph? Write an equation and solve.
9. This was Stan's first year for collecting baseball cards. If he continues to collect cards at the same rate each year, how many cards will he have at the end of the third year? Write a pattern that shows the solution.

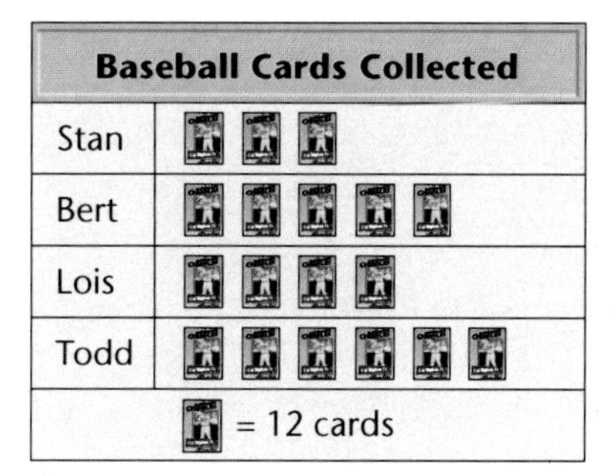

7.7 Understanding Percents

Objective: to find percents

The soccer teams from two middle schools are matched in a local soccer tournament. Sixty-five out of every one hundred spectators support the team wearing the yellow jerseys.
The ratio 65 to 100 can be written as the fraction $\frac{65}{100}$.

A **percent** is a ratio that compares a number to 100.

Percent means hundredths.

Since percent means hundredths, you can write $\frac{65}{100}$ as a percent.

Write: 65%
Say: *sixty-five percent*

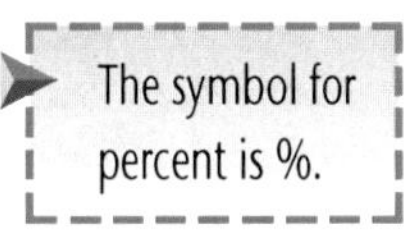

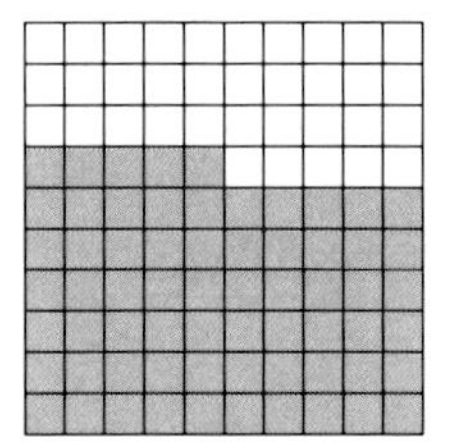

$\frac{65}{100}$ or 65% of the diagram is shaded.

Examples

A. Ms. Andrews' class distributed a sports survey to all sixth graders in Jackson Middle School. Out of the 100 surveys returned, 50 students said they were regularly involved in at least one sports activity. What percent of the students are active in a sport?

50 out of 100 students are in a sport.

$$\frac{50}{100} = 50\%$$

B. What percent of the diagram is shaded?

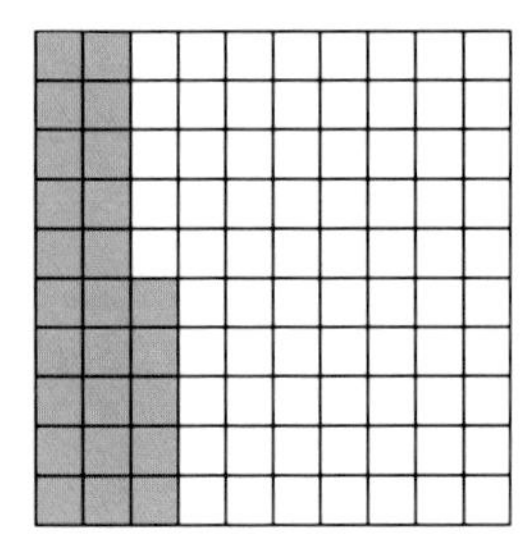

There are 100 squares, and 25 are shaded.

25 out of 100 = $\frac{25}{100} = 25\%$

Try THESE

Rename each fraction as a percent.

1. $\frac{6}{100}$
2. $\frac{78}{100}$
3. $\frac{150}{100}$

What percent of each diagram is shaded?

4.

5. 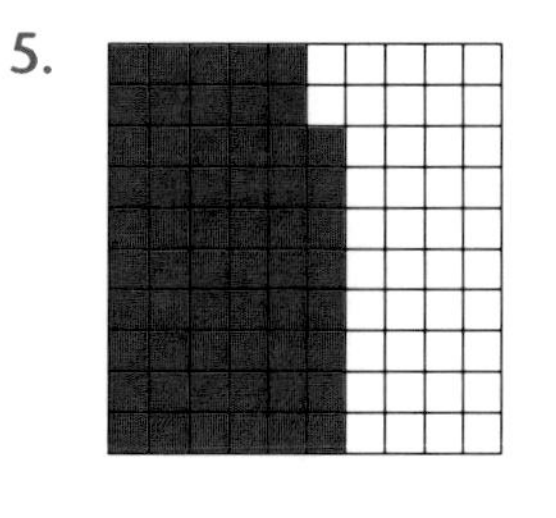

Exercises

Rename each fraction or ratio as a percent.

1. $\frac{8}{100}$
2. $\frac{23}{100}$
3. $\frac{32}{100}$
4. $\frac{85}{100}$
5. $\frac{98}{100}$
6. $\frac{1}{100}$
7. $\frac{50}{100}$
8. $\frac{100}{100}$
9. $\frac{125}{100}$
10. $\frac{205}{100}$
11. 35 to 100
12. 60 to 100
13. 9¢ to 100¢
14. 99¢ to 100¢
15. 170 to 100
16. 300 to 100

Match.

17. 15 to 100 — a. 87%
18. $\frac{87}{100}$ — b. $\frac{149}{100}$
19. one hundred one percent — c. twelve to one hundred
20. 12% — d. 101%
21. 149% — e. fifteen percent

Problem SOLVING

22. In a poll of 100 students, 36 chose red as their favorite color. What percent chose a color other than red?

23. The school baseball team won 60 percent of its home games. What percent of its home games did the team lose?

24. Suppose John correctly solves 87 problems out of 100 on a test. What percent did he solve correctly?

7.8 Percents as Fractions and Decimals

Objective: to write percents as decimals and fractions

The results of the Jackson Middle School sports survey indicate that 5% of the sixth graders play lacrosse. What fraction and decimal represent the number of students who play lacrosse?

To rename a percent as a fraction, write the percent as a fraction with a denominator of 100. Simplify if possible.

$$5\% = \frac{5}{100} = \frac{5 \div 5}{100 \div 5} = \frac{1}{20}$$

To rename a percent as a decimal, write the percent as a fraction with a denominator of 100. Write the decimal.

$$5\% = \frac{5}{100} = 0.05$$

Examples

A. Rename $33\frac{1}{3}\%$ as a fraction in simplest form.

$33\frac{1}{3}\% = \dfrac{33\frac{1}{3}}{100}$ Write the percent as a fraction with a denominator of 100.

$= 33\frac{1}{3} \div 100$ $\dfrac{33\frac{1}{3}}{100}$ means $33\frac{1}{3} \div 100$.

$= \frac{100}{3} \times \frac{1}{100}$ $33\frac{1}{3} = \frac{100}{3}$; $\frac{1}{100}$ is the recipocal of 100.

$= \frac{\overset{1}{\cancel{100}}}{3} \times \frac{1}{\underset{1}{\cancel{100}}}$ Simplify. Then multiply.

$= \frac{1}{3}$

B. Rename 12.5% as a decimal.

$12.5\% = \frac{12.5}{100} = 0.125$

$12.5\% = 0.125$

Move the decimal point two places to the left and drop the % symbol.

Copy and complete.

1. $25\% = \frac{■}{100} = \frac{■}{4}$
2. $30\% = \frac{■}{100} = \frac{■}{10}$
3. $18\% = \frac{■}{100} = \frac{■}{50}$
4. $94\% = \frac{■}{100} = 0.■$
5. $6\% = \frac{■}{100} = 0.■$
6. $80\% = \frac{■}{100} = 0.■$ or $0.■$

Exercises

Rename each percent as a fraction in simplest form.

1. 11%
2. 75%
3. 20%
4. 15%
5. 99%
6. 50%
7. 55%
8. 1%
9. 5%
10. 95%
11. 100%
12. 156%
13. $12\frac{1}{2}\%$
14. $37\frac{1}{2}\%$
15. $66\frac{2}{3}\%$
16. $16\frac{2}{3}\%$
17. $87\frac{1}{2}\%$
18. $83\frac{1}{3}\%$

Rename each percent as a decimal.

19. 16%
20. 28%
21. 36%
22. 98%
23. 8%
24. 7%
25. 3%
26. 1%
27. 70%
28. 10%
29. 54%
30. 12%
31. 40%
32. 125%
33. 250%
34. 37.5%
35. 12.8%
36. 0.5%
37. What whole number is equivalent to 100%? to 0%?

Problem SOLVING

38. Write as a decimal the part of the day spent at meals and recreation.
39. Write as a fraction the part of the day spent at school and doing homework.
40. What is the sum of the six percents? Write this sum as a fraction, a decimal, and a whole number.
41. The sum of the percents in what three categories equal the percent of time spent sleeping?

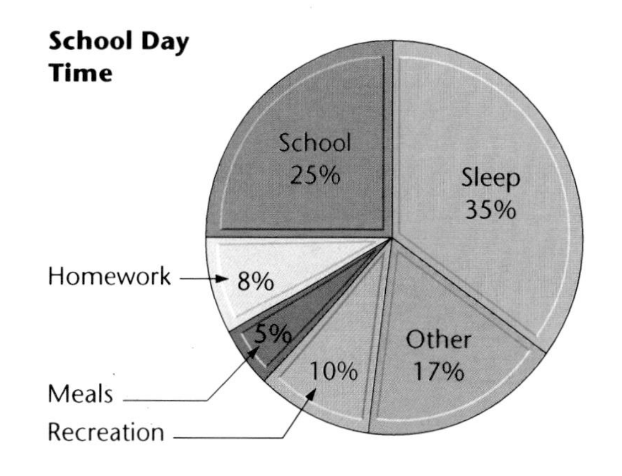

Test PREP

42. Which percent is greater than 0.6?

    a. 66% b. 59% c. 55% d. 54%

43. Four-fifths of the students drive. What percent of students drives?

    a. 15% b. 20% c. 25% d. 80%

7.9 Fractions and Decimals as Percents

Objective: to write decimals and fractions as percents

Nancy makes 5 out of 10 or $\frac{5}{10}$ of the field goals she attempts during a basketball game. What percent of the field goals does she make?

To rename $\frac{5}{10}$ as a percent, follow the steps shown below.

Step 1	Step 2
Write an equivalent fraction with a denominator of 100. $\frac{5}{10} \times \frac{10}{10} = \frac{50}{100}$	Rename that fraction as a percent. $\frac{50}{100} = 50\%$

Nancy makes 50% of her field goal attempts.

Examples

A. Rename $\frac{12}{25}$ as a percent.

$$\frac{12}{25} \times \frac{4}{4} = \frac{48}{100} = 48\%$$

B. Rename 0.4 as a percent.

$$0.4 = 0.40 = \frac{40}{100} = 40\%$$

Rename 0.4 as hundredths.

OR

$$0.4 = 0.40 \rightarrow 0.40 \rightarrow 40\%$$

Move the decimal point two places to the right and write the % symbol.

C. You can rename $\frac{7}{8}$ as a percent using division.

Remember $\frac{7}{8}$ means $7 \div 8$.

$$\begin{array}{r} 0.875 \\ 8\overline{)7.000} \\ -6\,4 \\ \hline 60 \\ -56 \\ \hline 40 \\ -40 \\ \hline 0 \end{array}$$

Add zeros as necessary to complete the division.

$0.875 = 87.5\%$

D. Rename $\frac{2}{3}$ as a percent.

$$\frac{2}{3} = 2 \div 3$$

$$\begin{array}{r} 0.6666 \\ 3\overline{)2.0000} \end{array} \rightarrow 0.\overline{6666} = 66.\overline{6}\%$$

Try THESE

Copy and complete.

1. $\frac{3}{10} \times \frac{10}{10} = \frac{\blacksquare}{100} = \blacksquare\%$

2. $\frac{7}{20} \times \frac{5}{5} = \frac{\blacksquare}{100} = \blacksquare\%$

3. $\frac{11}{25} \times \frac{4}{4} = \frac{\blacksquare}{100} = \blacksquare\%$

4. $\frac{3}{4} \times \frac{25}{25} = \frac{\blacksquare}{100} = \blacksquare\%$

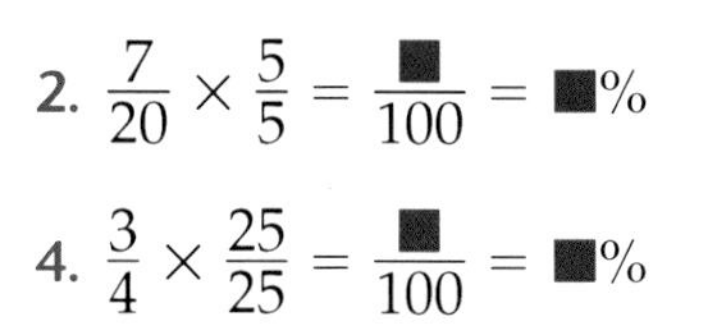

Exercises

Rename each fraction as a percent.

1. $\frac{7}{10}$
2. $\frac{3}{10}$
3. $\frac{2}{5}$
4. $\frac{3}{50}$
5. $\frac{1}{20}$
6. $\frac{1}{25}$
7. $\frac{19}{20}$
8. $\frac{3}{8}$
9. $\frac{4}{5}$
10. $\frac{1}{6}$
11. $\frac{3}{2}$
12. $\frac{1}{3}$

Rename each decimal as a percent.

13. 0.96
14. 0.05
15. 0.25
16. 0.72
17. 0.09
18. 0.64
19. 0.33
20. 0.01
21. 0.2
22. 1.725
23. 83.4

★24. $0.\overline{6}$

Write each fraction as a decimal. Then rename as a percent.

25. $\frac{6}{20}$
26. $\frac{1}{2}$
27. $\frac{4}{5}$
28. $\frac{7}{10}$
29. $\frac{2}{25}$
30. $\frac{9}{50}$
31. $\frac{5}{25}$
32. $\frac{9}{10}$
33. $\frac{5}{20}$

Copy and complete the table below. Write each fraction in simplest form.

	34.	35.	36.	37.	38.	39.
Fraction	$\frac{11}{50}$	$\frac{17}{25}$	$\frac{91}{100}$			$\frac{3}{5}$
Decimal				0.45		
Percent					32%	

Problem SOLVING

40. The enrollment of a school increased by $\frac{1}{10}$. By what percent did the enrollment increase?

41. Sam's Little League team won $\frac{4}{5}$ of its games. What percent of the games did Sam's team win?

42. Mr. Thomas has read 100 pages of a 400-page book. What percent of the book has he read?

43. Suppose you answer 25 out of 30 questions correctly on a test. What percent of the questions did you answer correctly?

Constructed RESPONSE

44. Explain how to write a decimal as a percent.

45. What percent of the numbers from 1–100 are prime numbers? Explain how you found your answer.

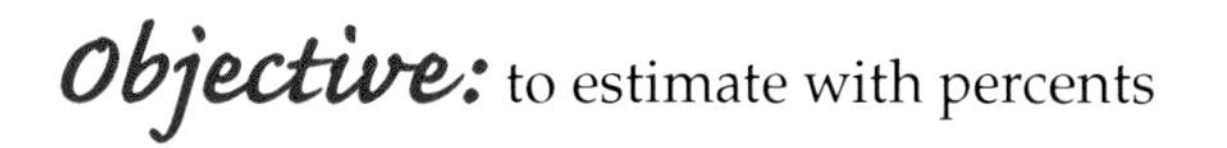

7.10 Estimating Percent of a Number

Objective: to estimate with percents

Each player on the Tiger team has to pay 45% of the cost of a uniform. *About* how much does each player pay for a $30 uniform?

To estimate the amount each player pays, think that 45% is about 50% or $\frac{1}{2}$.

$\frac{1}{2}$ of $30 is $15.

$$\frac{1}{2} \times 30 = 15$$

Each player must pay about $15.

Examples

Use the table of Fraction-Percent Equivalents to help you estimate.

A. Estimate 32% of 60.

32% is close to $33\frac{1}{3}\% = \frac{1}{3}$.

$$\frac{1}{3} \times 60 = 20$$

So, 32% of 60 is about 20.

B. Estimate 25% of 27.

25% is equal to $\frac{1}{4}$, and 27 is about 28.

$$\frac{1}{4} \times 28 = 7$$

So, 25% of 27 is about 7.

C. Estimate 36% of 78.

36% is close to $40\% = \frac{2}{5}$.

78 is almost 80.

$$\frac{2}{5} \times 80 = 32$$

So, 36% of 78 is about 32.
Is this estimate *high* or *low*? Explain.

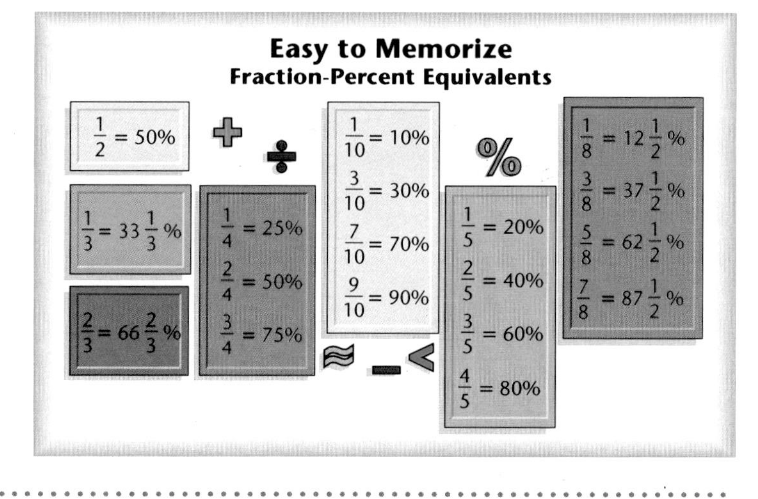

Try THESE

Name the percent you would use for estimating.

1. 11%
2. 24%
3. 65%
4. 52%
5. 77%
6. 82%

Exercises

Rewrite each expression so it will be easy to estimate.

1. 50% of 75
2. 13% of 24
3. 19% of 35
4. 12% of 18
5. 68% of 36
6. 31% of 58
7. 98% of 64
8. 74% of 25
9. 90% of 47

Estimate.

10. 25% of 35
11. 18% of 45
12. 14% of 16
13. 65% of 25
14. 52% of 67
15. 68% of 42
16. 32% of 29
17. 20% of 74
18. 82% of 22
19. 41% of 125
20. 36% of 72
21. $33\frac{1}{3}$% of 121
22. 12% of 18
23. $62\frac{1}{2}$% of 78

★24. 88% of 162

Problem SOLVING

25. A baseball glove is on sale for 20% off the regular price. About how much money is saved by buying the glove on sale?
26. Tickets to the baseball game are reduced 21 percent. If full-priced tickets cost $18, about how much is the reduced price?
27. A coach receives a special savings if he purchases all of the items listed. The sale package is 25% off the total price. About how much can he save by buying the package?

★28. Numbers can be compared using fractions, decimals, or percents. Which form is used most often for baseball averages? for finding interest on a savings account? for a sale?

★29. One number is two less than a second number. The ratio of the first number to the second number is 3 to 4. What are the two numbers?

Item	Regular Price
Bat	$ 6
Shoes	$12
Glove	$26
Uniform	$34
Baseball	$ 4

Constructed RESPONSE

30. To estimate 36% of 72, you could use $33\frac{1}{3}$% or 40%. Which one should you use to estimate? Explain your choice.

7.11 Finding Percent of a Number

Objective: to use proportions with percents; to use decimals and fractions with percents

Murphy's Department Store is having a sale. All merchandise is reduced 35%. Your mother wants to purchase a coat that was originally \$150, and she wants you to tell her what the discount will be.

Percent problems can be solved using proportions, decimals, or fractions.

- Method 1: Use a proportion
  Find 35% of 150.

$$\frac{35}{100} = \frac{x}{150} \begin{matrix} \leftarrow \text{part} \\ \leftarrow \text{whole} \end{matrix}$$

Remember, percents are ratios out of 100.

$$100x = 35 \times 150$$
$$100x = 5{,}250$$
$$x = 5{,}250 \div 100$$
$$x = 52.50$$

> When writing a percent as a fraction, remember percents will always have a denominator of 100.

- Method 2: Use a decimal or a fraction
  Find 25% of 200.

Write 25% as a decimal and multiply.
25% × 200

$$0.25 \times 200 = 50$$

Write 25% as a fraction and multiply.
25% × 200

$$\frac{1}{4} \times 200 = 50$$

Examples

A. What is 32% of 60?

- Method 1

$$\frac{32}{100} = \frac{x}{60}$$
$$100x = 32 \times 60$$
$$100x = 1{,}920$$
$$x = 19.2$$

- Method 2

$$32\% \times 60 = 0.32 \times 60$$
$$= 19.2$$

Try THESE

Write each percent as a fraction and as a decimal.

1. 50%
2. 35%
3. 70%
4. 125%

Write the proportion you would use to find each percentage.

5. 4% of 40
6. 12% of 3
7. 110% of 33
8. 5.5% of 10

Exercises

Use a proportion to find each percentage.

1. 60% of 20
2. 22% of 50
3. 2.5% of 30
4. 75% of 48

Change each percent to a decimal and find the percentage.

5. 42% of 450
6. 35% of 20
7. 12.5% of 80
8. 2% of 110

Change each percent to a fraction and find the percentage.

9. 20% of 25
10. 50% of 224
11. 60% of 90
12. $33\frac{1}{3}$% of 123

Solve. Use any method.

13. What number is 40% of 20?
14. Find 25% of 44.
15. 45% of 40 is what number?
16. What number is 62.5% of 24?
17. Find $12\frac{1}{2}$% of 16.
18. 73% of 55 is what number?
19. What number is 3.5% of 60?
20. 99% of 300 is what number?

Problem SOLVING

21. Claire saves 25% of her baby-sitting earnings. If she earned $150 babysitting, how much should she save?
22. Margaret works at a restaurant and receives a 15% tip from each table that she services. If the bill for one of her tables is $35, what should Margaret receive as her tip?
23. Rhett lives in the District of Columbia. The sales tax added to purchases in the District of Columbia is 5.75%. If Rhett purchases a television for $250, how much tax will be added to his purchase?

Constructed RESPONSE

24. Max was completing an assignment on finding percent of a number. He was not sure about when to use a proportion, a decimal, or a fraction. Explain to Max in which situations you feel it is best to use each method.

7.12 Using Percents

Objective: to use percents to calculate sales tax, tips, and discounts

David went to a sports equipment store to buy new basketball shoes. One pair of shoes costs \$65, but is on sale for 10% off. He also needs to pay sales tax of 5% on the sale price. David only has \$60 to spend. Does he have enough money to buy the shoes?

In life there are many applications for percents. The most common everyday uses for percents are **sales tax**, **tips**, and **discounts**.

Sales tax is a percentage of any purchase added by the local government. After calculating the sales tax, the amount is added to the cost of a purchase.

Tips are paid to a server at a restaurant. Some restaurants require mandatory percentages of the bill to be paid to the server. Generally speaking, 15% is a common percentage used to determine the amount of the tip.

A discount is an amount subtracted from the regular price of an item. If the discount is given as a percentage, first you calculate the percent of the regular price. Then subtract that amount from the regular price to calculate the sale price.

Examples

A. David wants to purchase a pair of shoes that are on sale. The regular price of the shoes is \$65, and the discount is 10%. What is the sale price of the shoes?

$\$65 \times 0.10 = \$6.50$	First, determine the amount of the discount.
$\$65 - 6.50 = \$58.50$	Then subtract the discount from the original price.

B. David has \$60 to purchase the shoes that cost \$58.50. Sales tax is 5%. What is the final price of the shoes including tax?

$\$58.50 \times 0.05 = \$2.925$	First, determine the amount of the sales tax.
$\$2.925 \approx \$2.93$	Then round to the nearest cent.
$\$58.50 + \$2.93 = \$61.43$	Finally add the tax to the sales price.

David does not have enough money. He needs an extra \$1.43 to purchase the shoes.

Exercises

Find the sales tax of the following items. Use 5% sales tax.

1. candy bar, \$0.60
2. water, \$1.29
3. lunch, \$5
4. TV, \$595
5. shoes, \$59.99
6. jacket, \$39

Find the amount of the discount of the following items.

7. donuts, \$4.99, 10% discount
8. gloves, \$6, 25% discount
9. jersey, \$80, 30% discount
10. flowers, \$20, 12.5% discount

Find the tip amount for the following bills. Use a 15% tip.

11. \$50
12. \$75
13. \$49.74

Find the final price for each given situation.

14. train ticket, \$30, 10% discount
15. amusement park admission, \$45, 5% sales tax
16. desk, \$250, 15% discount
17. restaurant bill, \$34.65, 15% tip
18. jeans, \$39.50, 5% sales tax
19. sweatshirt, \$24.49, 20% discount, 5% sales tax
20. restaurant bill, \$86.50, 25% discount, 15% tip after discount

Problem SOLVING

21. At a clearance sale, the price of each item was reduced by 40%. The original price of a blouse was \$35. What is the sale price?
22. A box of pencils costs \$1.49. There is a 12% discount on each box if you purchase 3 or more boxes. Find the price of 5 boxes of pencils.

Constructed RESPONSE

23. CarsRUs sells a popular car for \$15,000 with a discount of \$500. Another car dealer, SellUCars, sells the same car for \$16,000 with a discount of 8%. Which dealer sells the car at the lowest price? Explain how you know your answer is correct. Include the sale price for each car in your answer.

Objective: to solve a simpler problem

If a problem seems to have a large number of steps, it may be easier to solve a simpler problem first. Look for a pattern in the simpler problem to help you solve the complex one.

Example

When you simplify 3^{30}, what number is in the ones place?

You know that 3^{30} is a very large number, but you only need to find the ones digit. Maybe there is a pattern that you can use to help.

Simplify easier expressions, such as 3^2, 3^3, 3^4, and so on to see if there is a pattern in the number in the ones place.

$3^1 = 3$, $3^2 = 9$, $3^3 = 27$, $3^4 = 81$, $3^5 = 243$, $3^6 = 729$, $3^7 = 2187$, $3^8 = 6581$

Ones digit: 3, 9, 7, 1, 3, 9, 7, 1

There is a pattern. The pattern is 3-9-7-1. Every fourth power repeats the pattern. 3^4 and 3^8 end in the number 1. That means 3^{28} would end in 1.

Exponent:	28	29	30	31	32
Ones digit:	1	3	9	7	1

The number in the ones place would be 9.

Since you are using a pattern, you may want to find a few more powers of 3 by hand to make sure that the pattern continues.

$3^9 = 19{,}683$; $3^{10} = 59{,}049$; $3^{11} = 177{,}147$; $3^{12} = 531{,}441$

Try THESE

Solve.

1. Suppose you need to read pp. 115–134 in your Social Studies book. How many pages is this?
2. Find the sum of the first ten even numbers.

Solve

1. What is the pattern for the ones digit for powers of 5?
2. At a store a jacket costs $100. For the week, every day the store will take an additional 10% off. What is the price on the 7th day if the discount begins on the first day?
3. What is the pattern for the ones digit for powers of 7? What number is in the ones place for 7^{18}?
4. Mark agrees to mow a lawn for $7.50 every week. If he misses a week, he pays $8.50. At the end of 20 weeks, he earned $6. How many weeks did he mow the lawn?
5. There are ten different pairs of high-top shoes in the closet. Without looking, what is the greatest number of shoes that must be chosen to have a matching pair?
6. What is the pattern for the ones digit for powers of 2? What number is in the ones place for 2^{14}?
7. Find the sum of the first ten odd numbers.

Constructed RESPONSE

8. A new car comes in 7 different exterior colors and 2 different interior colors. How many different color combinations are possible? Explain how you determined your answer by solving a simpler problem.
9. Find the sum of the numbers 1–100. Explain how you determined your answer.

MIXED REVIEW

Evaluate each expression if $a = 3.45$, $b = 7$, $c = 2$, and $d = \frac{1}{4}$.

10. $a + b$
11. $b + c$
12. $b \times c + d$
13. $a + b - c$
14. $c^2 + d$
15. $b + c + a$
16. Mr. King is planning a 500-mile trip. His car gets 32 miles per gallon. Can he make the trip on 19 gallons of gas?
17. The product of two numbers is 1,340. The sum of the two numbers is 144. What are the two numbers?
18. What is the greatest sum you can get by adding two 2-digit numbers?
19. There are 32 teams in a tournament. The teams come from 4 states. The ratio of teams from Idaho to teams from Iowa is 2:5. What is the largest number of teams that could come from Iowa?

Chapter 7 Review

Language and Concepts

Write *true* or *false*. If false, replace the underlined word to make a true sentence.

1. If two ratios form a proportion, the cross products are <u>not</u> equal.
2. One way to write a <u>ratio</u> is 7 out of 10.
3. <u>Similar</u> figures have the same size and shape.
4. <u>Proportions</u> can be used to find actual distances on scale drawings.
5. A percent is a <u>proportion</u> that compares a number to 100.
6. The symbol for percent is <u>/</u>.

Skills and Problem Solving

Use a ratio to compare numbers of objects. (Section 7.1)

7. days in a year to weeks in a year
8. pints in a quart to pints in a gallon

Complete. (Section 7.1)

9.

Nickels	2	3	?	?	?
Pennies	10	15	?	?	?

10.

People	4	8	?	?	?
Cars	1	2	?	?	?

Use cross products to solve each proportion. (Section 7.3)

11. $\frac{p}{7} = \frac{9}{21}$
12. $\frac{2}{r} = \frac{9}{27}$
13. $\frac{4}{9} = \frac{8}{g}$
14. $\frac{2}{7} = \frac{s}{21}$

Rename each fraction as a percent. (Section 7.7)

15. $\frac{75}{100}$
16. $\frac{13}{25}$
17. $\frac{4}{50}$
18. $\frac{6}{8}$

Rename each decimal as a percent. (Section 7.9)

19. 0.35
20. 0.05
21. 0.3
22. 0.725
23. 1.49

Rename each percent as a fraction in simplest form and as a decimal. (Section 7.8)

24. 1% **25.** 81% **26.** 4% **27.** 20% **28.** 36%

Estimate. (Section 7.10)

29. 9% of 60 **30.** 20% of 47 **31.** 32% of 96

Solve. (Sections 7.4–7.5)

32. For the pair of similar triangles, use a proportion to find the missing length.

a

17 m

15 m

16 m

34 m

30 m

33. On a map, 1 centimeter stands for 75 kilometers. Find the map distance for an actual distance of 300 kilometers. Use a proportion.

Solve. (Sections 7.6, 7.12–7.13)

34. Jessie drove her car 130 miles and used 6.5 gallons of gas. How many miles per gallon does her car get?

35. The temperature of the room is 20°C. Use the formula $F = \frac{9}{5}C + 32$ to determine the temperature of the room in Fahrenheit.

36. Chris made 45% of his three-pointers during the season. If he shot 60 three-pointers, how many did he make?

37. Shelly purchased a crab cake platter for $12.49. Sales tax is 5%. What is the total price that she will have to pay including sales tax?

38. Three farmers can plow three fields in 12 hours. How many fields can six farmers plow in 24 hours?

39. Mrs. Cooke has 2 daughters and 1 son. Each daughter has 2 sons and 1 daughter. Her son has 3 daughters and 1 son. How many grandsons does Mrs. Cooke have? How many granddaughters does she have? Explain how you found each answer.

40. Mary is stacking cans for a display. She begins with the bottom row of seven cans. Each row has one less than the previous row. The last row will have one can in it. How many cans are in the display?

Chapter 7 Test

Write each ratio as a fraction in simplest form.

1. 3 to 7
2. 4 out of 6
3. 15:21
4. twelve to twenty

Solve each proportion.

5. $\frac{t}{5} = \frac{6}{80}$
6. $\frac{6}{w} = \frac{34}{51}$
7. $\frac{3}{2} = \frac{a}{98}$
8. $\frac{4.5}{9} = \frac{10}{t}$

Rename each fraction or decimal as a percent.

9. $\frac{4}{10}$
10. $\frac{12}{16}$
11. $\frac{23}{25}$
12. 0.61
13. 0.03

Rename each percent as a fraction in simplest form and as a decimal.

14. 17%
15. 3%
16. 20%
17. 1.5%
18. 135%

Estimate.

19. 40% of 38
20. 65% of 96
21. 13% of 48

Find each percentage.

22. 20% of 60
23. 45% of 10
24. 73% of 200

Solve.

25. For the pair of similar triangles, use a proportion to find the missing length.

26. Games4Cheap is having a special. A popular game system will be on sale. The original price is $400, and the store will take off 10% every day until they run out of game systems.
    a. Complete a chart showing the daily price over 6 days of the game system. The first day will be at regular price.
    b. Jimmy has $300. On which day will Jimmy be able to purchase a game system? Do not include tax.
    c. Play4Less is having a special, too. They will take off $35 each day. After 6 days, which store will have a cheaper price for the game system?

Percent Problems

Percent problems can be solved using proportions and a calculator. The **percentage** is the number compared to another number called the **base.** The **rate** is a percent expressed as a fraction.

$\frac{\text{percentage}}{\text{base}} = \frac{r}{100}$ $\frac{r}{100}$ is the rate.

Examples

A. What percent of 20 is 4?

$\frac{\text{percentage}}{\text{base}} = \frac{r}{100} \rightarrow \frac{4}{20} = \frac{r}{100}$ Use cross products.

$100 \times 4 = 20 \times r$

$100 \times 4 \div 20 = r$ 4 is 20% of 20.

$20 = r$

B. 50% of what number is 200?

$\frac{\text{percentage}}{\text{base}} = \frac{r}{100} \rightarrow \frac{200}{b} = \frac{50}{100}$ Use cross products.

$100 \times 200 = b \times 50$

$100 \times 200 \div 50 = b$ 50% of 400 is 200.

$200 = b$

C. 23% of 50 is what number?

$\frac{\text{percentage}}{\text{base}} = \frac{r}{100} \rightarrow \frac{p}{50} = \frac{23}{100}$ Use cross products.

$100 \times p = 50 \times 23$

$50 \times 23 \div 100 = p$ 23% of 50 is 11.5.

$11.5 = p$

Solve. Use proportions and a calculator.

1. 16 is what percent of 20?

2. What number is 2% of 10?

3. 3% of what number is 36?

4. 12.5 is what percent of 50?

★**5.** What percent of 50 is 150?

★**6.** 540 is 120% of what number?

7. Murphy's Department Store is having a sale. All merchandise that was reduced 50% is reduced another 15%. If the original price of a coat is $85, how much does it cost now?

Cumulative Test

1. $\frac{3}{7} =$ ____
   a. $\frac{1}{5}$
   b. $\frac{2}{6}$
   c. $\frac{6}{14}$
   d. $\frac{4}{8}$

2. $\frac{3}{4} \times 2\frac{2}{5} =$ ____
   a. $1\frac{1}{5}$
   b. $2\frac{3}{10}$
   c. $2\frac{5}{9}$
   d. none of the above

3. $33\frac{1}{3}\% =$ ____
   a. $\frac{1}{6}$
   b. $\frac{1}{3}$
   c. $\frac{2}{3}$
   d. $\frac{3}{3}$

4. Which of the following is equivalent to $\frac{2}{5}$?
   a. 0.4%
   b. 4%
   c. 40%
   d. 400%

5. What is the LCM for 6 and 8?
   a. 16
   b. 18
   c. 24
   d. none of the above

6. Find the rule.

Input	4	6	8
Output	12	18	24

   a. add 4
   b. add 8
   c. multiply by 2
   d. multiply by 3

7. The sum of three numbers is 18. The sum of two numbers equals the third number. The product of the three numbers is 162. What are the three numbers?
   a. 4, 6, and 10
   b. 3, 6, and 9
   c. 3, 5, and 11
   d. none of the above

8. What is the GCF of 24 and 36?
   a. 4
   b. 6
   c. 12
   d. none of the above

9. Ling-Ling bought $2\frac{1}{2}$ yards of material for a new dress. Then she realized that she needed 4 yards for her dress. How many more yards does she need?
   a. $1\frac{1}{2}$
   b. $2\frac{1}{2}$
   c. 4
   d. $6\frac{1}{2}$

10. Roger gave Toya 8 more seashells to add to her collection. How many seashells does she have now?

    Which expression represents this situation?
    a. $s - 8$
    b. $s + 8$
    c. $s \times 8$
    d. $s \div 8$

CHAPTER 8

Probability

Brittany Garza
Florida

8.1 Introduction to Probability

Objective: to find the probability of an event

What is the **probability** (chance) that a tossed coin will come up tails? Probability is used to describe the likeliness that an event will happen.

There are two *equally* likely outcomes. Since $\frac{1}{2}$ of the outcomes are tails, the probability (chance) of tails is $\frac{1}{2}$.

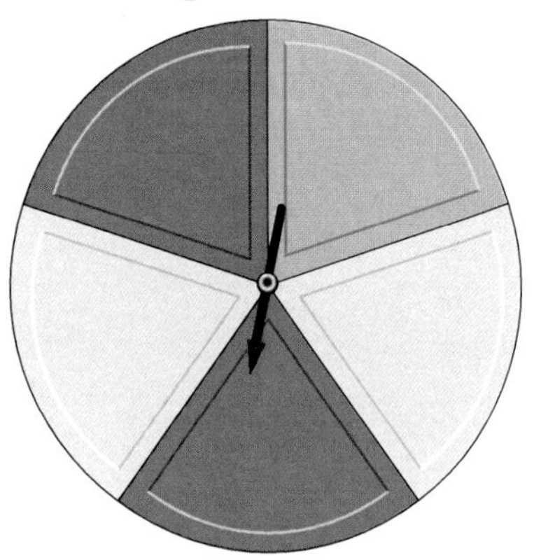

On the spinner at the left, there are 5 equally likely outcomes. There are 2 ways to spin red. Find the probability of spinning red as follows.

$\mathbf{2}$ — number of ways to spin red
$\mathbf{5}$ — number of possible outcomes

The probability of spinning red is $\frac{2}{5}$.
Write $\mathbf{P(red)} = \frac{\mathbf{2}}{\mathbf{5}}$.

You can also read a probability as a ratio.
Two out of 5 spins land on red.

More Examples

A. What is the probability of spinning either red, yellow, or blue?

Every spin is either red, yellow, or blue. This is called a **certainty**, and the probability is 1.

ways to spin red, yellow, or blue — $\frac{5}{5} \rightarrow 1$ — possible outcomes

***P*(red, yellow, or blue) = 1**

B. What is the probability of spinning orange?

Orange is not on the spinner, so it is **impossible**. The probability is 0.

ways to spin orange — $\frac{0}{5} \rightarrow 0$ — possible outcomes

***P*(orange) = 0**

Try THESE

Use the spinner above to write a fraction for each outcome.

1. yellow, $\frac{■}{5}$
2. blue, $\frac{■}{5}$ 
3. red or blue, $\frac{■}{5}$
4. What does the numerator represent?
5. Why is 5 the denominator?

Exercises

A number cube is rolled. Find the probability of each outcome.

1. 1
2. 3 or 5
3. an even number
4. a multiple of 3
5. a multiple of 1
6. 0
7. a number greater than 4
8. a prime number
9. Tell how you read each answer for problems 1–8 as a ratio.

Suppose you can choose from a red rose, a red tulip, a red carnation, a yellow rose, a yellow tulip, and a yellow carnation. How would you interpret each probability?

10. P(yellow rose)
11. P(carnation)
12. P(yellow or red)
13. P(white)
14. P(red)
15. P(tulip or rose)

Problem SOLVING

16. There are 8 boys in Kuan's diving class. They drew straws to see who would dive first. If there was 1 short straw and 7 long straws, what was the probability of having to dive first?

Mind BUILDER

Adding Probabilities

Another way to find the probability of two or more outcomes is to add the probabilities of each outcome.

Spin the spinner once. What is the probability of a 3 or a 5?

$P(3) = \frac{1}{8}$ $\qquad P(5) = \frac{1}{8}$

$P(3 \text{ or } 5) = \frac{1}{8} + \frac{1}{8} = \frac{2}{8} \text{ or } \frac{1}{4}$

Spin the spinner once. Find each probability.

1. P(3 or 2)
2. P(even number)
3. P(prime number)
4. P(odd or even number)
5. P(9 or 10)

★6. P(2 or a factor of 8)

★7. P(3 or an odd number)

8.2 Counting Outcomes

Objective: to use probability to find outcomes

Angie has just won a stuffed animal at one of the games on the midway. She may choose a dog or a bear. The color choices are white, brown, or pink. List the possible choices and find how many choices are possible.

Each choice that Angie has is called an **outcome.** There are two outcomes for animal and three outcomes for color.

You can use a **tree diagram** to list and count the possible outcomes.

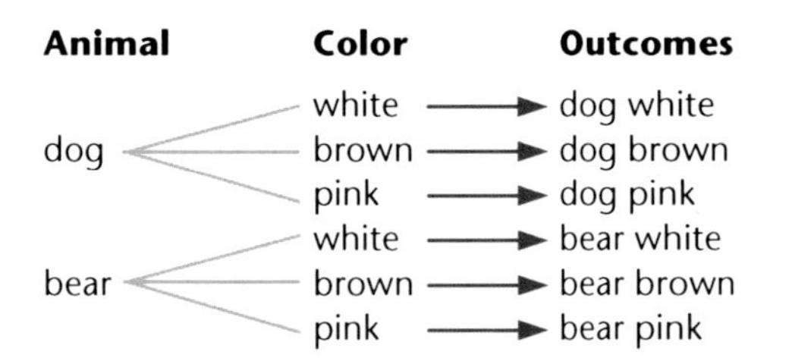

There are six possible outcomes. Name them.

Another Example

You can also use multiplication to count outcomes. This is called the **Counting Principle**. If there are m ways of making one choice, and n ways of making a second choice, then there are $m \times n$ ways of making the first choice followed by the second.

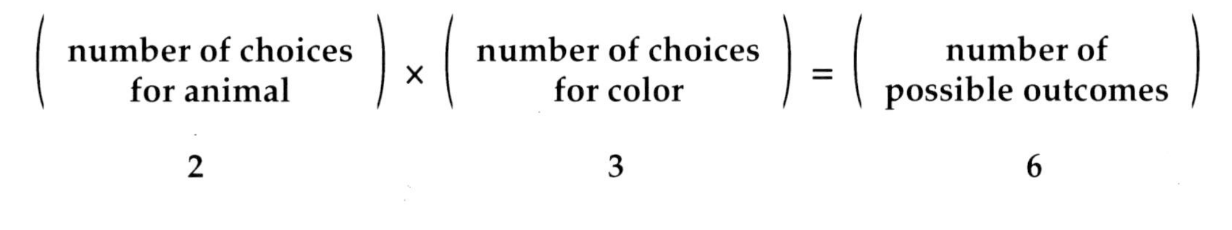

Try THESE

List the possible outcomes. Then tell how many outcomes are possible.

1. rolling a number cube

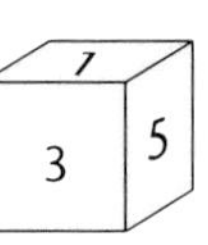

2. spinning the spinner

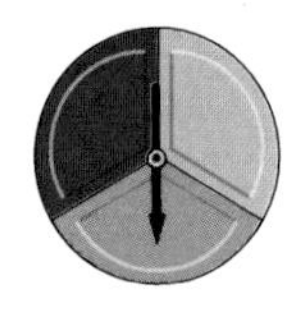

3. choosing a date in April

S	M	T	W	T	F	S
					1	2
3	4	5	6	7	8	9
10	11	12	13	14	15	16
17	18	19	20	21	22	23
24	25	26	27	28	29	30

Exercises

Suppose a nickel and a dime are tossed at the same time. Answer the following questions using the tree diagram. (H is heads and T is tails.)

1. How many outcomes are possible?
2. List the possible outcomes.
3. What does HT mean?
4. How do HT and TH differ?

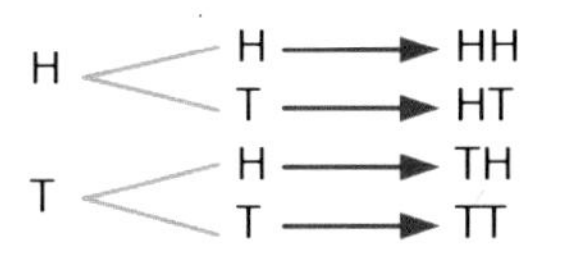

Copy and complete the tree diagram.

5.

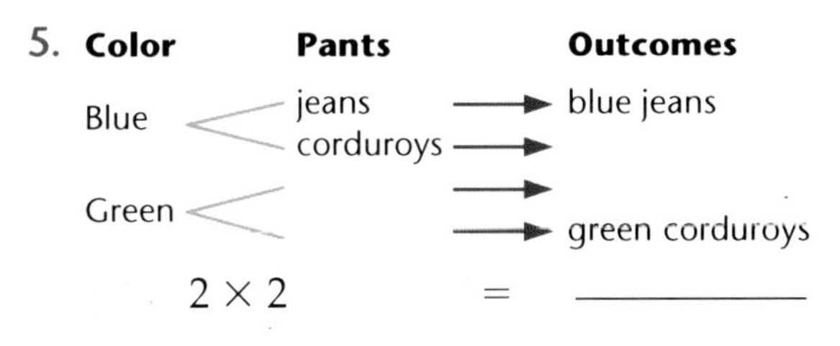

Use a tree diagram to list the possible outcomes.

6. spinning the spinner and rolling the number cube

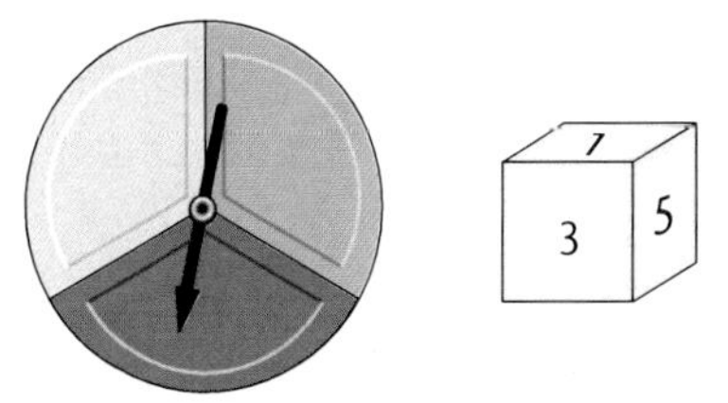

Use the Counting Principle to find the number of possible outcomes.

7. 5 flavors of ice cream with 3 possible toppings
8. rolling two number cubes
9. 3 sandwich choices, French fries or chips, and 7 flavors of soda

Problem SOLVING

10. Zach was trying to decide what to wear to school. He has a white shirt, a blue shirt, and a yellow shirt. He also has two pairs of pants to wear with the shirt he chooses. How many different outfits can he make from the shirts and pants?
11. Elaine has a red skirt, a red blouse, a white skirt, a white blouse, a tan skirt, and a tan blouse. Use a tree diagram to list the possible outcomes.

Constructed RESPONSE

12. Charles tosses a coin and rolls a number cube. What is the probability of Charles flipping heads and rolling a 3? Explain how you found your answer.

8.3 Theoretical Probability

Objective: to find the probability of an event

Suppose you select a number at random from the numbers 1–10. Find the probability of selecting a factor of 8. Remember probability is used to describe the likeliness that an event will happen. One type of probability is **theoretical probability**.

$$\textbf{theoretical probability} = P\text{ (event)} = \frac{\text{number of favorable outcomes}}{\text{total number of possible outcomes}}$$

You can express probability as a fraction, a decimal, or a percent.

Examples

A. Find the probability of selecting a factor of 8 from numbers 1–10.

The factors of 8 are 1, 2, 4, and 8. There are four factors of 8.

$$P(\textbf{factor of 8}) = \frac{4}{10}$$

4 ← number of factors of 8
10 ← total number of possible outcomes

$$P(\textbf{factor of 8}) = \frac{4}{10} = \frac{2}{5} = 0.4 = 40\%$$

B. Find the probability of selecting a multiple of 3 from numbers 1–10.

The first multiples of 3 are 3, 6, and 9. There are three multiples of 3 in the numbers 1–10.

$$P(\textbf{multiple of 3}) = \frac{3}{10}$$

3 ← number of multiples of 3
10 ← total number of possible outcomes

$$P(\textbf{multiple of 3}) = \frac{3}{10} = 0.3 = 30\%$$

All probabilities range from 0 to 1. The probability of something that is *impossible* is 0. The probability of a **certain event** is 1. The **complement** of an event is all of the outcomes not included in a given event. The probability of the complement of an event could be written as P(not event). For example, P(girl) + P(not girl) = 1.

C. You roll a number cube once. Find the probability of not rolling a 6.

$$P(\textbf{not 6}) = \frac{5}{6} = 0.8\overline{3} = 83.\overline{3}\%$$

Try THESE

You have cards with the letters M, A, T, H, E, M, A, T, I, C, and S written on them. You mix the cards thoroughly and place them into a bag. Without looking, you draw a letter. Find the probability of each event as a fraction, a decimal, and a percent.

1. *P*(**E**)
2. *P*(**M**)
3. *P*(**A**)
4. *P*(vowel)
5. *P*(consonant)
6. *P*(not **T**)

Exercises

Write each fraction as a decimal and as a percent. Round to the nearest hundredth if necessary.

1. $\frac{1}{10}$
2. $\frac{3}{4}$
3. $\frac{51}{100}$
4. $\frac{11}{40}$
5. $\frac{17}{20}$
6. $\frac{4}{7}$

Suppose you spin the spinner once. Find each probability. Write each probability as a fraction, a decimal, and a percent.

7. *P*(7)
8. *P*(12)
9. *P*(even)
10. *P*(multiple of 5)
11. *P*(number less than 3)
12. *P*(factor of 24)

Suppose you place 3 black marbles, 2 yellow marbles, and 4 red marbles in a bag. Find the probability of each event as a fraction, a decimal, and a percent.

13. *P*(black)
14. *P*(yellow)
15. *P*(not yellow)
16. *P*(black or yellow)
17. *P*(purple)
18. *P*(not red)

Problem SOLVING

19. Mercury, Venus, Earth, and Mars are the inner planets. Jupiter, Saturn, Uranus, and Neptune are the outer planets. If you randomly selected a planet from these eight planets, what is the probability of selecting an outer planet? Write your answer as a decimal. Round your answer to the nearest hundredth if necessary.

Constructed RESPONSE

20. A bag contains marbles. You know that the *P*(blue) = 25% and the *P*(green) = 50%.
    a. Are all of the marbles blue or green? Explain how you know.
    b. How many marbles might be in the bag? Are there other correct answers to the number of marbles in the bag? How do you know?

8.4 Experimental and Theoretical Probability

Objective: to find the experimental and theoretical probability of events

Suppose you are flipping two coins. To calculate the theoretical probability of both coins landing heads, you could use a tree diagram to determine the number of possible outcomes and how many of the outcomes are both heads.

Coin 1	Coin 2	Outcomes
H	H	→ HH
	T	→ HT
T	H	→ TH
	T	→ TT

There are four outcomes in all, and only one way to get both coins to be heads.

$$P(\text{HH}) = \frac{1}{4} = 0.25 = 25\%$$

Theoretical probability is a prediction of what is likely to happen. Different from theoretical probability is **experimental probability**. Experimental probability is probability based on experimental data or observations. If you tossed two coins 100 times, you would *predict* both coins should land heads 25 times. However, if you actually tossed two coins 100 times, the results may not match exactly with the prediction.

Example

Matt tossed two coins 10 times. His results are shown in the table below:

Toss	1	2	3	4	5	6	7	8	9	10
Result	HH	HT	TT	HT	TT	HT	HH	HH	TT	HT

A. What was the experimental probability of tossing two heads?

HH appears three times out of ten trials. $P(\text{HH}) = \frac{3}{10} = 0.3 = 30\%$

B. What is the most likely event when tossing two coins according to Matt's data?

HT appears four times, HH appears three times, and TT appears three times. HT is the most likely event.

Try THESE

Toss two coins 10 times and record your findings in a table like the one that Matt used. Find the experimental probabilities. Write each probability as a fraction, a decimal, and a percent.

1. P(HH)
2. P(HT)
3. P(TT)
4. P(HH or TT)
5. P(not HT)
6. P(not HH)

Exercises

Find each experimental probability. Write each probability as a fraction, a decimal, and a percent.

1. 50 coin tosses, 32 heads

   P(heads) =

2. 200 coin tosses, 104 tails

   P(tails) =

Problem SOLVING

3. A potato chip factory finds 10 green chips out of every 75 chips.
   a. What is the experimental probability that a potato chip may be green?
   b. If the factory produces 3 million potato chips each day, how many would you expect to be green?
4. Nate plays on a basketball team, and makes 50% of his shots.
   a. How could you make an experiment to simulate how many of his next 20 shots he will make? (*Hint:* What else has a probability of 50%?)
   b. Conduct the experiment.
   c. How many of his next 20 shots did your experiment show he would make?
   d. Find your experimental probability. Write the probability as a fraction, a decimal, and a percent.

Constructed RESPONSE

5. When you flip a coin there are two possible outcomes, heads or tails. After flipping a coin ten times and getting heads every time, what is the probability that heads will come up on the eleventh flip? Explain your answer.

How Many Prizes?

Cathy Contestant is waiting backstage to appear on "Guess Your Prize!" The stage is filled with prizes ranging in price from $500 to $1,000. Each prize has a different price. She is allowed to keep each prize with a price tag whose digits add up to 10. Since Cathy is only given 1 minute to find those prizes, she is allowed to make a list of possible price tags. How many prizes is it possible for Cathy Contestant to take home?

Extension

What will be her most expensive prize? What will be her least expensive prize?

Cumulative Review

Order from greatest to least.

1. 503; 3,500; 305; 3,501
2. 44; 404; 4,004; 4,000; 400

Round to the underlined place-value position.

3. 5$\underline{2}$4
4. 7,$\underline{0}$53
5. 1$\underline{6}$,897
6. 5$\underline{4}$9,000

Estimate.

7. 343 + 591
8. 2,306 + 7,119
9. 35,210 + 9,431
10. 757 − 376
11. 6,940 − 5,399
12. 48,500 − 23,904

Compute.

13. $\begin{array}{r} 32 \\ +\ 7 \\ \hline \end{array}$
14. $\begin{array}{r} \$45 \\ +\ 23 \\ \hline \end{array}$
15. $\begin{array}{r} 63 \\ +\ 28 \\ \hline \end{array}$
16. $\begin{array}{r} 24 \\ +\ 98 \\ \hline \end{array}$
17. $\begin{array}{r} 352 \\ +\ 479 \\ \hline \end{array}$
18. $\begin{array}{r} 733 \\ +\ 197 \\ \hline \end{array}$
19. $\begin{array}{r} 4{,}097 \\ +\ 552 \\ \hline \end{array}$
20. $\begin{array}{r} 5{,}617 \\ +\ 8{,}704 \\ \hline \end{array}$
21. $\begin{array}{r} 12{,}365 \\ +\ 9{,}788 \\ \hline \end{array}$
22. $\begin{array}{r} \$37{,}975 \\ +\ 43{,}709 \\ \hline \end{array}$
23. $\begin{array}{r} 89 \\ -\ 47 \\ \hline \end{array}$
24. $\begin{array}{r} \$93 \\ -\ 54 \\ \hline \end{array}$
25. $\begin{array}{r} 128 \\ -\ 79 \\ \hline \end{array}$
26. $\begin{array}{r} 50 \\ -\ 33 \\ \hline \end{array}$
27. $\begin{array}{r} 500 \\ -\ 256 \\ \hline \end{array}$
28. $\begin{array}{r} 3{,}754 \\ -\ 2{,}909 \\ \hline \end{array}$
29. $\begin{array}{r} 62{,}327 \\ -\ 34{,}758 \\ \hline \end{array}$
30. $\begin{array}{r} \$8{,}000 \\ -\ 6{,}783 \\ \hline \end{array}$
31. $\begin{array}{r} 72{,}800 \\ -\ 29{,}900 \\ \hline \end{array}$
32. $\begin{array}{r} 60{,}200 \\ -\ 54{,}736 \\ \hline \end{array}$
33. $\begin{array}{r} 70 \\ \times\ 3 \\ \hline \end{array}$
34. $\begin{array}{r} \$600 \\ \times\ 4 \\ \hline \end{array}$
35. $\begin{array}{r} 304 \\ \times\ 7 \\ \hline \end{array}$
36. $\begin{array}{r} 526 \\ \times\ 8 \\ \hline \end{array}$
37. $\begin{array}{r} 7{,}324 \\ \times\ 9 \\ \hline \end{array}$
38. 50 × 70
39. 800 × 30
40. 600 × 500

Write an equation and solve.

41. A bus seats 66 people and gets 18 miles per gallon. How many people do 10 of these buses seat?
42. Mary Timmons pays $8,378 for her car. Tom Yost pays $9,109 for his car. How much more does Tom pay?
43. The Boones spend $87 on food, $29 on supplies, and $18 on a boat rental. How much do they spend in all?
44. There are 12,304 people at a concert. Tickets cost $12 each. How much do 100 tickets cost?

8.5 Probability of Independent Events

Objective: to find the probability of independent events

Toss a coin and roll a number cube. What is the probability that the coin comes up heads and the number cube comes up 3?

A **compound event** consists of two or more events. In the example there are two events, tossing a coin and rolling a number cube. In compound events, the events may or may not depend on each other.

Two events are **independent events** if one event does not affect the outcome of the other event. Does tossing a coin affect rolling a number cube? No, the two events do not affect each other, so they are independent events.

If you have two independent events, and you know both probabilities, you can also multiply the probabilities.

$$P(\text{A, then B}) = P(\text{A}) \times P(\text{B})$$

Examples

A. A coin is tossed and a number cube is rolled. Find the probability of flipping heads and rolling a 3.

$P(\text{heads, then } 3) = P(\text{heads}) \times P(3) = \frac{1}{2} \times \frac{1}{6} = \frac{1}{12}$

B. You flip a coin three times. What is the probability of flipping three tails in a row?

Each coin toss has a probability of $\frac{1}{2}$.

You can multiply to find the probability of three heads in a row:

$P(\text{three heads in a row}) = \frac{1}{2} \times \frac{1}{2} \times \frac{1}{2} = \frac{1}{8}$

Try THESE

A coin is tossed and a number cube is rolled. Find the probability for each outcome. Write each probability as a fraction is simplest form.

1. *P*(heads, 6)
2. *P*(tails, 2)
3. *P*(heads, even number)
4. *P*(tails, prime number)
5. *P*(tails, number less than 4)
6. *P*(heads, number 5 or greater)

Exercises

There are two jars containing marbles. Jar A has 3 marbles: 1 red marble and 2 blue marbles. Jar B has 7 marbles: 3 red marbles, 2 yellow marbles, and 2 blue marbles.

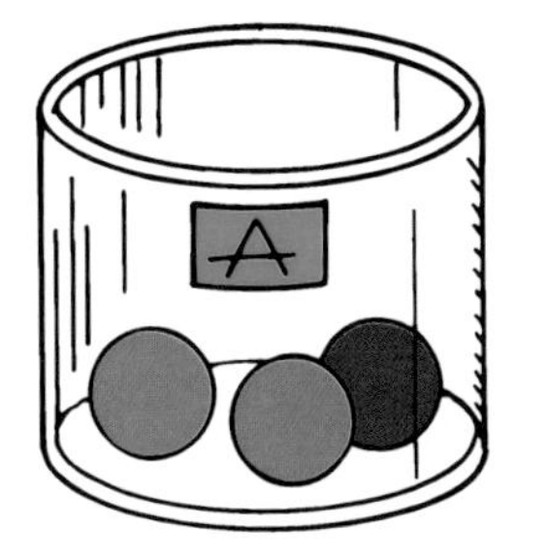

a. Use a tree diagram to find the possible outcomes of taking one marble out of each jar.

b. Find the probability for each outcome. Write each probability as a fraction in simplest form.

1. What is the probability of selecting a red marble from jar A?
2. What is the probability of selecting a red marble from jar B?
3. What is the probability of selecting a red marble from both jars?
4. *P*(blue from A, yellow from B)
5. *P*(red from A, blue from B)
6. *P*(blue from both)
7. *P*(yellow from both)

You roll a number cube twice. Find each probability.

8. *P*(1, then 1)
9. *P*(2, then even number)
10. *P*(less than 3, then 6)
11. *P*(greater than 2, then odd)

Problem SOLVING

12. You toss a coin four times. What is the probability of getting heads all four times?

Mid-Chapter REVIEW

The spinner to the right is spun. Find the probability of each outcome. Write each probability as a fraction, a decimal, and a percent.

1. *P*(red)
2. *P*(3)
3. *P*(odd number)
4. *P*(yellow)
5. *P*(multiple of 2)

Use a tree diagram to list the outcomes.

6. 5 flavors of ice cream, 3 types of toppings, cherry or no cherry

8.6 Problem-Solving Strategy: Act It Out

Objective: to solve problems by acting them out

Trudy bought a painting for $15, sold it for $20, bought it back for $25, and then sold it again for $30. How much did Trudy gain or lose in buying and selling her painting?

You need to know how much money Trudy gained or lost by buying and selling her painting.

This problem can be solved by acting out all purchases and sales. Suppose Trudy started with $40. Use $40 in play money and a card marked "painting."

Act out all steps with a friend.

Step 1	Step 2	Step 3	Step 4
Buy for $15. $40 – $15 = $25	Sell for $20. $25 + $20 = $45	Buy for $25. $45 – $25 = $20	Sell for $30. $20 + $30 = $50

Trudy's profit is $50 – 40 = $10. Trudy gained $10.

Trudy started with $40, and ended with $50. She gained $10.

Example

On a table, there are 6 pennies placed in a horizontal line and 8 pennies placed in a vertical line. What is the least number of pennies placed on the table?

You need to find out the least number of pennies that could be used to make one row of 6 pennies in a horizontal line and 8 pennies in a vertical line.

You could use pennies, counters, or a diagram to show how to arrange the pennies on a table. The highest number of pennies that you would need is 14, since 8 + 6 is 14.

Overlapping the horizontal and vertical lines gives you the smallest number of pennies needed to place the pennies on the table.

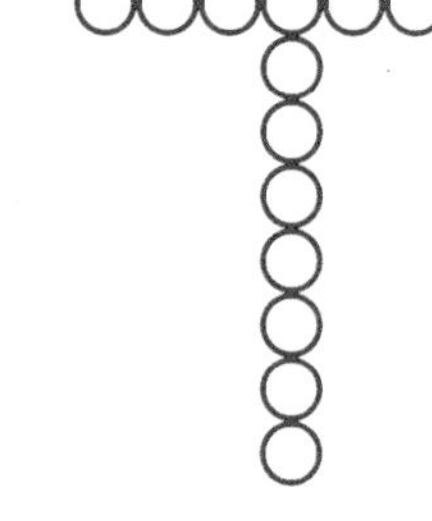

Altogether, this means 13 pennies are needed.

Any other ways to overlap the pennies in horizontal and vertical rows result in 13 pennies altogether.

Try THESE

1. A group of 32 people counted by ones, beginning with the number 1. Each person who counted a number that was a multiple of 2 stood up. Then the people sitting down counted by ones again. This time each person who counted a number that was a multiple of 2 stood up. How many people are still seated?
2. There is a paper clip in front of 2 paper clips, a paper clip behind 2 paper clips, and a paper clip between 2 paper clips. What is the least number of paper clips that could be in this group?

Solve

1. Two boys weighing 75 pounds each and one man weighing 150 pounds want to cross a river. The boat they have is safe for only 150 pounds at a time. Using only the boat, how can they cross the river?
2. Find each girl's full name. Wendy lives next door to the Abbotts. Candy baby-sits for the Costello children. Mandy is an only child. The Buccis have two children.
3. Viola and Edna bowl together. Their combined score for the first game was 280. Viola's score was 40 more than Edna's. What is each of their scores?
4. There are two jars that contain marbles. In the first jar, there are 3 marbles: 1 red, 1 purple, and 1 black. The second jar contains 10 marbles: 1 red, 4 yellow, 1 black, and 4 blue. What is the probability of selecting a black marble from both jars?
5. Suppose you are ordering a pizza. You have the choice of three toppings: sausage, pepperoni, or mushrooms. It is possible to have cheese only, one topping, two toppings, or three toppings. If you randomly select the pizza, what is the probability that the pizza would have sausage and pepperoni?

Constructed RESPONSE

6. Susan has six coins equaling $1.15. What coins does she have? Explain how you found your answer.

MIXED REVIEW

Compute.

7. $\frac{3}{4} + \frac{5}{6}$

8. $1\frac{2}{5} - \frac{7}{8}$

9. $4\frac{3}{5} \times 1\frac{1}{8}$

10. $\frac{1}{2} \div \frac{1}{4}$

Chapter 8 Review

Language and Concepts

Choose the correct word or phrase to complete each sentence. Write the letter of the word or phrase that best matches each description.

1. Another word for this term is *chance.*
2. You can use this to list all of the possible outcomes.
3. This is a way to find the total number of outcomes without actually listing every possible outcome.
4. If two outcomes are equally likely, their probabilities are this.
5. When the probability of a given outcome is 1, it is called a(n)______.
6. When the probability of a given outcome is 0, it is ______.
7. The predicted probability of an event is called the ______.

a. equal
b. Counting Principle
c. probability
d. certainty
e. theoretical probability
f. impossible
g. tree diagram

Skills and Problem Solving

A number cube is rolled. Find the probability of each outcome. (Section 8.1)

8. *P*(2)
9. *P*(multiple of 2)
10. *P*(odd number)

Suppose you are ordering a pizza and you have the choice of regular or thick crust and one choice of meat—sausage, pepperoni, or ham. (Section 8.2)

11. Copy and complete the tree diagram for the possible outcomes.
12. How many outcomes are possible?
13. List the possible outcomes.
14. What does RH mean?
15. How does RH differ from TH?
16. What is the probability of choosing a pizza with sausage as a topping?

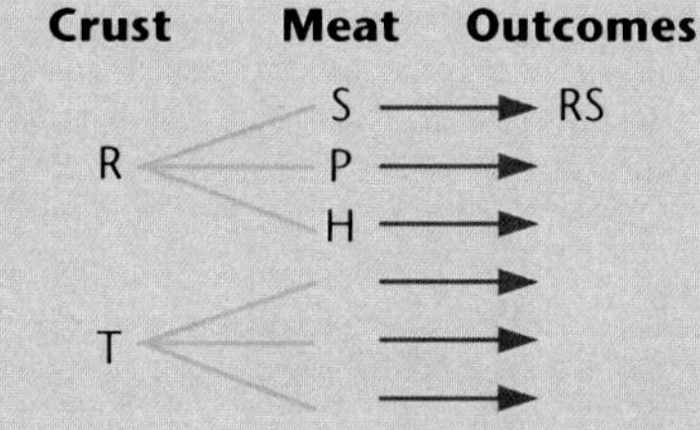

You have cards with the letters B, A, S, K, E, T, B, A, L, and L written on them. You mix the cards thoroughly and place them into a bag. Without looking, you draw a letter. Find the probability of each event as a fraction, a decimal, and a percent. (Section 8.3)

17. $P(\mathbf{K})$
18. $P(\mathbf{A})$
19. $P(\text{not } \mathbf{S})$
20. $P(\mathbf{B} \text{ or } \mathbf{L})$

Murray tossed three coins 10 times. His results are shown in the table below. Use your knowledge of experimental and theoretical probability to answer the following questions. (Section 8.4)

Toss	1	2	3	4	5	6	7	8	9	10
Result	THT	HTT	TTT	HHT	TTH	HTT	HHH	HHH	HTT	HTT

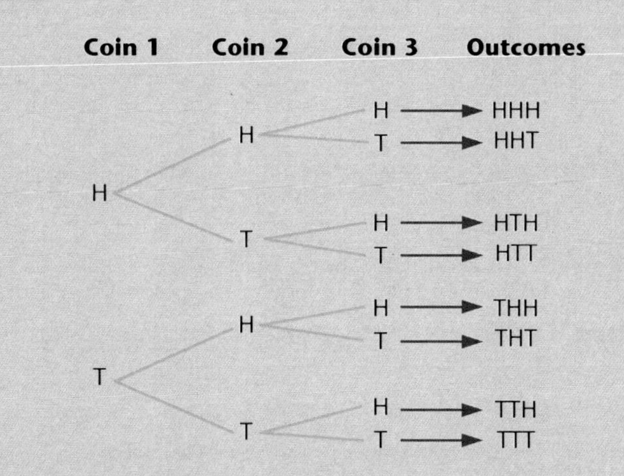

21. What is theoretical probability of flipping three heads?
22. What is the experimental probability of flipping three heads?
23. How many ways can you flip two heads and one tails?

A coin is tossed and a number cube is rolled. Find the probability for each outcome. Write each probability as a fraction in simplest form. (Section 8.5)

24. $P(\text{heads, } 1)$
25. $P(\text{tails, } 3 \text{ or } 4)$
26. $P(\text{tails, even number})$

You roll a number cube twice. Find each probability. (Section 8.5)

27. $P(4, \text{ then } 5)$
28. $P(\text{less than } 2, \text{ then } 5)$

Solve. (Section 8.6)

29. Samantha has a white blouse, a red sweater, and a black turtleneck that she likes to wear with either blue jeans or black jeans. If Samantha randomly selected an outfit, what is the probability that she would select the red sweater and the blue jeans? Explain how you found your answer.

Chapter 8 Test

A number cube is rolled. Find the probability of each outcome.

1. $P(1)$
2. P(prime number)
3. P(not even)

A number cube is rolled, and the spinner to the right is spun. Find the probability of each outcome.

A B D C

4. P(2, then **A**)
5. P(even, then **B**)

Molly tossed a coin, then spun the spinner above ten times. Her results are shown in the table below. Use your knowledge of experimental and theoretical probability to answer the following questions.

Toss	1	2	3	4	5	6	7	8	9	10
Result	H-B	H-C	T-A	H-B	T-D	T-C	T-A	H-A	H-D	T-A

6. What is the theoretical probability of flipping heads and spinning an A?
7. What is the experimental probability of flipping heads and spinning an A?

You have cards with the letters C, U, P, C, A, K, E, and S written on them. You mix the cards thoroughly and place them into a bag. Without looking, you draw a letter. Find the probability of each event as a fraction, a decimal, and a percent.

8. P(**U**)
9. P(vowel)
10. P(not **C**)

You roll a number cube twice. Find each probability.

11. P(both 2 and 5)
12. P(less than 4, then 2)

Solve.

13. Find each boy's full name. Leo lives next door to the Smiths. Tom plays basketball with the Newton children, but he does not live in their neighborhood. Kurt and Nicolas are brothers. The Littles and Newtons attend the same school.
14. Stacey has 4 knit shirts that she likes to wear. They are white, black, blue, and light blue. Stacey also has 4 pairs of jeans that she wears with her knit shirts. The jeans are dark blue, black, light blue, and red.
    a. Make a tree diagram to show the possible outfits Stacey can make by selecting one knit shirt and one pair of jeans.
    b. Use the Counting Principle to determine the number of possible outfits Stacey can make by selecting one knit shirt and one pair of jeans.
    c. If Stacey randomly selected an outfit, what is the probability that she would select the light blue knit shirt and the black jeans? Explain how you found your answer.

Change of Pace

Arrangements

Suppose the letters **B**, **E**, **F**, and **H** are to be used together on license plates. How many different ways can the letters be *arranged* if each letter is used only once?

One way to find **arrangements** is to make a list of all the possible combinations. Notice that order is important.

BEFH BEHF BFEH BFHE BHEF BHFE

EBFH EBHF EFBH EFHB EHBF EHFB

FBEH FBHE FHEB FHBE FEBH FEHB

HBEF HBFE HEBF HEFB HFEB HFBE

Do you notice a pattern in the arrangement?

There are 24 different ways the letters can be arranged.

Arrangements such as these, in which order is important, are called **permutations**.

A quicker way to find the number of arrangements is to multiply the number of choices. Use each letter only once.

If you have 1 letter, you will have 1 arrangement.

If you have 2 letters, you will have 2 arrangements. $2 \times 1 = 2$

If you have 3 letters, you will have 6 arrangements. $3 \times 2 \times 1 = 6$

If you have 4 letters, you will have 24 arrangements. $4 \times 3 \times 2 \times 1 = 24$

The symbol for $3 \times 2 \times 1$ is 3! It is read "3 factorial."

Solve.

1. Make a list of the different ways your initials can be arranged. If two of your initials are the same, change one of them. How many arrangements are there?
2. How many different ways can 1, 2, 3, 4, 5, and 6 be arranged on license plates?
3. How many different ways can 5 people be seated in 5 chairs?
4. How many different ways can 10 people stand in line?

Cumulative Test

1. Find $t - 13$, if $t = 16.9$.
   a. 3.9
   b. 13.9
   c. 29.9
   d. 39.9

2. What is the probability of rolling a number cube and getting a 5?
   a. $\frac{1}{8}$
   b. $\frac{1}{6}$
   c. $\frac{1}{4}$
   d. $\frac{1}{2}$

3. What is the remainder?
   $6\overline{)512{,}892}$
   a. 0
   b. 3
   c. 1
   d. 6

4. Sonya has three pairs of shorts and five shirts. How many outfits are possible?
   a. 3
   b. 5
   c. 8
   d. none of the above

5. Rename 32% as a decimal.
   a. 0.032
   b. 0.32
   c. 3.2
   d. 32

6. $\frac{2}{4} = \frac{5}{m}$
   a. $2\frac{1}{2}$
   b. 5
   c. 10
   d. none of the above

7. What is the LCM of 25 and 75?
   a. 75
   b. 5
   c. 25
   d. 3

8. What is the probability that the spinner will stop in the red area if you spin it once?
   a. $\frac{51}{100}$
   b. $\frac{1}{4}$
   c. $\frac{1}{2}$
   d. 1

9. How much more does Dorothy spend on housing than on food?
   a. $177
   b. $386
   c. $600
   d. none of the above

   Dorothy's Expenses
   $15,000
   Other Expenses 27%
   Food 24%
   Clothing 8%
   Housing 28%
   13% Transportation

10. Which of the following is the decimal equivalent for $\frac{14}{15}$?
    a. 0.9
    b. 0.933
    c. 0.93
    d. $0.9\overline{3}$

CHAPTER

Data Analysis and Graphs

Taylor Begnoche
New York

9.1 Frequency Tables and Histograms

Objective: to create and interpret frequency tables and histograms

The Canton Cycling Club has 52 members. The ages of the members are listed below. You can use the ages to make a frequency table and a histogram.

15, 28, 38, 43, 50, 53, 63, 69, 23, 20, 20, 31, 36,
31, 49, 55, 66, 60, 22, 20, 34, 18, 42, 52, 34, 21,
41, 50, 32, 61, 30, 46, 23, 59, 31, 81, 32, 16, 25,
47, 52, 29, 20, 32, 68, 43, 40, 29, 39, 26, 32, 33

To create a frequency table…

1. Create a table with three columns.
2. In the first, list the items in the data set, sometimes in intervals.
3. In the second column, mark the tallies.
4. In the third, write the frequency.

To create a histogram…

1. Label the horizontal axis with the intervals.
2. Label the vertical axis with the frequency.
3. Draw the bars.
4. Give the graph a title.

Ages of Canton Cycling Club Members

Age	Tally	Frequency
15–25	~~IIII~~ ~~IIII~~ II	12
26–35	~~IIII~~ ~~IIII~~ ~~IIII~~	15
36–45	~~IIII~~ III	8
46–55	~~IIII~~ IIII	9
56–65	IIII	4
66–75	III	3
76–85	I	1

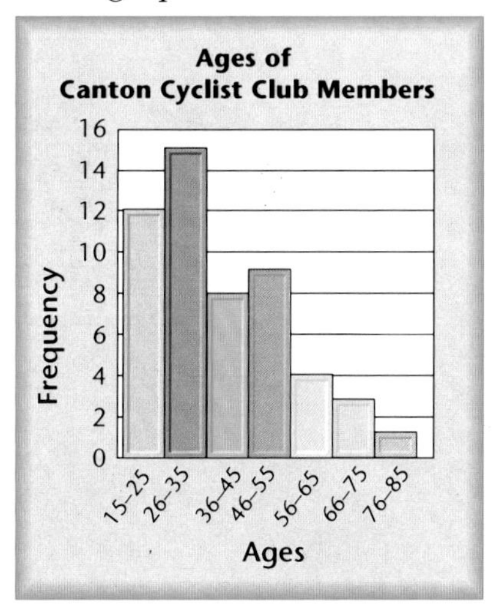

The frequency table and histogram above are divided into 10-year intervals. You can also see that the 26–35-year-old interval has the highest frequency and the 76–85-year-old interval has the lowest frequency.

Use the frequency table and histogram above to answer questions 1–3.

1. How many members of the cycling club are in the 46–55 year age range?
2. Are more cyclists younger than 26 or older than 35?
3. How many cyclists are at least 26?

Exercises

Make a frequency table and a histogram for each set of data in questions 1–2.

1. scores on a science test

85	62	88	78	74	90	92	74	92
87	85	55	70	76	80	85	94	68
74	79	80	74	97	82	84	86	58
75	69	70	71	99				

2. money spent on tennis shoes in dollars

85	26	66	84	54	30	15	18	64
28	95	47	87	40	22	38	62	90
70	30	40	62	41	36	27	29	45

Problem SOLVING

Use the histogram on the right, displaying the number of hours sixth grade students watch TV in a week, to answer questions 3–6.

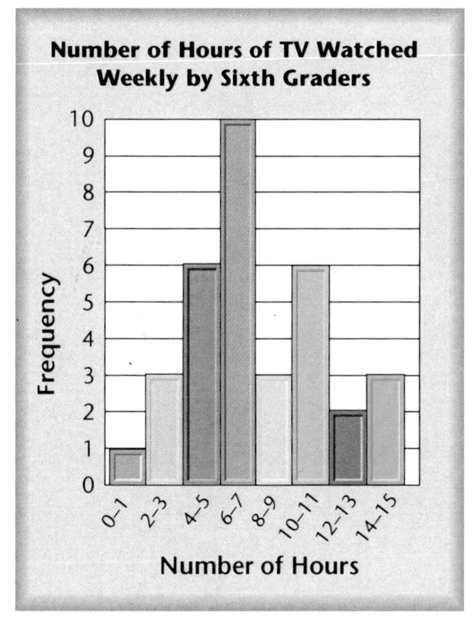

3. Which interval has the lowest frequency?
4. How many students spend 4–5 hours a week watching TV?
5. How many more students watch 6–7 hours of TV a week than watch 2–3 hours of TV a week?
6. Which is greater: the number of students watching more than 6–7 hours of TV a week, or the number of students watching less than 6–7 hours of TV a week?

Test PREP

7. Refer to the frequency table on the right. Twenty-five students were surveyed. How many students like chocolate ice cream?

   a. 6 b. 7

   c. 8 d. 9

8. Which flavor is the least favorite?

   a. chocolate b. vanilla

   c. strawberry d. peach

Ice Cream Flavor	Tally	Frequency
Chocolate	?	?
Vanilla	𝍸 IIII	9
Strawberry	𝍸 I	6
Peach	II	2

9.2 Double-Bar Graphs

Objective: to create and interpret double-bar graphs

Megan surveyed all of the sixth graders in her class to find out their favorite ice cream flavor. Megan decided that she wanted to compare the data using a graph. She chose to make a **double-bar graph** of the data.

Favorite Ice Cream Flavor	Boys	Girls
Vanilla	5	9
Chocolate	7	4
Strawberry	3	2
Chocolate chip	5	3

Use the steps listed below to make the graph.

1. Label the graph with a title.
2. Draw and label the vertical axis and the horizontal axis.
3. Mark off the equal spaces on the horizontal and vertical axis and label them with the best scale to represent the data.
4. Draw two bars for each flavor. Use different colors to represent the two categories.
5. Make a key to show what each bar represents.

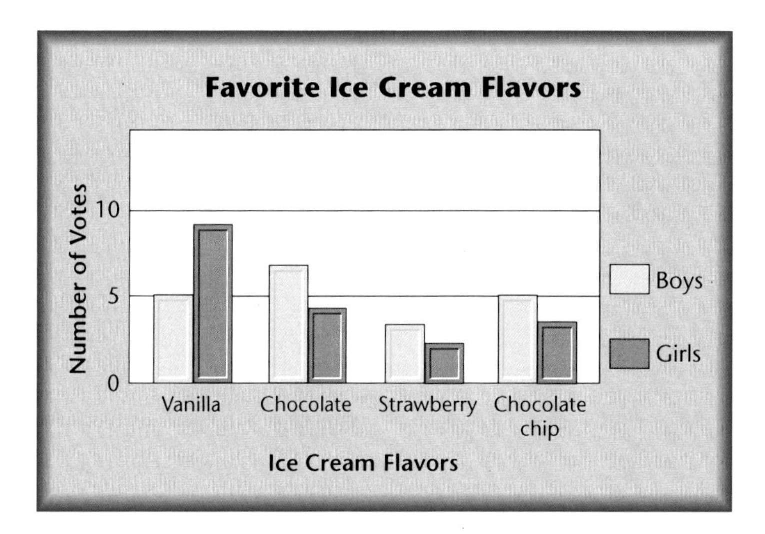

Try THESE

Use the double-bar graph.

1. How many students were surveyed?
2. Which ice cream received the most votes by the boys? by the girls?
3. How many students chose vanilla, chocolate, or chocolate chip?
4. What percentage of the sixth graders chose strawberry?

Exercises

Use the double-bar graph to answer questions 1–3.

1. Which player scored the most goals overall?
2. Which player scored the most goals in soccer?
3. Who scored the least goals overall?

Choose the best representation for the vertical scale on a bar graph.

4. the data ranges from 3 to 32
   a. 0 to 30
   b. 0 to 35
   c. 3 to 32
   d. 0 to 100

5. the data ranges from 122 to 355
   a. 100 to 400
   b. 0 to 350
   c. 122 to 355
   d. 0 to 400

Make a double-bar graph from the information in each table below.

6.

Color of Dog and Cat Fur		
Color	**Dog**	**Cat**
Black	14	9
Brown	5	4
Yellow	8	7

7.

DVD and CD Collections		
Name	**DVD**	**CD**
Joan	12	8
Tom	5	3
Sara	6	15
Paul	9	10

9.3 Stem-and-Leaf Plots

Objective: to create and interpret stem-and-leaf plots

Susan is the statistician for her school's baseball team. At the right is a table showing the number of hits for each player after fifteen games. Susan wants to organize the table in such a way that the individual data will be displayed.

Player	Hits
Rivera	15
Anderson	12
Webster	22
Palmer	32
Carroll	25
Jackson	17
Chang	15
Martinez	28
Larkin	19
King	29
Simon	9
Cruz	11

One way to organize this data is to use a **stem-and-leaf plot.** The greatest place value of the data can be used for the stem. The next greatest place value forms the leaves. Follow these steps to construct a stem-and-leaf plot from the table.

Example

Look at the number 12. Think about it as a stem (1) and a leaf (2). So, 12 becomes 1 | 2, writing the stem, then a bar, and then the leaf. Since 15 shares the same stem (1), you can think of 12 and 15 as two leaves on the same stem: 1 | 2 5.

- **Find the stems.**

  The stems are in the tens place value.

  The least number, 9, has a 0 in the tens place.

  The greatest number, 32, has a 3 in the tens place. Therefore, the stems are digits from 0 to 3.

  Make a vertical list of the stems with a line to their right.

Stem	Leaves
0	
1	
2	
3	

- **Put the leaves on the plot.**

  The leaves are in the ones place value.

  Record each of the data on the graph by pairing the units digit, or leaf, with the correct stem. For example, 25 is plotted by placing the units digit, 5, to the right of the stem 2.

Stem	Leaves
0	9
1	5 2 7 5 9 1
2	2 5 8 9
3	2

- **Arrange the leaves so they are ordered from least to greatest.**

- **Include an explanation of the data.**

  2|5 represents 25 hits

Stem	Leaves
0	9
1	1 2 5 5 7 9
2	2 5 8 9
3	2

Notice that the data is organized from least to greatest. The **range** of a set of data is the difference between the least and greatest numbers.

In this example, the range is 32 – 9 = 23.

Try THESE

Use the stem-and-leaf plot on the previous page to answer each question.

1. How many players are represented on the stem-and-leaf plot?
2. What does 2 | 9 represent?
3. Write a sentence that describes the information from the plot.

Exercises

Write the stems that would be used to plot each set of data.

1. 12, 19, 22, 38, 41, 52, 25, 47
2. 9, 22, 45, 38, 11, 19, 32, 22
3. 230, 456, 784, 235, 745, 357
4. 4.5, 6.1, 5.8, 9.8, 4.1, 3.2

The table at the right shows the height in inches of 18 students.

59, 62, 60, 58, 67, 64,
58, 65, 61, 58, 63, 51,
65, 66, 56, 68, 70, 71

5. What is the range of the data?
6. Which stems would you use to plot the data?
7. Make a stem-and-leaf plot.
8. Into which interval (stem) does most of the data fall?

Use the stem-and-leaf plot at the right for exercises 9–12.

9. What is the highest test score?
10. What is the range?
11. How many students took the test?
12. How many students received a score of 80 or better?

History Test Scores	
6	1 1 4 6 7 8
7	2 3 5 7 9
8	1 3 5 6 6 7 7 8 9
9	0 0 3 4 6 8 9 9
10	0 0

9|3 = 93

13. Write a sentence that describes the similarities and differences of the data in the two stem-and-leaf plots.

1	2 3 5 5
2	6 8 9
3	1 5 8 9
4	0 7 9 9
5	3 3 4 8 9

1	3 4
2	5 6 7 7
3	3 4 4 7 8 8 9 9
4	1 2 6 6
5	5 8

9.4 Mean, Median, and Mode

Objective: to find the mean, median, and mode of a data set

The **mode, median,** and **mean** are all measures of central tendency. It is important to know which measure to use or which measure is being used in a given situation. The following information may help you choose the best statistical measure.

mode The mode is the number or item that appears most often in a set of data.

median The median is the middle number in a set of data arranged from smallest to largest.

mean The mean is found by adding all the numbers and dividing by the number of addends. It is often referred to as the average.

Example

The ages of the workers at a concession stand are shown at the right. What is the central tendency of the age of the workers?

16 19 66 27
20 66 18

1. Choose the measure that best describes their ages. The mode is not the best choice because of the older ages. The mean is not the best choice because it will be affected by the older ages. The median will not be affected by the older ages.
2. List the ages in order.

   16 18 19 **20** 27 66 66 The middle number is 20.

   The median age of the workers is 20.
3. Choosing which measure of central tendency to use depends on the situation. In this case, the median represents the ages the best.

Try THESE

Answer the following questions using the workers' ages.

1. Which statistic would you use to attract retired people to work at the concession stand? Why?
2. Find the mean. How is the mean affected by the two older ages?
3. Which statistic would you use to attract teenagers to work at the stand? Why?

Exercises

Find the mean, median, mode, and range of each data set.

1. 12, 8, 7, 10, 8, 8, 10
2. 1, 2, 0, 1, 1, 1, 3, 4, 6
3. 86, 90, 88, 84, 102, 95, 8
4. 1.8, 3.2, 4.1, 3.2, 0.7

Problem SOLVING

5. The class sizes of the sixth-grade classes are 28, 20, 30, and 25. Find the mean.
6. The predicted high temperatures for the first week of March in Orlando were 76, 78, 81, 77, 80, 79, and 77. Find the mean, median, mode, and range.
7. Mr. McKenzie announced that the test scores were as follows: 100, 100, 97, 96, 95, 93, 92, 92, 84, 81, 80, 77, 75, 73, 70, 65, 62, 58, and 49. Find the mean, median, and mode. Name the best measure. Explain.
8. The mean of four numbers is 30. If three of the numbers are 12, 50, and 48, what is the missing number?

Use the data to complete the following.

9. Make a stem-and-leaf plot.
10. Find the mean, median, and mode.
11. Name the best measure to use. Explain.

Passengers on Amusement Park Rides				
8	9	10	10	7
12	13	14	13	13
10	13	11	8	6
13	9	7	10	13
11	8	13	9	0

Constructed RESPONSE

12. Create a problem that involves choosing the best measure of central tendency.

Test PREP

13. In the data set 84, 96, 94, 92, 94, which measure of central tendency is the least?

    a. mode b. median c. median d. median and mode

14. Find the median of 25, 30, 25, 15, 27, 26, and 28.

    a. 24 b. 25 c. 26 d. 30

9.5 Box-and-Whisker Plots

Objective: to create and interpret a box-and-whisker plot

We have learned different ways to organize and display data. Another way to describe a set of data is by using a **box-and-whisker plot**. A box-and-whisker plot is a graph that shows five main values from a set of data: the median, the highest and lowest numbers, and the values that separate the numbers into fourths. The values that separate the data into four equal parts are called **quartiles**.

The **middle quartile** is the median. The **lower quartile** is the median of the lower half of the data, and the **upper quartile** is the median of the upper half of the data.

Example

A. Find the middle, lower, and upper quartiles for the following set of data.

Points scored by a basketball player in a month.
26, 29, 31, 22, 35, 18, 33, 44, 43, 26, 33, 25, 22

- Step 1: Find the median of the data.

18, 22, 22, 25, 26, 26, 29, 31, 33, 33, 35, 43, 44 — Order the data.

29 is the median or middle quartile.

- Step 2: Find the median of the lower half of the data.

18, 22, 22, 25, 26, 26 — List the lower half of the data. Find the median.

$$\frac{22 + 25}{2} = 23.5$$

When there are two middle numbers, add them and divide by two.

- Step 3: Find the median of the upper half of the data.

31, 33, 33, 35, 44, 43 — List the lower half of the data. Find the median.

$$\frac{33 + 35}{2} = 34$$

The middle quartile is 29, the lower quartile is 23.5, and the upper quartile is 34.

B. Make a box-and-whisker for the data in example A.

- Step 1: Find the least and greatest values.

  18, 22, 22, 25, 26, 26, 29, 31, 33, 33, 35, 43, 44 Order the data.

  18 is the least value and 44 is the greatest value.

- Step 2: Draw a number line that includes the range of data. Use the least and greatest numbers as your guide. Remember, when you draw a number line, the interval between each mark should be the same.

18 20 22 24 26 28 30 32 34 36 38 40 42 44

- Step 3: Plot the least value, the greatest value, and the quartiles.

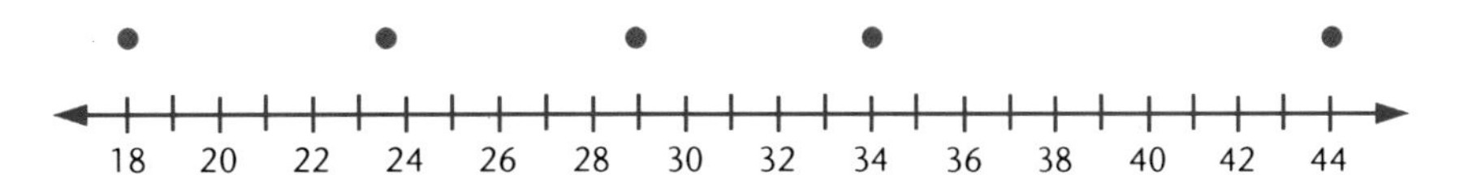

- Step 4: Draw a box through each quartile as shown.

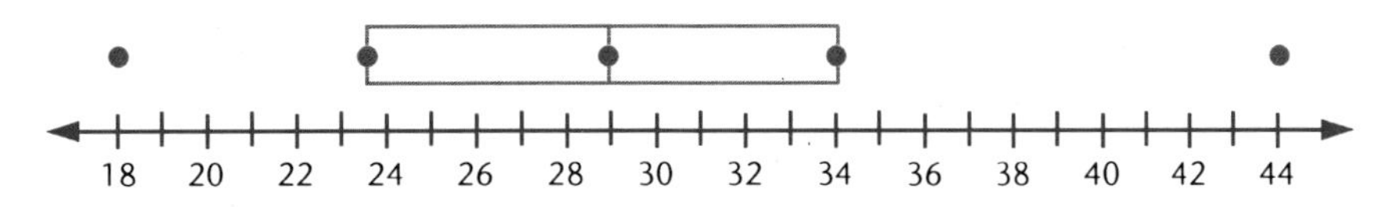

- Step 5: Connect the least and greatest values to the box to make the whiskers.

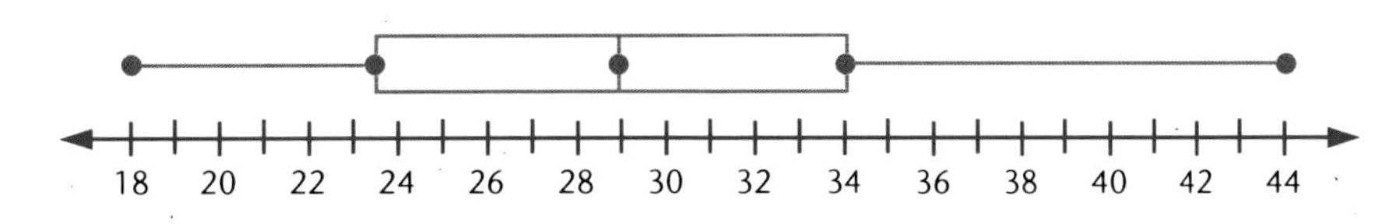

Find the middle, lower, and upper quartiles for each set of data.

1. 6, 8, 2, 1, 3, 10, 15, 12, 12, 1, 8
2. 12, 24, 13, 15, 20, 30, 48, 33, 54, 10, 11, 11, 11, 1, 10, 16

Exercises

Make a box-and-whisker for each set of data.

1. 6, 8, 2, 1, 3, 10, 15, 12, 12, 1, 8
2. 12, 24, 13, 15, 20, 30, 48, 33, 54, 10, 11, 11, 11, 1, 10, 16
3. 58, 67, 63, 60, 61, 60, 64, 66
4. 46, 37, 33, 42, 40, 37, 34, 49, 73, 46

Problem SOLVING

5. What fraction of the data is represented by each quartile?
6. Use the points scored by a basketball player from Example B to answer the following questions.
   a. 50% of the points that the basketball player scored are greater than what value?
   b. According to the box-and-whisker plot, what percentage of games did the basketball player score 34 or more points?
7. George earned the following scores on his math tests: 86, 72, 43, 78, 91, 88, 75.
   a. Make a box-and-whisker plot for his test score.
   b. How does the box-and-whisker plot change if the teacher drops the lowest test score?

Mid-Chapter REVIEW

Use the double-bar graph.

1. Which brand of shampoo was preferred most often?
2. Did more or less people prefer Brand B the second year compared to year 1?

Complete. Use the test scores.

3. Make a frequency table.
4. Find the mean, median, and mode. Name the best measure. Explain.

Test Scores				
87	94	100	72	64
85	61	94	99	72
94	87	100	87	94

Cumulative Review

Estimate.

1. $\begin{array}{r} 468 \\ +\ 801 \\ \hline \end{array}$

2. $\begin{array}{r} 9{,}320 \\ -\ 4{,}105 \\ \hline \end{array}$

3. $\begin{array}{r} 91 \\ \times\ \ 6 \\ \hline \end{array}$

4. $\begin{array}{r} 83 \\ \times\ 39 \\ \hline \end{array}$

5. $4\overline{)1{,}217}$

Compute.

6. $\begin{array}{r} 956 \\ +\ 623 \\ \hline \end{array}$

7. $\begin{array}{r} \$807 \\ -\ \ 494 \\ \hline \end{array}$

8. $\begin{array}{r} 10.8 \\ \times\ \ 6.5 \\ \hline \end{array}$

9. $0.9\overline{)51.3}$

10. $23\overline{)1{,}863}$

11. $16.25 + $9.62

12. 17.26 − 9.08

13. $128.92 − $29.87

14. 46 × 10

15. 5.4 × 1,000

16. 890 ÷ 100

17. 1.6 ÷ 10

Compute. Write each answer in simplest form.

18. $\frac{2}{9} + \frac{5}{9}$

19. $\frac{1}{4} + \frac{3}{8}$

20. $\frac{2}{3} + \frac{5}{6}$

21. $1\frac{3}{4} + \frac{5}{8}$

22. $\frac{5}{7} - \frac{4}{7}$

23. $\frac{5}{6} - \frac{1}{6}$

24. $1\frac{2}{3} - \frac{1}{2}$

25. $1\frac{5}{8} - \frac{15}{16}$

26. $\frac{2}{3} \times 6$

27. $\frac{1}{3} \times \frac{4}{5}$

28. $2\frac{1}{3} \times 3$

29. $6\frac{1}{2} \times 1\frac{1}{2}$

30. $\frac{7}{8} \div \frac{3}{4}$

31. $4 \div \frac{2}{3}$

32. $\frac{3}{8} \div 3$

33. $2\frac{1}{2} \div 4$

Solve.

34. Ramon buys 2 cans of tennis balls. How much change does he receive from $10?

35. Julia drives $3\frac{3}{5}$ miles to work each day. Jenn drives $3\frac{5}{9}$ miles to work each day. Who drives farther?

36. Steve buys $1\frac{1}{4}$ pounds of turkey, $2\frac{2}{3}$ pounds of roast beef, and $\frac{1}{8}$ pound of ham. How many pounds of deli meat did he buy?

37. Tanya bought a silver bracelet for $14.25 and a pair of silver earrings for $16.50. How much change did she receive from $31.00?

9.6 Problem-Solving Strategy: Use a Variety of Strategies

Objective: to solve problems using a variety of strategies

Dora is planning a biking trip in Wisconsin near Green Bay. The Mountain Bay trail is 80 miles long. She plans to bike 25% of the trail on the first day, then 40% of the trail the second day. How many miles of the trail will she have to bike in order to complete the trail the third day?

You have been given many problem-solving strategies to use. You can choose from the list for ways to solve word problems.

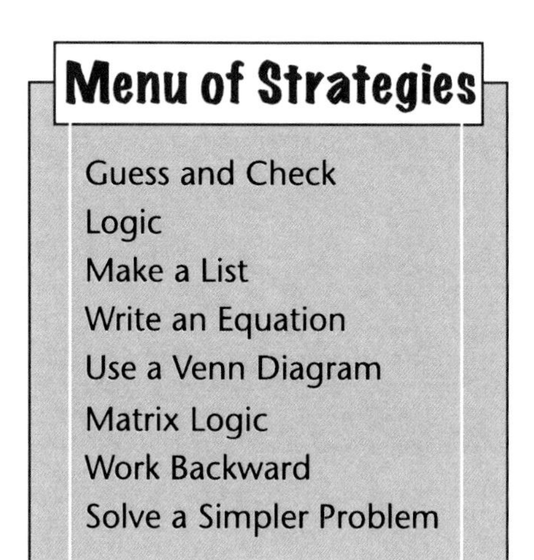

You can use logic to draw a diagram to help you understand a problem better. You can solve a simpler problem and use the information to help you solve the more complex problem.

Remember, *there is usually more than one correct way to solve a word problem.* Try to choose the best method of solving a given problem using the skills that you have.

In the problem, Dora is planning to bike an 80-mile long trail. She will bike 25% the first day and 40% the second day. You need to determine how many miles are remaining on the third day.

You could use logic to solve the problem. If she bikes 25% the first day, and 40% the second day, you can determine what percentage is remaining. The total must add up to 100%. Then you can use that percentage to determine how many miles are remaining.

You could also act it out. You can use your knowledge of percents to determine how many miles Dora plans to bike the first two days. Then you can subtract from 80 miles to find the remaining miles.

- Method 1:

25% + 40% = 65%
100% – 65% = 35%
35% of 80 =
$0.35 \times 80 = 28$
28 miles remain

- Method 2:

25% of 80 = 20
40% of 80 = 32
80 – (32 + 20) = 28
28 miles remain

Both methods check. **25% + 40% + 35% = 100%** and **20 + 32 + 28 = 80**

Solve

For each problem, list the problem-solving strategy that you would use. Explain why you chose that problem-solving strategy. Then solve.

1. A runner is planning on increasing the number of miles that he runs each week. The first week he will run 2 miles. The second week he will run 5 miles, and the third week he will run 8 miles and so on. How many miles will he run the eighth week?
2. Suppose a lunch table can fit 4 people on each side and 2 people on each end. What is the largest number of people that could sit at 4 tables if the tables are all pushed together?
3. Find one way to add any of the numbers 3, 5, 7, 9, 11, and 12 to get a sum of 20.
4. In a basketball tournament, teams are eliminated from competition after a loss. If the champion won three games, how many teams were in the tournament?
5. A pizza costs $9.00. Each additional topping costs 75 cents. If the total cost of the pizza is $12.00, how many additional toppings are on the pizza?
6. Maxine buys 3 wristbands. Each wristband is identical. She uses a $2 off coupon and pays $7 to the cashier. How much does each wristband cost?
7. The mean of a set of 5 numbers is 25. The median is 12. Two of the numbers are 8 and 80. The smallest number is 5. What are the five numbers?
8. The difference between two whole numbers is 14. Their product is 1,800. Find the two numbers.
9. You have a cup of quarters and dimes. You counted $6.75. The number of dimes is one more than the number of quarters. How many quarters are there?

Mixed Review

10. $\begin{array}{r} 517.421 \\ +\ 2.474 \\ \hline \end{array}$

11. $\begin{array}{r} \$6.24 \\ -\ 2.42 \\ \hline \end{array}$

12. $\begin{array}{r} 5.0 \\ -\ 3.9 \\ \hline \end{array}$

13. $\begin{array}{r} 378.51 \\ +\ 952.99 \\ \hline \end{array}$

14. $\begin{array}{r} 8.540 \\ \times\ 1.65 \\ \hline \end{array}$

15. $\begin{array}{r} 0.008 \\ \times\ 0.012 \\ \hline \end{array}$

16. $0.37\overline{)1.887}$

17. $0.53\overline{)16.43}$

9.7 Constructing Line Graphs

Objective: to create and interpret line graphs to show change

Temperatures Throughout the Day	
Time	**Temperature**
8 A.M.	31°F
9 A.M.	32°F
10 A.M.	32°F
11 A.M.	33°F
12 P.M.	35°F
1 P.M.	37°F
2 P.M.	38°F
3 P.M.	39°F
4 P.M.	39°F
5 P.M.	39°F
6 P.M.	38°F
7 P.M.	37°F
8 P.M.	36°F

Students in Mrs. Merwin's science class were asked to record the temperature at different times of the day. Their data is shown in the table.

The next day, Mrs. Merwin told the class that they would be constructing **line graphs**. A line graph uses a group of connected line segments to show changes in data. Usually, a line graph shows changes over a period of time.

Just like when you constructed a bar graph, you need to determine the scale that you will use for the vertical axis, label the vertical and horizontal axes, and then make the graph.

Example

Construct a line graph for the temperature data.

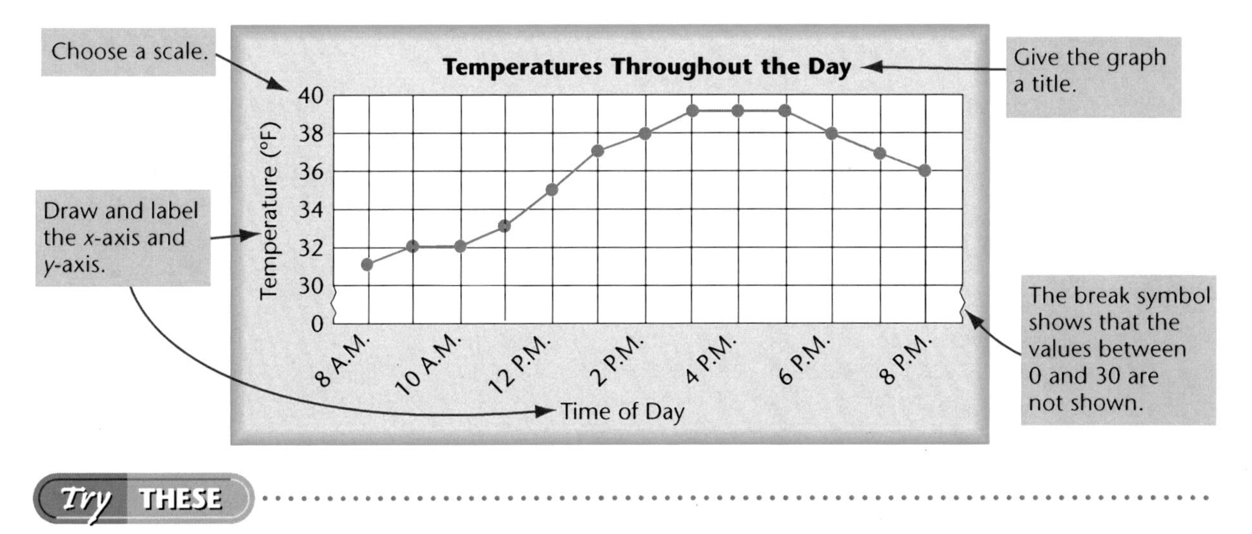

Try THESE

Make a line graph for the data in the following table.

1.

Fundraiser Contributions Collected				
Week	1	2	3	4
Amount	$50	$75	$80	$30

Exercises

Make a line graph of the data in each table. Then answer the questions that follow.

1.

Books Read Each Month				
Month	Sept.	Oct.	Nov.	Dec.
Number	50	300	450	350

a. Make a line graph to represent the data in the table.

b. Why do you think only 50 books were read in September?

2.

Temperature of Heated Water											
Time (minutes)	0	2	4	6	8	10	12	14	16	18	20
Temperature (°C)	25	27	30	35	45	60	80	95	100	100	100

a. Make a line graph to represent the data in the table.

b. What is the range of the temperature?

c. Why do you think the temperature stayed at 100°C?

3.

Test Scores							
Test Number	1	2	3	4	5	6	7
Rachel's scores (%)	85	87	99	75	100	99	95
Emily's scores (%)	95	92	97	100	100	99	100

a. Make a double-line graph to represent both sets of data on the same graph. Include a key.

b. What differences do you see in the girls' scores?

c. Which girl do you think has the highest mean? median? mode?

d. Calculate the mean, median, and mode for each girl.

Mixed Review

Use the line graph to match each time with the day. The times shown on the graph are rounded to the nearest tenth of a second.

4. 29.545
5. 29.365
6. 28.95
7. 29.22
8. 29.34
9. 29.08

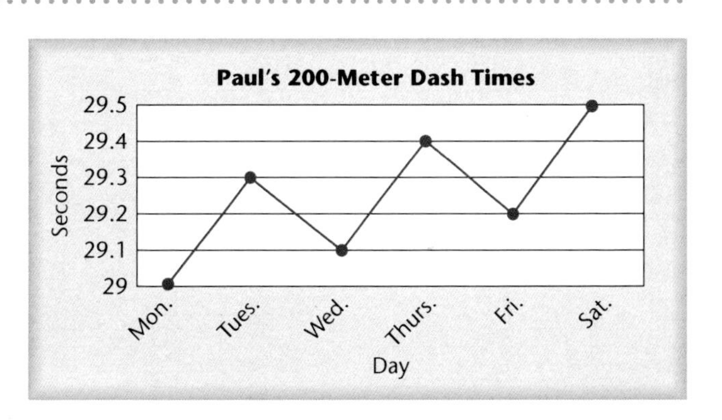

9.8 Reading Graphs

Objective: to create and interpret a variety of graphs

There are many different types of graphs. You should be able to read and use each type of graph. **Bar graphs** are used to compare related quantities. **Line graphs** show changes in data, usually over a period of time. **Circle graphs** are used to compare parts of a whole. Circle graphs often contain percents. A **box-and-whisker plot** divides the data into four equal parts and can be used to show trends in data. Two other kinds of graphs are line plots and histograms. A **line plot** is a graph that shows the frequency of data along a number line by stacking **X**s above each data value on a number line. This type of graph shows data that could be represented in a frequency table. A **histogram** is a bar graph that shows the frequency of a range of data. Usually, the data is represented in equal-sized intervals. This type of graph also shows data that could be represented in a frequency table.

Examples

A. Make a line plot for the following set of data.

Lengths of shoes in inches:
10, 8, 6, 6, 8, 7, 8, 8, 9, 12, 8

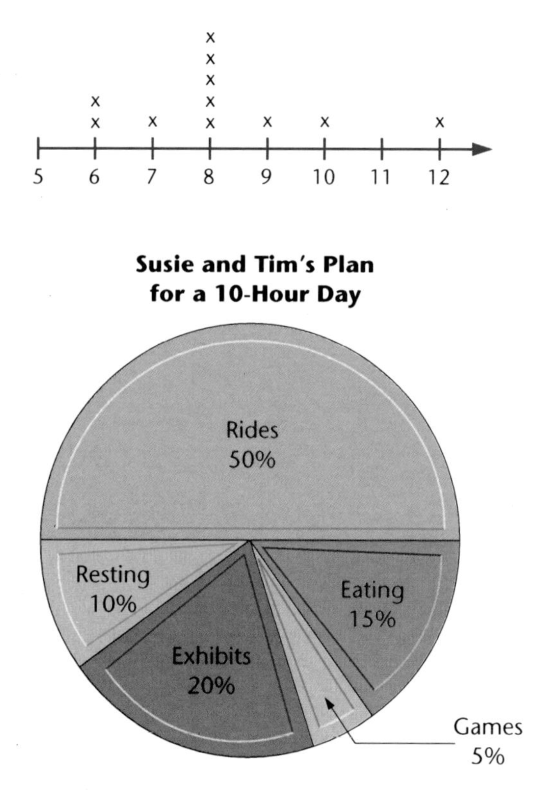

B. Use the circle graph to find how many hours Susie and Tim planned for exhibits.

Use a fraction.

$20\% = \frac{20}{100} \rightarrow \frac{1}{5}$

$\frac{1}{5} \times 10 = 2$

Use a decimal.

$20\% = 0.20$

$0.20 \times 10 = 2.00$

Susie and Tim plan to spend 2 hours at the exhibits.

Exercises

Make a line plot of each data set.

1. 4, 5, 8, 9, 10, 10, 10, 4, 5, 3, 7, 10, 8, 9, 10, 15
2. 6, 6, 6, 7, 7, 7, 8, 8, 8, 8, 8, 8, 8, 9, 9, 9, 9, 9, 9, 10, 10, 10, 11, 11, 12, 12, 13, 14

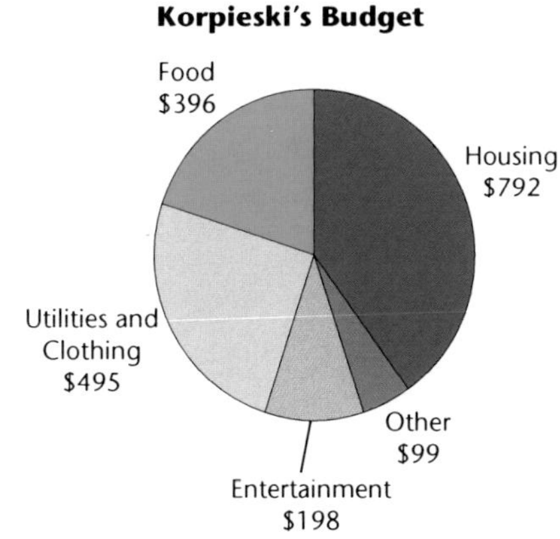

The Korpieskis earned $1,980 last month. Answer questions 3–5 using the circle graph.

3. What percentage of the Korpieskis' budget was used for food?
4. What percentage of the Korpieskis' budget was used for entertainment?
5. Suppose the food costs more than they budgeted. How can they still balance their budget?

Use the line plot to answer questions 6–8.

6. What is the mode for ages of students in the class?
7. What is the median age of the students in the class?
8. Which is easiest to determine from the line plot: mean, median, or mode? Why?

Student's Ages

15	16	17	18	19	20
x	x	x		x	x
x	x			x	x
	x			x	
	x				
	x				

Use the double-line graph to answer questions 9–12.

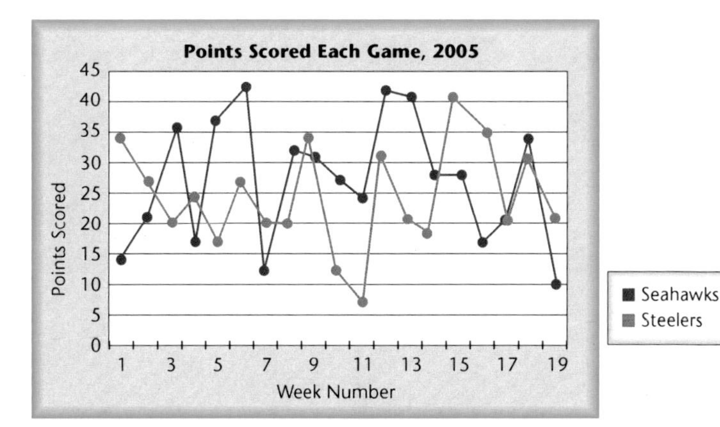

9. Which team had the most games with 40 or more points scored?
10. Which team had the most games with less than 20 points?
11. How many times did the Steelers score exactly 20 points?
12. The two teams played against each other in the final game. What was the score?

9.9 Choose the Appropriate Graph

Objective: to choose an appropriate graph for a set of data

Math	90 minutes
Reading	120 minutes
Science	30 minutes
Recess	15 minutes
Social Studies	30 minutes

Miss Tuscany wants to display how she uses her time in a day for every subject that she teaches. What type of graph would best show this?

When choosing a graph you should ask yourself a few questions. What kind of data is given? Can the data be counted? Is the data in multiples or in intervals? Is the data given for a time span?

A line graph is not appropriate because you are not showing change over time. A bar graph is not appropriate because the data cannot be counted. A histogram is not appropriate because the data does not show frequency. A circle graph is appropriate because you are showing parts of a whole.

The circle graph to the right shows how Miss Tuscany spends her day teaching.

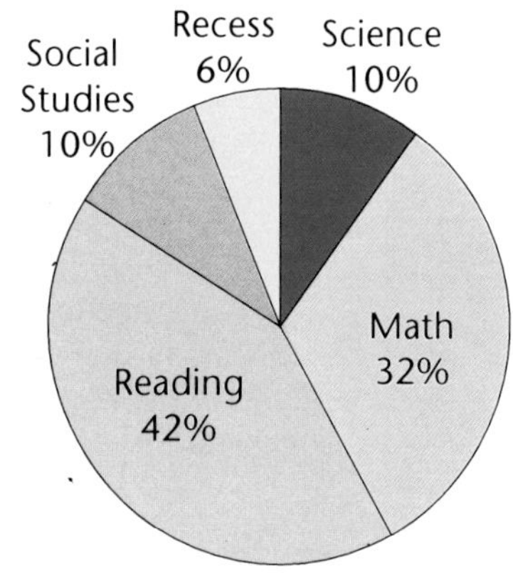

Try THESE

Each graph contains the same information about a 5-day temperature range. Tell which graph is the appropriate graph for the data. Explain why.

a.

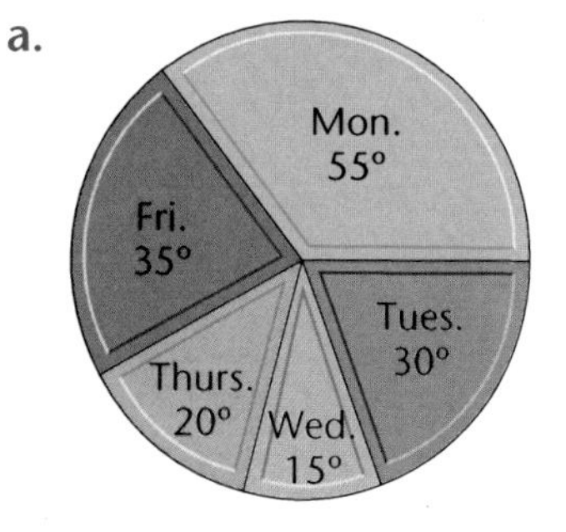

b.

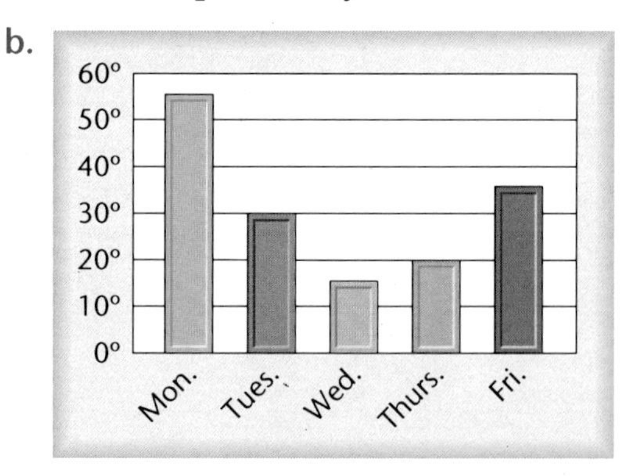

Exercises

Tell which kind of graph you would use for each set of data and why you would use it.

1. height of waterfalls
2. changes in your height
3. how you spend your money
4. daily fair attendance
5. number of television stations in various cities
6. what part of your class likes various types of music

Solve. Use the circle graph.

7. King's Land used a circle graph to show how a new ride had affected the attendance after it was installed in July. This graph is not appropriate for this data. Explain why.
8. Which graph would be more appropriate for this data?
9. Use the data from the circle graph to make the appropriate graph for this data. Include a title, labels, and a scale.

Monthly Attendance

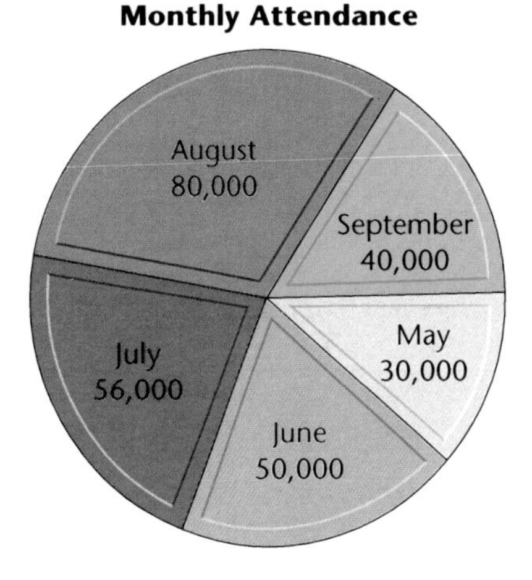

Problem SOLVING

10. What type of data should be included in a circle graph? Give an example of data that you could collect that would be best represented in a circle graph.
11. Ask your friends or classmates a survey question. Here are some examples.
    - What is your favorite type of music?
    - How many hours do you spend studying each week?
    - How many pets do you have?

    Then create a graph to represent your data. Choose the most appropriate type of graph. Explain why you chose that type of graph to represent your data.
12. Collect data that could be represented in a double-bar graph. Then, make the double-bar graph. Include a title, labels, and a scale.

9.10 Misleading Graphs and Statistics

Objective: to identify misleading graphs; to identify misleading statistics

Often, data is represented to influence you. Sometimes graphs can be distorted to mislead you. Sometimes an inappropriate measure of central tendency may be used. As you look at graphs, and when you interpret the mean, median, and mode of a set of data ask, "Is the data represented accurately?" "Which measure of central tendency best represents the given data?"

Examples

The growth of the U.S. population is represented in the line graphs below. Which graph is misleading?

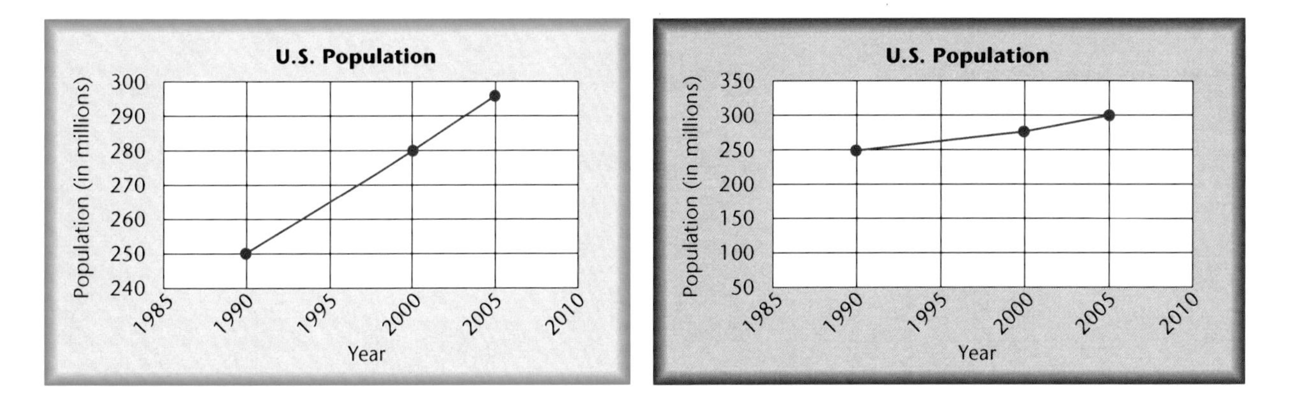

A. What impression is given by the first graph?

The graph suggests that the population is rapidly increasing.

B. Why is the first graph misleading?

The vertical scale increases by ten on the first graph and by fifty on the second graph. The vertical scale on the first graph starts at 240 million.

Try THESE

1. Which graph would be used by product A? Why is that graph misleading?
2. Which graph would be used by product B? Why is that graph misleading?
3. Soda A received 80 votes and Soda B received 98 votes. Redraw the bar graph so that it is not misleading.

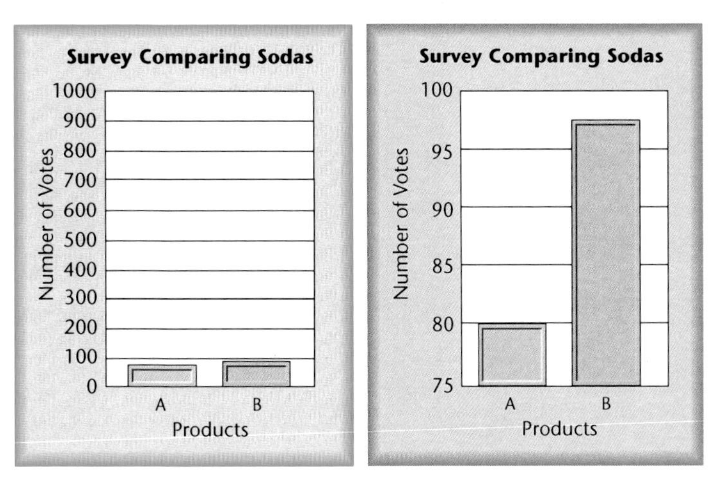

Exercises

Decide whether each graph is misleading. Answer the following questions for each graph.

a. Is the graph misleading? If the graph is misleading, explain why.

b. What impression is made by the graph?

c. Redraw any misleading graph so that it is no longer misleading.

1.

2.

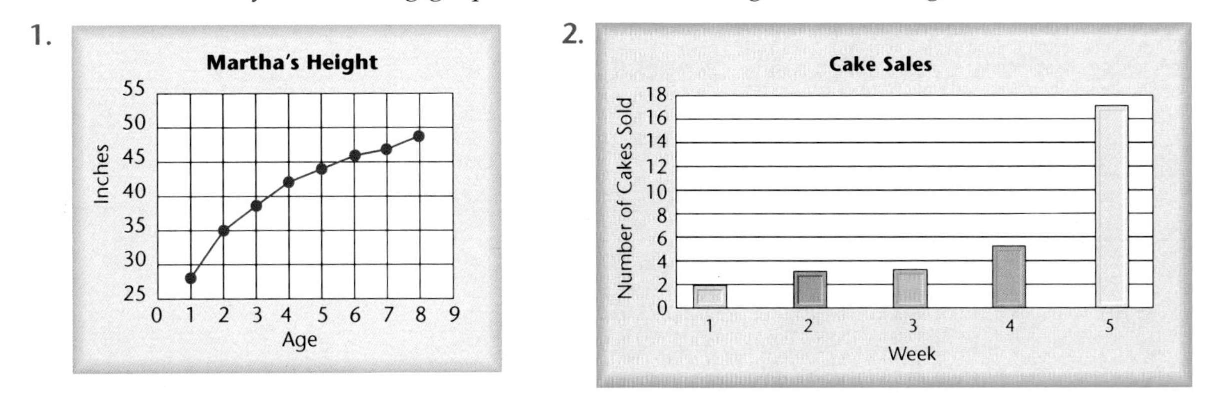

Solve.

3. Robert scored 100%, 100%, 90%, 75%, and 60% on five quizzes.
   a. Which makes his grades look best, the mean, the median, or the mode?
   b. Which measure would his teacher use to convince Robert to study harder for the next quiz?

Solve.

4. Willie and Robyn are competing in a bowling game. Both claim to have won the match. Their scores are shown in the table.

Game	Willie	Robyn
1	267	150
2	155	156
3	190	200

   a. Willie says that he won the match. How can he support his claim?

   b. Robyn claims that she won the match. How can she support her claim?

   c. The league gives 1 point for winning a game, and 1 point for the highest total score. Who is the winner, Willie or Robyn?

Constructed **RESPONSE**

5. Celeste has worked as a biologist for five years. Her salaries are shown in the table.

Year	Salary
1	$50,000
2	$52,000
3	$55,000
4	$60,000
5	$75,000

   a. Draw a line graph showing that her annual salary has increased greatly each year.

   b. Draw a line graph showing that her annual salary has increased slowly each year.

   c. Explain how you drew the graphs in parts (a) and (b) to get the desired results.

Bart owns a small business. He made two different graphs to show his monthly sales. Notice that both graphs show the same data. Compare the vertical scales.

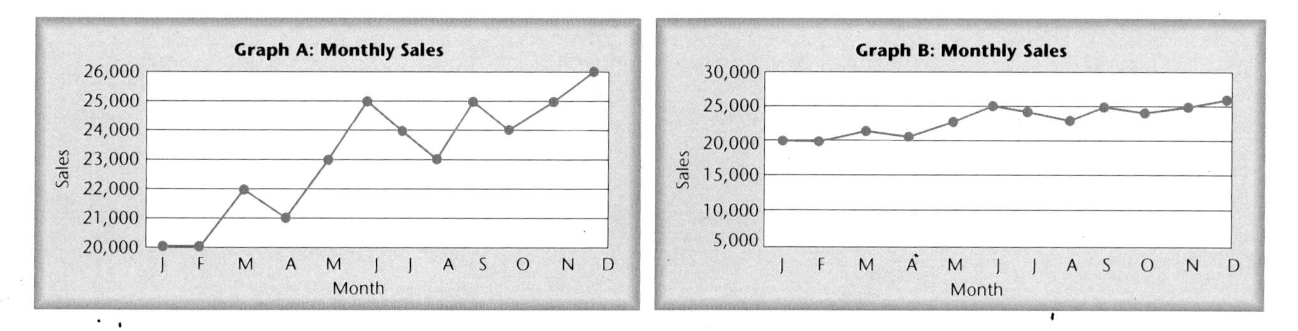

6. Which graph would Bart show someone interested in buying his business? Why?

7. Which graph would Bart show an employee who asked for a big raise? Why?

Akeem's Cube

For a class project, Akeem painted a large cube so that opposite faces were the same color. There were a pair of red faces, a pair of yellow faces, and a pair of blue faces. Then he cut the cube into 64 smaller cubes, all the same size. What is the greatest number of sides that are painted? Make a list of the different number of sides that are painted. How many of each kind did Akeem have?

Sides Painted	Number of Cubes
3	8

Extension

Suppose Akeem cut the cube in half and painted the sides that were cut. Then he cut each rectangular prism into 32 cubes. Make a list of the different number of sides that could be painted. How many of each kind does Akeem have now? Did the total number of cubes with paint on them increase or decrease from what he has in the problem above?

Chapter 9 Review

Language and Concepts

Choose the correct word or phrase to complete each sentence.

1. A (double-bar graph, double-line graph) is used to compare and contrast related quantities.
2. A broken line on a scale of a graph means that the scale (begins, does not begin) at zero.
3. A (chart, frequency table) tells you how often an item was chosen or how many times something occured.
4. A (circle graph, line graph) is used to show parts of a whole.
5. To choose the appropriate graph, you must consider the (frequency, type of data) that you want to graph.
6. The middle number in a set of data arranged from smallest to largest is called the (mode, median).
7. You can use a (box-and-whisker plot, stem-and-leaf plot) to separate data into fourths.

Skills and Problem Solving

Use the data in the table to complete exercises 8–9. (Sections 9.1–9.2)

Scores				
10	8	7	5	9
6	10	9	9	8
3	9	7	8	9
8	10	10	7	

8. Make a frequency table of the data in the table.
9. Make a bar graph of the data in the frequency table.

Ages of Chess Team Members			
30	33	25	36
36	36	30	24
25	10	45	

Use the following data table to complete exercises 10–12. (Sections 9.3–9.5)

10. Make a stem-and-leaf plot of the data in the table.
11. Calculate the mean, median, mode, and range of the data.
12. Make a box-and-whisker plot of the data in the table.

Use the two box-and-whisker plots to answer exercises 13–14. (Section 9.5)

13. Which class appears to have scored better on the test? How can you tell?
14. Why do you think the box for Class A is smaller than the box for Class B?

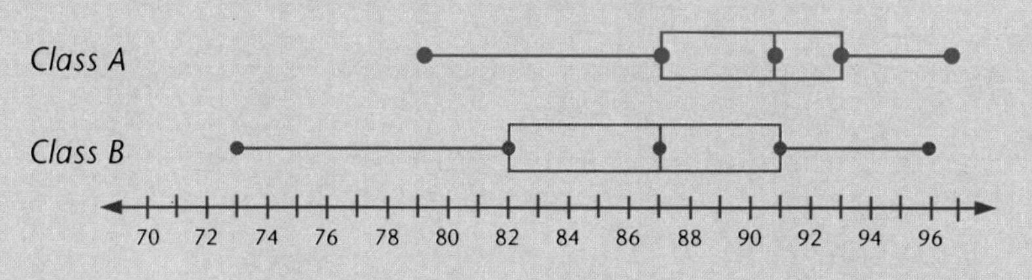

Use the following data table to complete exercises 15–16. (Section 9.7)

Average High Temperature in Baltimore												
Month	Jan.	Feb.	Mar.	Apr.	May	Jun.	Jul.	Aug.	Sep.	Oct.	Nov.	Dec.
High Temp. (°F)	44	47	57	68	77	86	91	88	81	70	59	49

15. Make a line graph of the data in the table.
16. Find the range of temperatures.

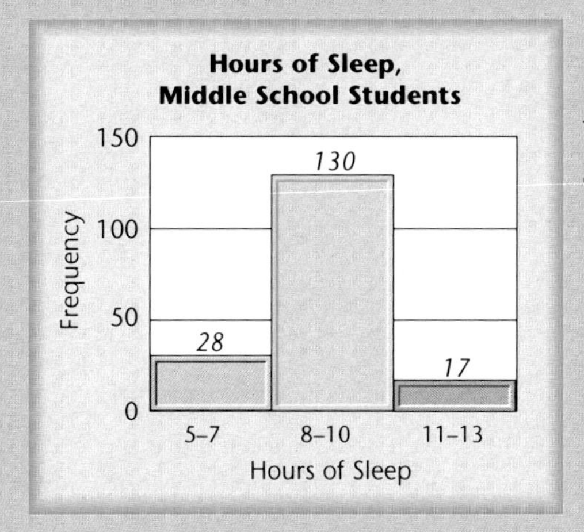

Use the histogram to answer the following questions. (Section 9.8)

17. How many students in the school get 8 or more hours of sleep?
18. How many students were surveyed?
19. What percentage of the students get between 5 and 7 hours of sleep?

Books Found in the School Library

Reference Books 300
Autobiography 225
Science Fiction 125
Young Adult Fiction 300
Romance Novels 50

Use the circle graph to answer the following questions. (Section 9.8)

20. How many books are in the library?
21. What percentage of the books in the collection are science fiction?
22. Would most of the books be classified as fiction or nonfiction?

For each problem, list the problem-solving strategy that you would use. Explain why you chose that problem-solving strategy. Then solve. (Section 9.6)

23. Suppose you toss a coin and roll a number cube. Find the number of possible outcomes, and then find the probability of getting heads and rolling a 4.

24. The mean of five numbers is 54. Four of the numbers are 25, 32, 50, and 98. What is the median of the set of five numbers?

Chapter 9 Test

Use the circle graph to the right to answer the following questions.

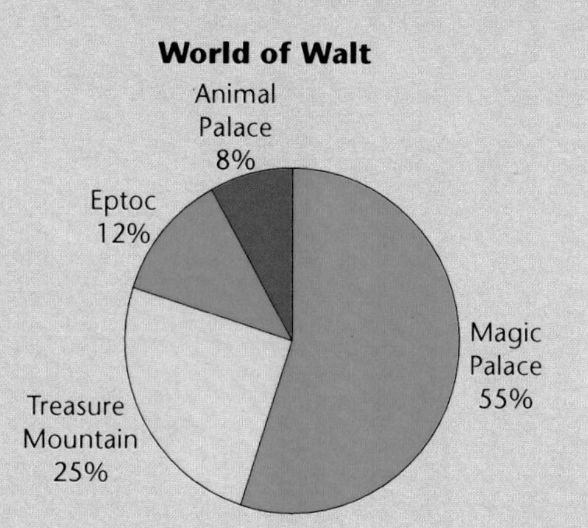

1. If the Small family spent 40 hours at the World of Walt, how many hours did they spend at Magic Palace?
2. How many hours did they spend at Eptoc and Animal Palace combined?

Scores				
9	8	10	4	9
6	9	9	9	8
6	9	7	8	9
5	10	8	7	10

Use the scores on a ten-point quiz to complete the following.

3. Make a frequency table of the data.
4. Make a line plot of the data.
5. Find the mean, median, mode, and range of the data.

Use the ages of the teachers to complete the following.

Teachers' Ages				
29	38	50	24	39
26	29	29	39	48
46	59	77	38	29

6. Find the mean, median, and mode of the teachers' ages.
7. Which measure of central tendency is most appropriate for representing the data? Explain.
8. Why could the mean be misleading?
9. Make a stem-and-leaf plot of the data.

Use the box-and-whisker plot to answer the following questions.

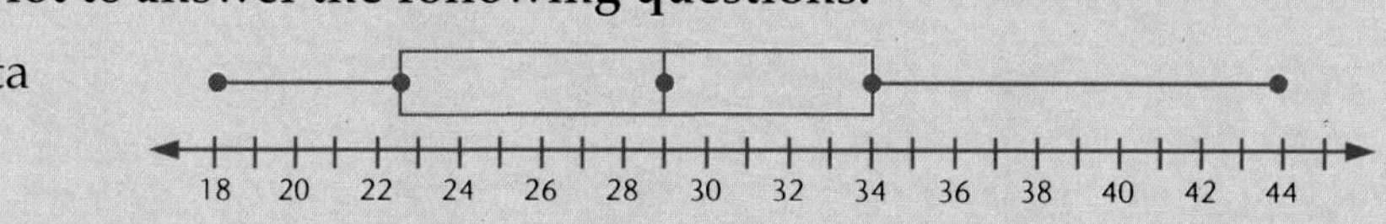

10. What fraction of the data is represented by each quartile?
11. What is the median of the data?
12. The graph represents the number of points scored by a basketball player. The player scored higher than what number 25% of the time?

Solve.

13. Summer is making cookies, brownies, and cake to celebrate her birthday. There are 48 students in her grade that will be eating desserts. Each cake makes 20 slices and she will make 3 cakes. There are 24 cookies per batch, and 18 brownies per batch. How many batches of cookies and brownies should she make so that each person gets exactly three items? She must make at least one batch of each dessert.
    a. List the problem-solving strategy that you would use.
    b. Explain why you chose that problem-solving strategy.
    c. Solve.

Change of Pace

Egyptian Multiplication

The ancient Egyptians developed their own method of multiplication long before 1650 B.C. They based the multiplication on successive doubling or adding a number to itself.

Follow these steps to multiply 12 and 23.

Step 1		**1 group of 23 = 23**	
Step 2	double 1 →	**2 groups of 23 = 46**	← double 23
Step 3	double 2 →	**4 groups of 23 = 92**	← double 46
Step 4	double 4 →	**8 groups of 23 = 184**	← double 92

Stop when you can add to get 12 groups.

Notice that 4 groups of 23 and 8 groups of 23 is 12 groups of 23. Therefore, add 92 and 184 to get the product 276. Check this result. $12 \times 23 = 276$ ✓

Another Example

Now follow this shorter version of the method to find 27×61.

1 — 61
2 — 122
4 — 244
8 — 488
16 — 976

List 1 group of 61 and continue to double both numbers.

Use the example for 27 × 61 to answer the following.

1. Which numbers in the first column add to 27?
2. List the numbers in the second column to be added.
3. Find the product of 27 and 61 by adding. Check your result.

List the numbers from the set 1, 2, 4, 8, and 16 that add to each number.

4. 23 5. 17 6. 3 7. 26 8. 13 9. 25

Multiply. Use Egyptian multiplication.

10. 22×42 11. 17×65 12. 31×53 13. 55×67

14. Relate Egyptian multiplication to the Distributive Property to explain why the method works.

Cumulative Test

1. In the equation $650 \div 20 = 32 \text{ R}10$, what is the number 650 called?
   a. dividend
   b. divisor
   c. quotient
   d. remainder

2. In the equation $230 \div 10 = 23$, what is the number 10 called?
   a. dividend
   b. divisor
   c. quotient
   d. remainder

3. What number comes next?
   240, 245, 255, 270, ?
   a. 260
   b. 265
   c. 275
   d. 290

4. Write 2 trillion in standard form.
   a. 2,000
   b. 2,000,000
   c. 2,000,000,000
   d. 2,000,000,000,000

5. $65 \times 10{,}000 = \_\_\_\_$
   a. 65,000
   b. 650,000
   c. 6,500,000
   d. none of the above

6. $45\overline{)1{,}354}$
   a. 3 R4
   b. 30
   c. 30 R4
   d. 34

7. Five softball players scored the following runs.

Angie	8
Bill	14
Sophia	15
Tanya	7
Ted	11

   What is the average number of runs scored by the five players?
   a. 5 c. 11
   b. 8 d. 56

8. Which of the following is the product of 524×130?
   a. 6,812
   b. 68,020
   c. 16,144
   d. 68,120

9. Which of the following equations helps to solve this problem?

   Bruce mows 32 lawns in 6 days. He mows 7 lawns the first day. He mows the same number of lawns each day for the next 5 days. How many lawns did he mow each day?
   a. $32 \div 5 = 6 \text{ R}2$
   b. $25 \div 5 = 5$
   c. $7 \div 5 = 1 \text{ R}2$
   d. none of the above

10. Which facts are given in this problem?

    Michelle built a bookshelf in her room. She paid $17 for supplies. She gave the clerk a $20 bill. How much change did she get?
    a. cost of supplies, change
    b. change, amount given clerk
    c. cost of supplies, amount given clerk
    d. cost of supplies, number of books on the bookshelf

CHAPTER

Geometric Figures

James Garey
Georgia

10.1 Introduction to Geometry

Objective: to understand, identify, and name basic geometry terms

Tourists driving or walking around an unfamiliar city will need a street map to help them find points of interest. A map uses lines to show buildings, streets, and locations. The lines form rectangles, squares, and other geometric shapes.

Geometry is about size, shape, and location.

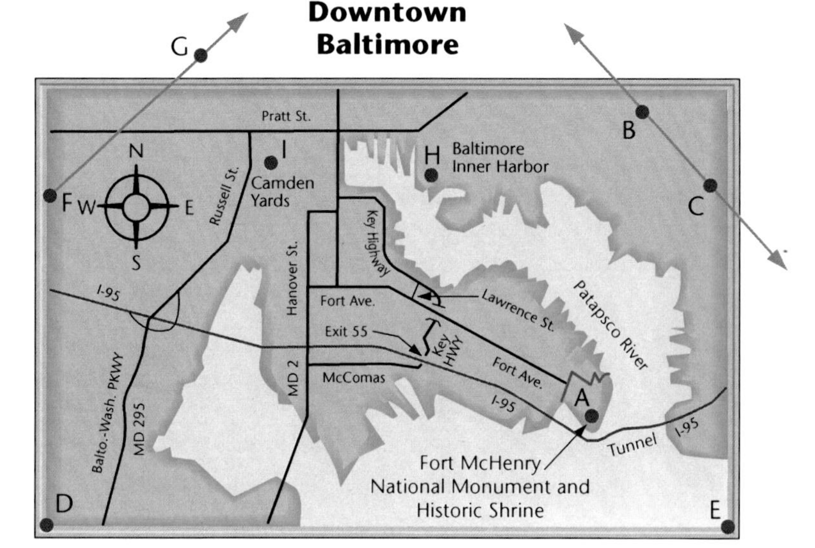

A **point** is an exact location. On the map, Fort McHenry is shown with a point labeled A.

A **plane** is a never-ending flat surface. A map might be considered a model of a plane.

A **line** is a never-ending straight path. It goes in both directions. On the map a line goes through points B and C, so the line is called BC. Write it as $\overleftrightarrow{BC}$ or $\overleftrightarrow{CB}$.

A **line segment** is a straight path with two endpoints. The lower border of the map is formed by line segment DE. Write it as $\overline{DE}$.

A **ray** is a never-ending straight path in one direction. The ray FG has its endpoint at point F and extends through point G. Write it as $\overrightarrow{FG}$. Name the endpoint first.

Points on the same line are **collinear**. Points that cannot be connected by one line are **noncollinear**.

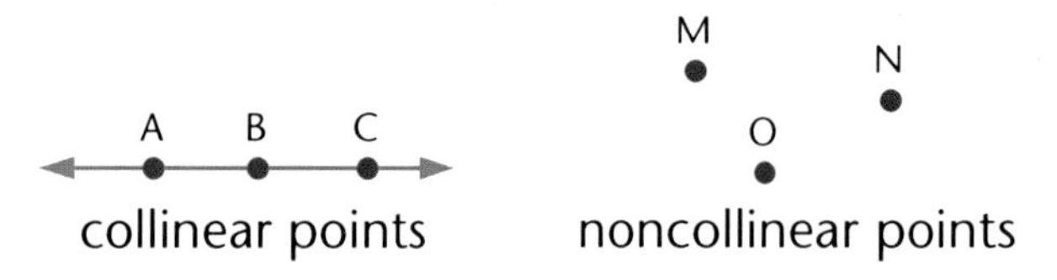

Lines that cross are called **intersecting lines**. Two lines that intersect or cross to form right angles are called **perpendicular lines**. In the figure below, line AB is perpendicular to line CD.

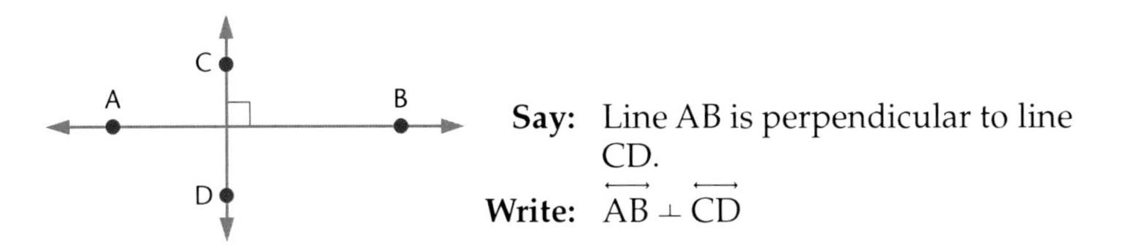

Say: Line AB is perpendicular to line CD.
Write: $\overleftrightarrow{AB} \perp \overleftrightarrow{CD}$

Some lines do *not* intersect. Lines in a plane that *never* cross are called **parallel lines**. In the figure below, line EF is parallel to line GH.

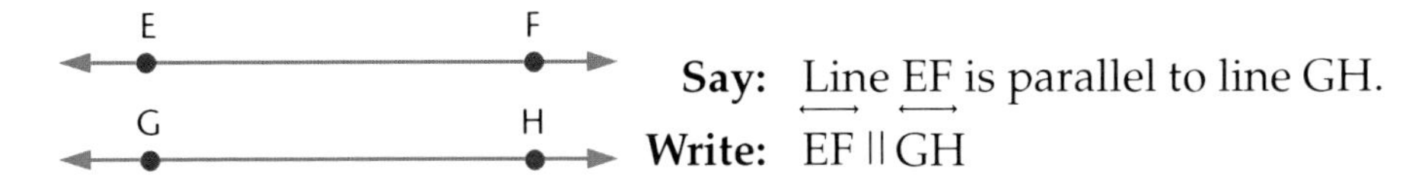

Say: Line EF is parallel to line GH.
Write: $\overleftrightarrow{EF} \parallel \overleftrightarrow{GH}$

Lines that do *not* intersect and are *not* parallel are called **skew lines**. In the figure shown at the right, line MN and line ST are skew lines.

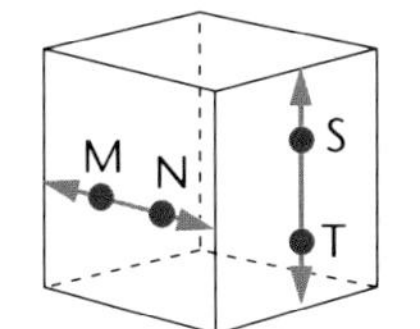

Line segments can also be intersecting, perpendicular, parallel, or skew. A line segment is part of a line.

Use symbols to name the following.

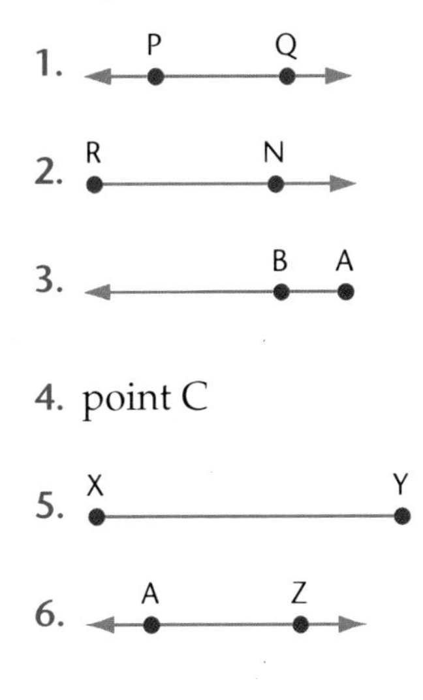

Exercises

Refer to the map of Baltimore on p. 258 to answer these questions.

1. Name two points on the map.
2. How many line segments form Fort Avenue?
3. Lawrence Street is a good example of what geometric term?
4. How many segments form the border of the map?
5. Key Highway cannot be considered a line or a line segment. Why?

Draw the following on your paper.

6. point C
7. line segment $\overline{EF}$
8. $\overrightarrow{HM}$
9. line $\overleftrightarrow{EB}$
10. ray $\overrightarrow{AB}$

Use symbols to name the following.

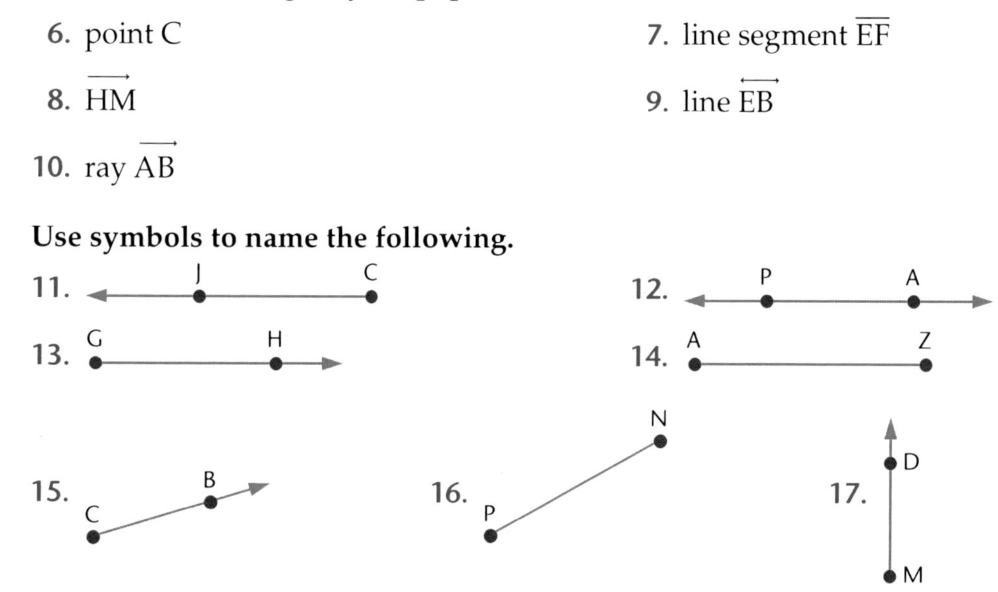

Use symbols to name all pairs of perpendicular and parallel lines and line segments.

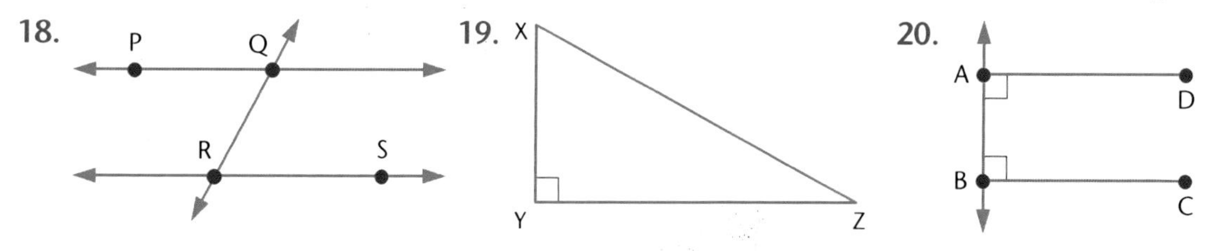

Complete. Use the figure at the right.

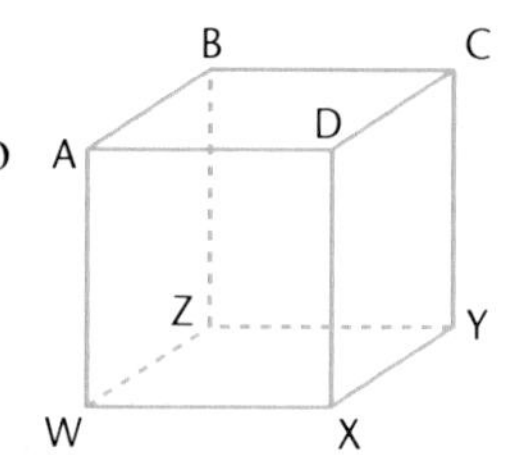

21. Name all the line segments that intersect and are perpendicular to line segment AD.
22. Name all the line segments that are parallel to line segment AD.
23. Name all the line segments that are skew to line segment AD.

Use *sometimes, always,* or *never* to complete each sentence.

24. A line ________________________ has two points.

25. Parallel lines ________________________ intersect.

26. Three points are ________________________ collinear.

27. Skew lines ________________________ intersect.

28. A line segment ________________________ has two endpoints.

Problem SOLVING

Draw as many line segments as you can.

29. Connect these points.

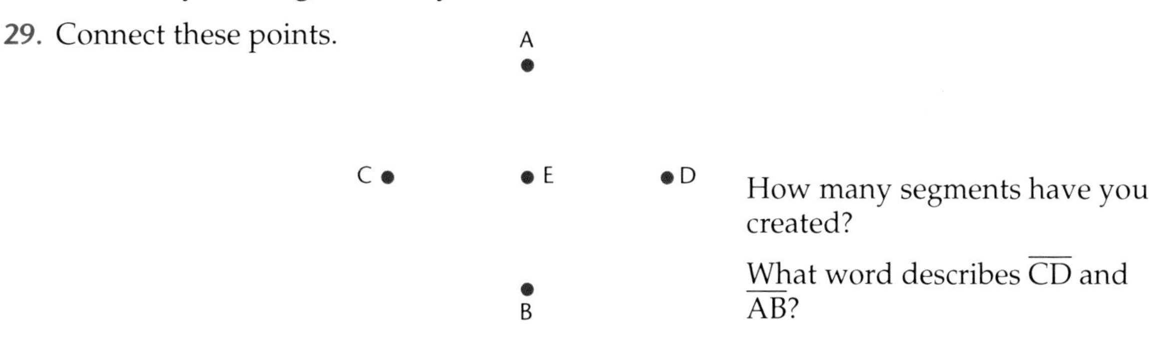

How many segments have you created?

What word describes $\overline{CD}$ and $\overline{AB}$?

Solve.

★30. Fold a sheet of paper in half. Into how many parts is the paper divided? Fold it again. How many parts are there? If it were possible, how many parts would there be after ten folds?

MIXED REVIEW

31. $\frac{3}{5} + \frac{1}{4}$

32. $\frac{2}{9} + \frac{1}{3}$

33. $\frac{3}{4} - \frac{1}{2}$

34. $\frac{5}{8} - \frac{11}{20}$

35. $\frac{1}{3} + \frac{4}{5}$

36. $\frac{7}{8} - \frac{5}{12}$

37. $7\frac{4}{5} + 9\frac{3}{10}$

38. $5\frac{2}{3} - 3\frac{1}{2}$

39. $9\frac{4}{7} - 3\frac{5}{14}$

40. $6\frac{5}{12} + 10\frac{5}{8}$

41. $8\frac{3}{5} - 4\frac{1}{5}$

42. $15 - 7\frac{2}{5}$

10.2 Angles

Objective: to measure, draw, and classify angles

The city recreation department is planning a new park. The designer of the park measures and draws many angles as the plan develops.

An **angle** has two sides, which are called **rays**, and a common endpoint called a **vertex**.

The most common unit used in measuring angles is the **degree.**

The angle below can be named three ways.

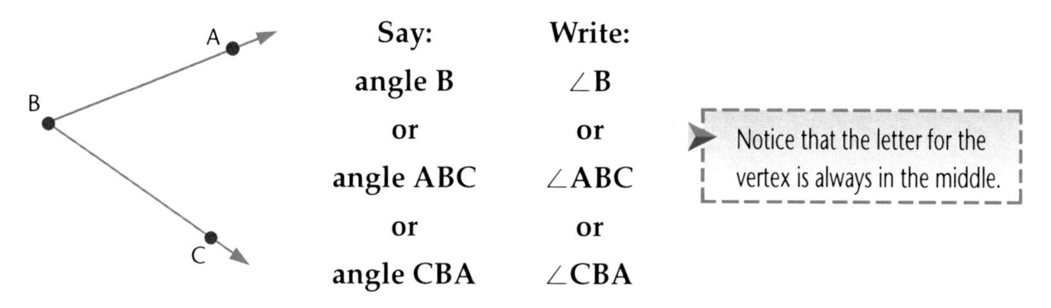

The angle shown in red *must* be named using three letters in either of the following ways.

∠JPK

or

∠KPJ

The angle is not named ∠P because ∠P could refer to any of the three angles having vertex P.

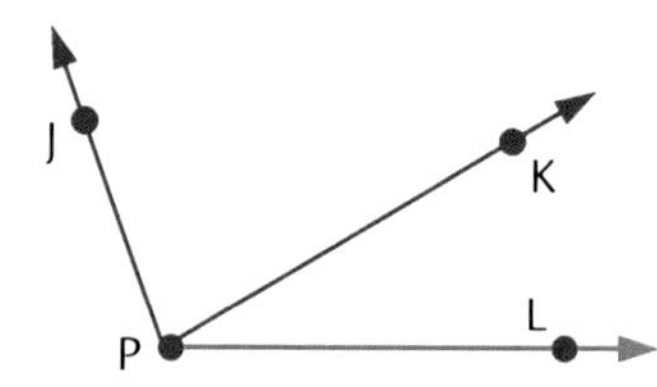

A **protractor**, such as the one shown to the right, is used to measure angles.

- Place the center of the protractor on the vertex.
- Place the 0° mark on one side of the angle.
- Read the measure of the angle where the other side crosses the scale. The angle is less than 90°.

The angle measures 50°.

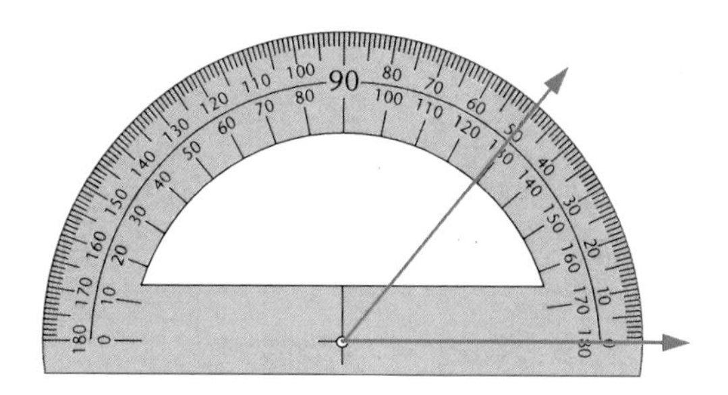

Will the angle measure less than or greater than 90°?

A special angle called a **right angle** is used to classify other angles. A right angle measures 90°.

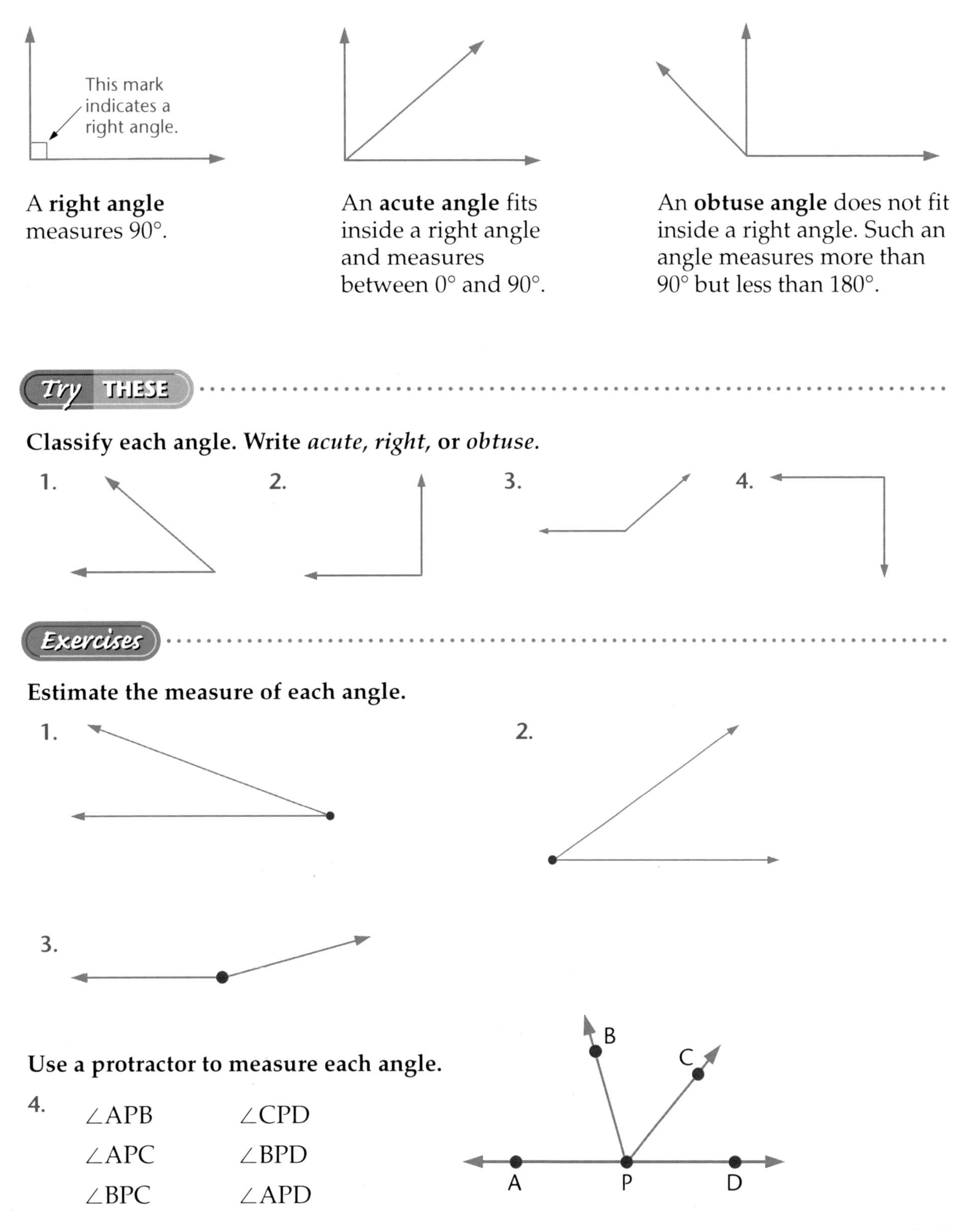

A **right angle** measures 90°.

An **acute angle** fits inside a right angle and measures between 0° and 90°.

An **obtuse angle** does not fit inside a right angle. Such an angle measures more than 90° but less than 180°.

Try THESE

Classify each angle. Write *acute, right,* or *obtuse.*

1.

2.

3.

4.

Exercises

Estimate the measure of each angle.

1.

2.

3.

Use a protractor to measure each angle.

4.

∠APB	∠CPD
∠APC	∠BPD
∠BPC	∠APD

Without using a protractor, sketch an angle for each measurement. Then use a protractor to check your work.

5. 20° 6. 90° 7. 105° 8. 180° 9. 60° 10. 120° 11. 150°

Use symbols to name each red angle. Then classify the red angle.

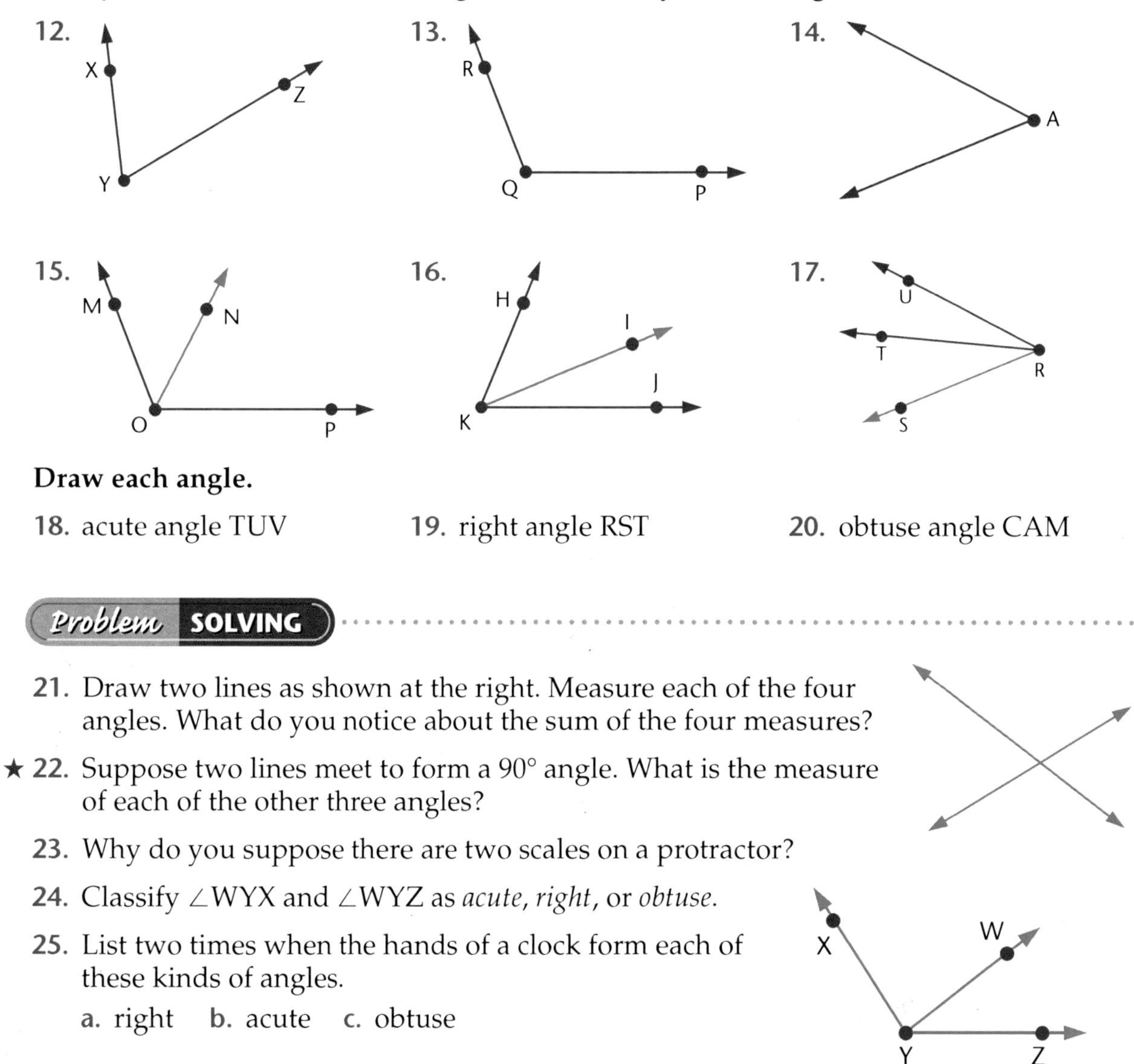

Draw each angle.

18. acute angle TUV 19. right angle RST 20. obtuse angle CAM

Problem SOLVING

21. Draw two lines as shown at the right. Measure each of the four angles. What do you notice about the sum of the four measures?

★ 22. Suppose two lines meet to form a 90° angle. What is the measure of each of the other three angles?

23. Why do you suppose there are two scales on a protractor?

24. Classify ∠WYX and ∠WYZ as *acute, right,* or *obtuse.*

25. List two times when the hands of a clock form each of these kinds of angles.

   a. right b. acute c. obtuse

Cumulative Review

Compare using $<$, $>$, or $=$.

1. 27,000 ● 270,000
2. 350,649 ● 350,649
3. 47,354 ● 47,345

Estimate.

4. $\begin{array}{r} 1{,}038 \\ +\ 874 \\ \hline \end{array}$
5. $\begin{array}{r} 9{,}127 \\ -\ 428 \\ \hline \end{array}$
6. $\begin{array}{r} 461 \\ \times\ 18 \\ \hline \end{array}$
7. $28\overline{)239}$
8. $57\overline{)4{,}798}$

Compute.

9. $\begin{array}{r} 54 \\ +\ 19 \\ \hline \end{array}$
10. $\begin{array}{r} 2{,}408 \\ +\ 4{,}207 \\ \hline \end{array}$
11. $\begin{array}{r} 39{,}037 \\ +\ 9{,}063 \\ \hline \end{array}$
12. $\begin{array}{r} \$148 \\ 62 \\ +\ 45 \\ \hline \end{array}$
13. $\begin{array}{r} 3{,}300 \\ 2{,}409 \\ +\ 466 \\ \hline \end{array}$
14. $\begin{array}{r} 88 \\ -\ 17 \\ \hline \end{array}$
15. $\begin{array}{r} 763 \\ -\ 159 \\ \hline \end{array}$
16. $\begin{array}{r} 1{,}009 \\ -\ 603 \\ \hline \end{array}$
17. $\begin{array}{r} \$16{,}789 \\ -\ 11{,}590 \\ \hline \end{array}$
18. $\begin{array}{r} 49{,}500 \\ -\ 24{,}175 \\ \hline \end{array}$
19. $\begin{array}{r} 465 \\ \times\ 8 \\ \hline \end{array}$
20. $\begin{array}{r} 7{,}000 \\ \times\ 90 \\ \hline \end{array}$
21. $\begin{array}{r} \$399 \\ \times\ 25 \\ \hline \end{array}$
22. $\begin{array}{r} 6{,}210 \\ \times\ 100 \\ \hline \end{array}$
23. $\begin{array}{r} 371 \\ \times\ 256 \\ \hline \end{array}$
24. $3\overline{)60}$
25. $9\overline{)191}$
26. $24\overline{)288}$
27. $33\overline{)499}$
28. $170\overline{)2{,}720}$

Find the average of each set of numbers.

29. 84, 71, 66, 99
30. 855; 1,033; 809

Write using exponents.

31. $7 \times 7 \times 7 \times 7 \times 7$
32. 100,000,000
33. 81×81

If the problem has enough facts, solve it. If not, write any missing facts.

34. Linda bought a doll for $24.98. Her change was $5.02. How much money did she give the clerk?

35. Felicia spent of $\frac{1}{5}$ her allowance on Saturday and $\frac{4}{8}$ of her allowance on Sunday. How much of her allowance did she spend?

36. The students at Parke Middle School either walk to school or ride the bus. There are 226 students who ride the bus. How many students walk?

37. Jason catches 5 fish on Monday, 6 fish on Tuesday, 2 fish on Wednesday, and 7 fish on Thursday. What is the average number of fish caught each day?

10.3 Special Angles

Objective: to identify special angles

Some pairs of angles have special relationships. Two angles whose sum is 90° are **complementary angles**. Two angles whose sum is 180° are **supplementary angles**.

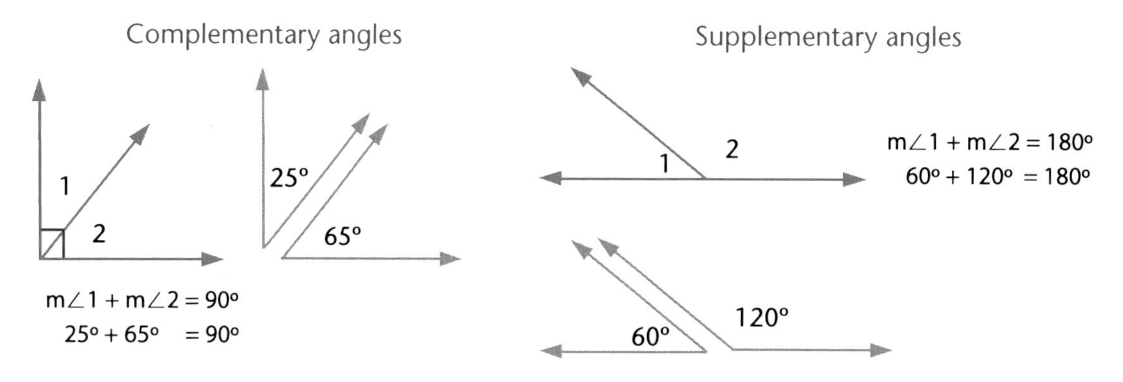

There are other special pairs of angles. **Vertical angles** are angles formed by two intersecting lines. Vertical angles have equal measures. Equal angles are **congruent angles**. Angles 1 and 2 are opposite each other and are therefore vertical angles. Can you name another pair of vertical angles?

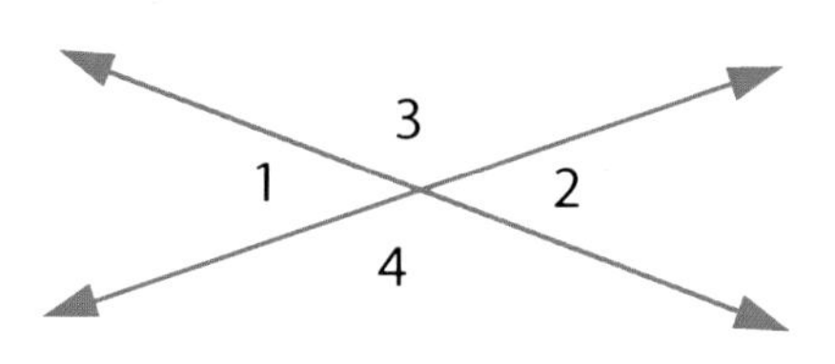

Adjacent angles share a side and have a common vertex. Angles 1 and 3 are an example of adjacent angles. Can you name the other adjacent angles?

Examples

A. Find the complement of the angle.

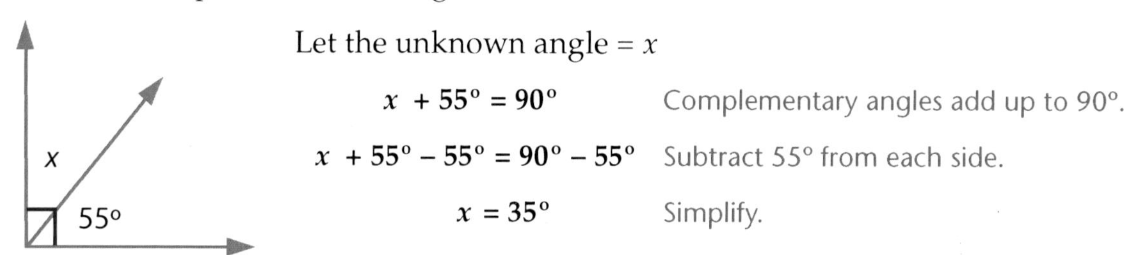

Let the unknown angle = x

$x + 55° = 90°$	Complementary angles add up to 90°.
$x + 55° - 55° = 90° - 55°$	Subtract 55° from each side.
$x = 35°$	Simplify.

The measure of the complement of the angle is 35°.

B. Angles P and T are supplementary. If $m\angle T = 110°$, find the $m\angle P$.

$\angle P + \angle T = 180°$	Supplementary angles add up to 180°.
$\angle P + 110° = 180°$	Replace $\angle T$ with 110°.
$\angle P + 110° - 110° = 180° - 110°$	Subtract 110° from each side.
$\angle P = 70°$	

The measure of the supplement of the angle is 70°.

Try THESE

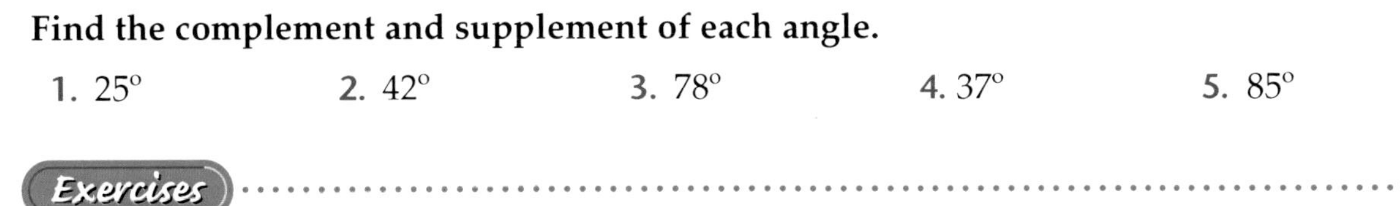

Find the complement and supplement of each angle.

1. 25°
2. 42°
3. 78°
4. 37°
5. 85°

Exercises

Find the value of x in each figure.

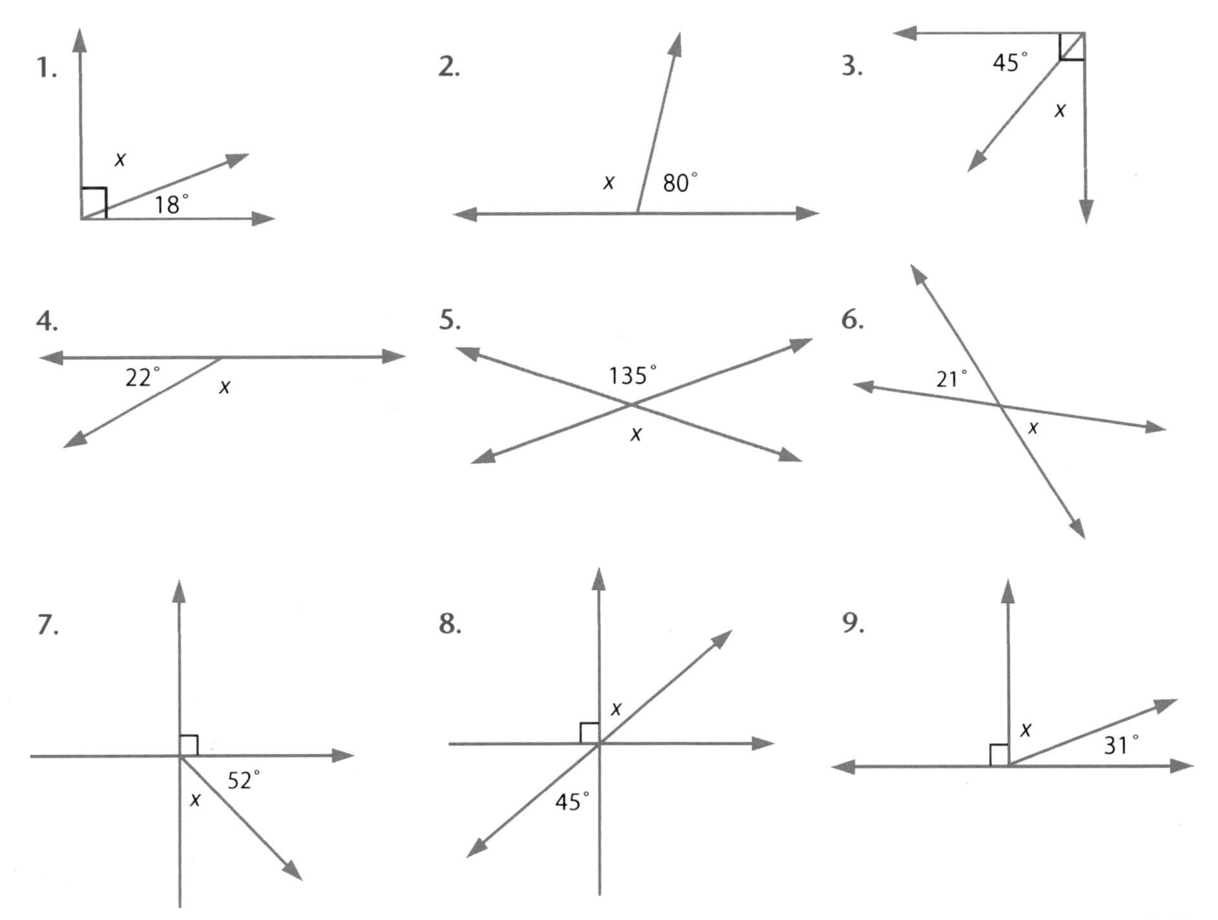

Use *sometimes*, *always*, or *never* to complete each sentence.

10. Two acute angles are ____________________ supplementary.
11. Two right angles are ____________________ complementary.
12. Two obtuse angles are ____________________ supplementary.
13. Two acute angles are ____________________ complementary.
14. Two obtuse angles are ____________________ complementary.
15. Two right angles are ____________________ supplementary.

Problem SOLVING

16. Can an angle that measures 125° have a complement and a supplement? Explain.

★ 17. A circle is divided into two equal sections. How many degrees are in each section? The circle is then divided into two more sections. How many degrees are in each section?

Test PREP

18. An angle measures 64°. Find its complement.

    a. 36° b. 116° c. 126° d. 26°

19. All vertical angles are equal.

    a. True b. False

Mind BUILDER

Mental Images

How many blocks do you see?

Turn the book around. Now how many blocks do you see?

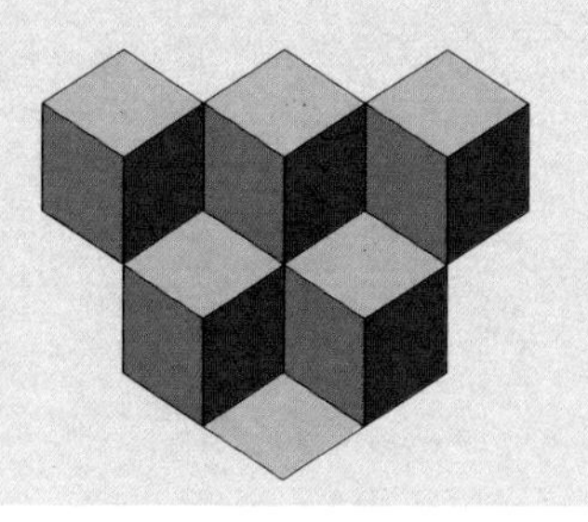

Mrs. Crosby's Puzzle

Mrs. Crosby, the geometry teacher, was cutting up some geometric shapes (star, triangle, circle, square, rectangle, hexagon) for her students to put together as puzzles. Each piece was cut into three pieces. When she smelled something burning in her kitchen, she ran to turn off the stove. In the meantime, her kitten Polygon mixed up all of the shapes. Help Mrs. Crosby put the shapes back together.

Extension

Sort the six shapes any way you choose.

10.4 Constructions: Angle and Angle Bisectors

Objective: to construct angles and angle bisectors

An architect uses a compass and straightedge to draw the design for a new building. You can use a compass and straightedge to construct angles and angle bisectors.

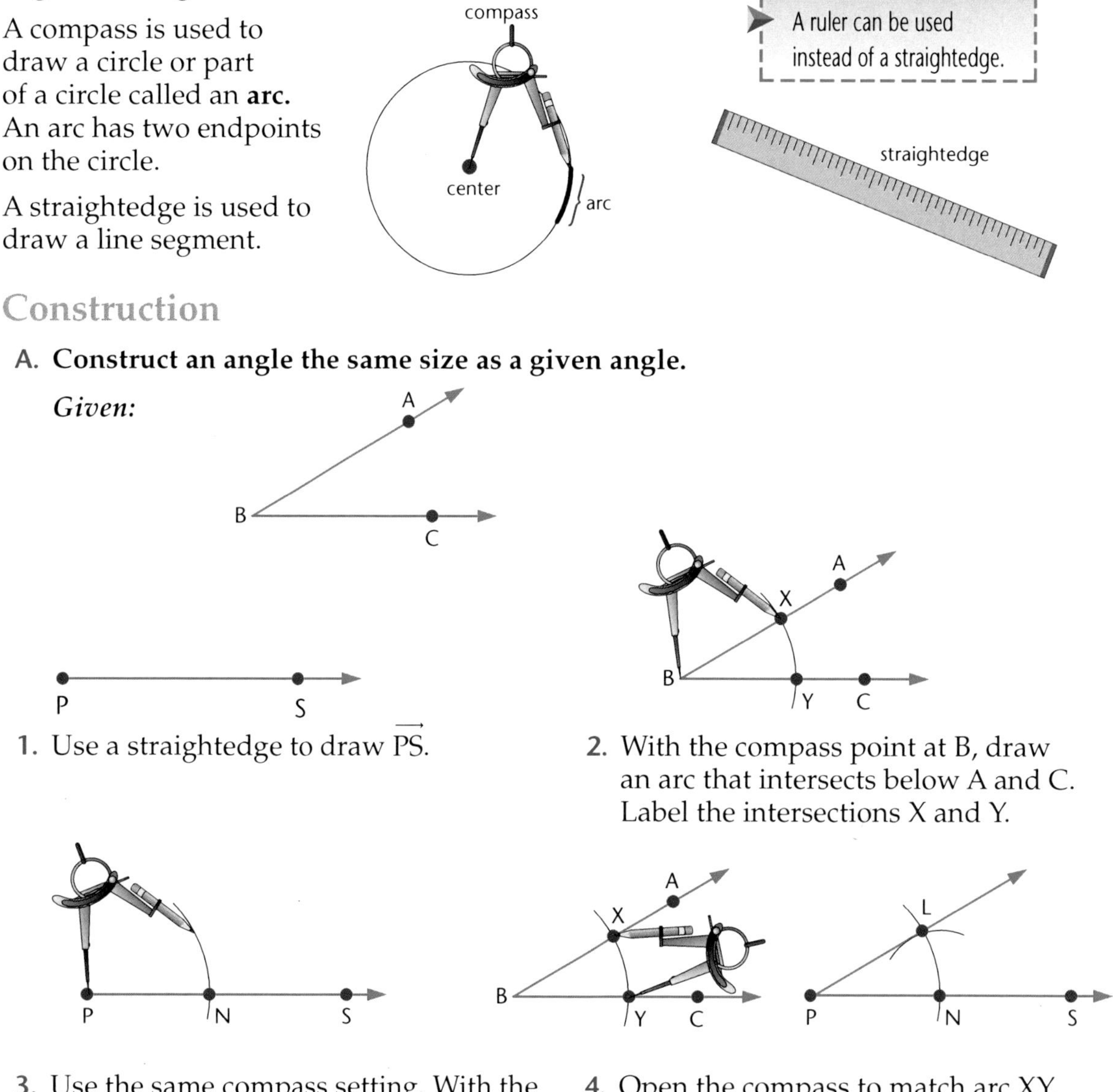

A compass is used to draw a circle or part of a circle called an **arc.** An arc has two endpoints on the circle.

A straightedge is used to draw a line segment.

A ruler can be used instead of a straightedge.

Construction

A. Construct an angle the same size as a given angle.

Given:

1. Use a straightedge to draw $\overrightarrow{PS}$.
2. With the compass point at B, draw an arc that intersects below A and C. Label the intersections X and Y.
3. Use the same compass setting. With the compass point at P, draw an arc. Label that intersection N.
4. Open the compass to match arc XY. With the compass point at N, draw an arc. Label the intersection of the arcs L. Use a straightedge to draw $\overrightarrow{PL}$. $\angle LPS$ is equal to $\angle ABC$.

B. Bisect a given angle.

Given:

> *Bisect* means to separate into two equal parts.

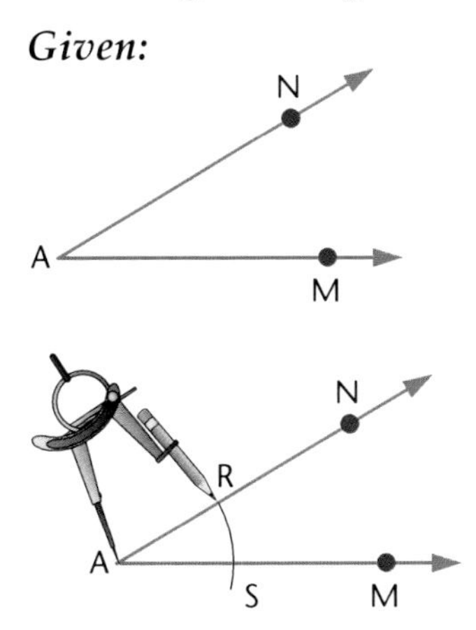

Draw an arc from A. Label intersection points R and S.

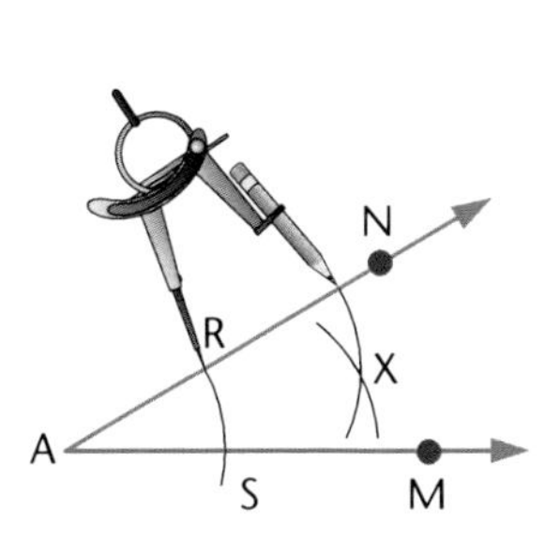

Draw arcs from points R and S intersecting at X.

Draw $\overrightarrow{AX}$. This ray bisects ∠NAM. ∠NAX is congruent to ∠XAM.

Exercises

Trace each angle. Construct an angle equal to each angle given.

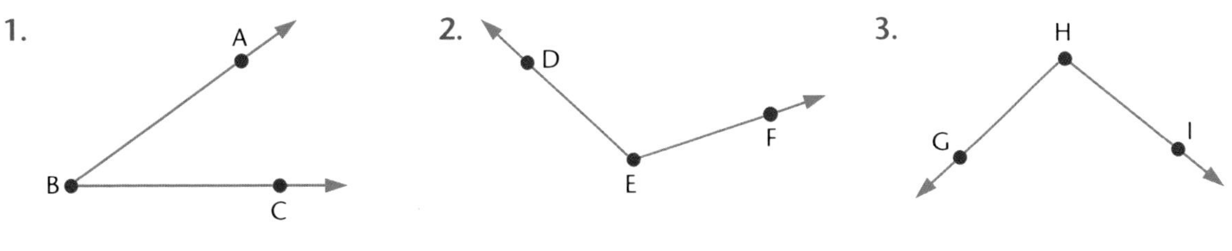

Trace each angle. Then bisect each angle given.

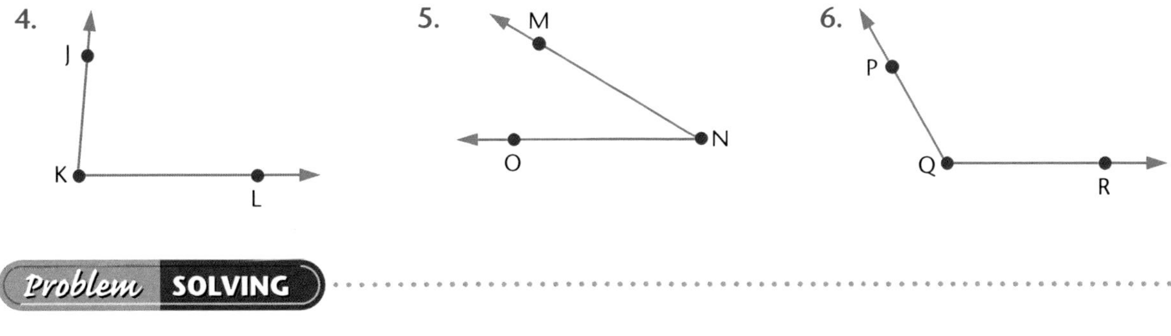

Problem SOLVING

7. Draw an acute angle and construct its bisector.

8. Draw an obtuse angle. Separate it into four congruent angles.

10.5 Classifying Triangles

Objective: to classify triangles

Triangles can be classified by the number of congruent sides.

A **scalene triangle** has no congruent sides.

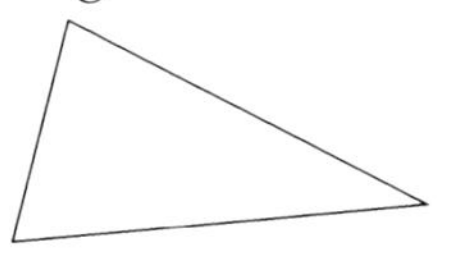

An **isosceles triangle** has two congruent sides.

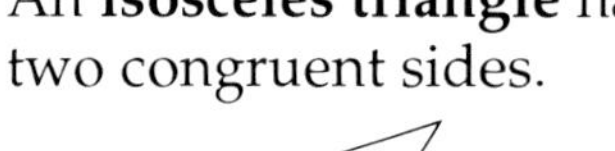

An **equilateral triangle** has three congruent sides.

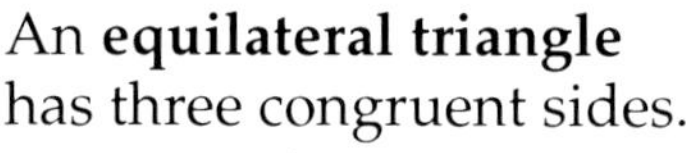

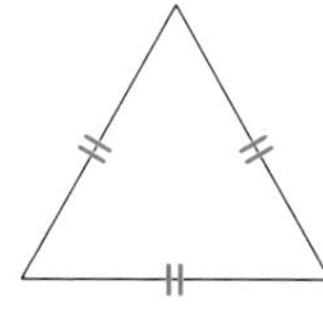

Triangles may also be classified by their angles.

The blue marks indicate which sides and angles are congruent.

An **acute triangle** has three acute angles.

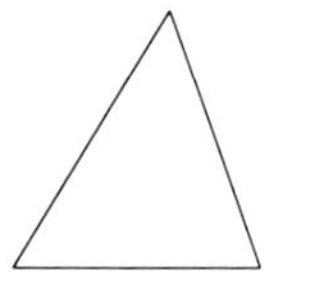

A **right triangle** has one right angle.

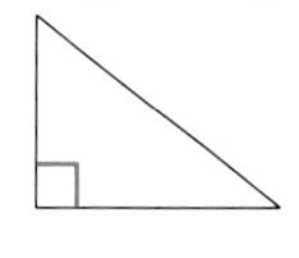

An **obtuse triangle** has one obtuse angle.

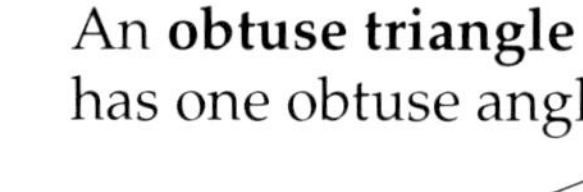

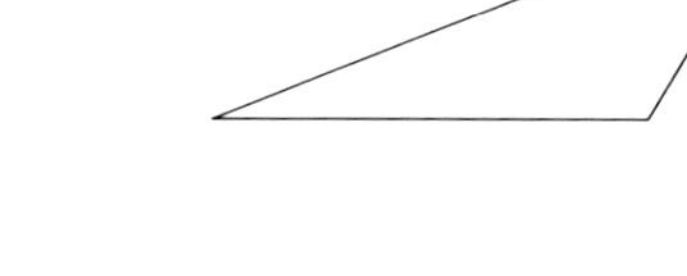

Draw any triangle. Label the angles 1, 2, and 3. Cut off the corners of the triangle and arrange them as shown.

This shows that the angles of a triangle form a straight angle whose measure is 180°. So, the sum of the measures of the angles of a triangle is 180°.

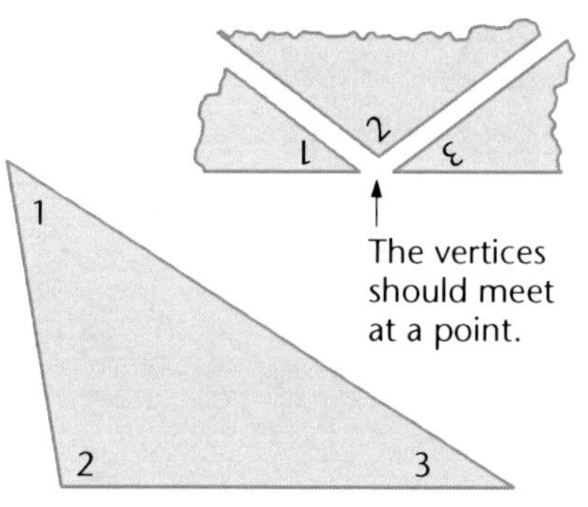

Try THESE

Classify each triangle by its sides.

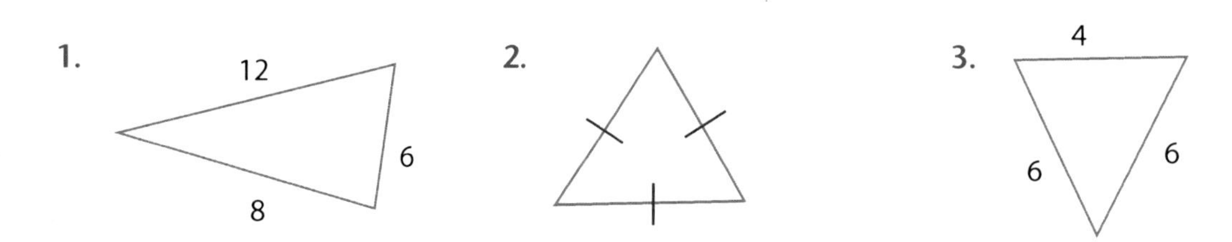

Exercises

Classify each triangle by its angles.

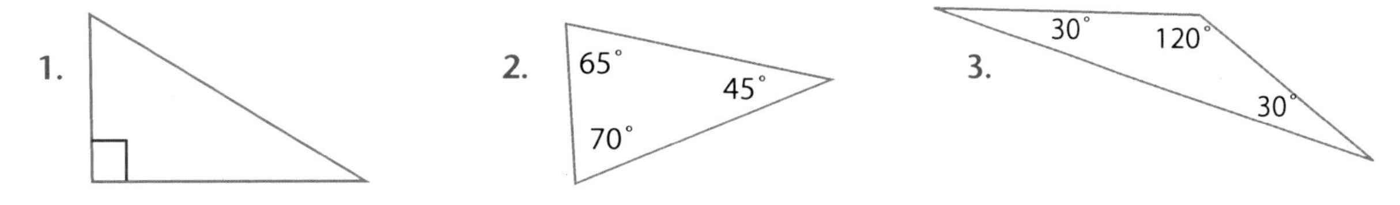

Use a ruler to measure the sides of each triangle and a protractor to measure the angles of each triangle. Classify each triangle by its angles and sides.

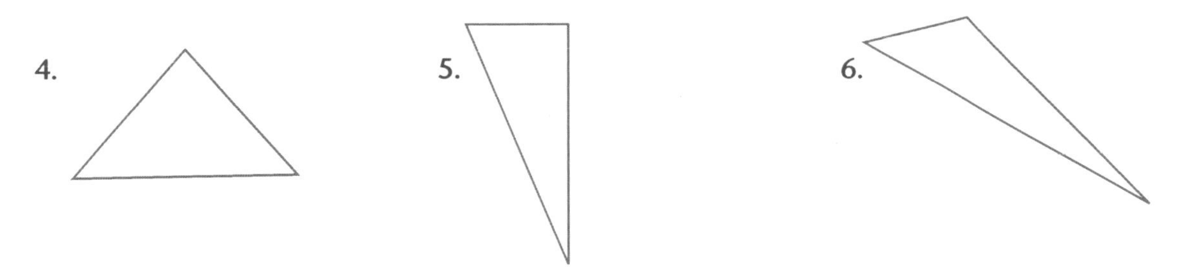

Find the value of x for triangles with the given angle measures.

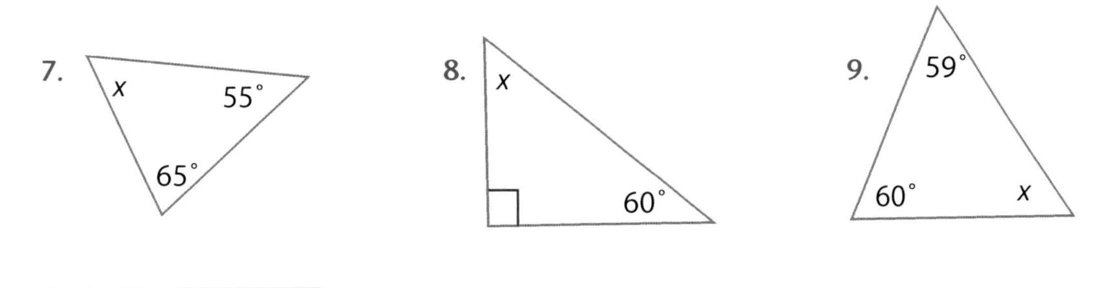

Problem SOLVING

10. A triangle has angles that measure 20° and 71°. Find the measure of the third angle.

11. A right triangle has an angle that measures 42°. Find the measure of the third angle.

12. What is the measure of each of the angles in an equilateral triangle?

★13. Is it possible to make a triangle with sides of length 3 cm, 15 cm, and 19 cm? Explain. You may want to cut strips of paper to demonstrate.

10.6 Classifying Quadrilaterals

Objective: to classify quadrilaterals

Quadrilaterals are classified according to the following.

- parallel sides
- congruent angles
- congruent sides

A **trapezoid** is a quadrilateral with only one pair of parallel sides.

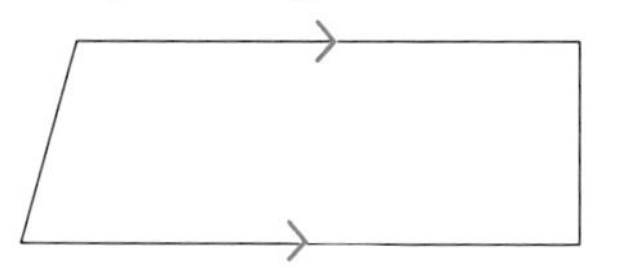

A **parallelogram** is a quadrilateral with two pairs of parallel sides.

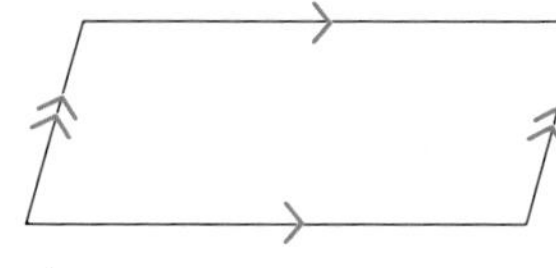

A **rectangle** is a parallelogram with all angles congruent.

A **rhombus** is a parallelogram with all sides congruent.

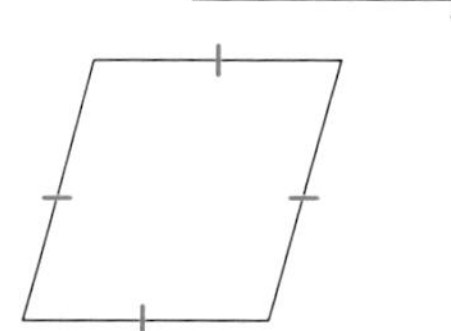

A **square** is a parallelogram with all sides and all angles congruent.

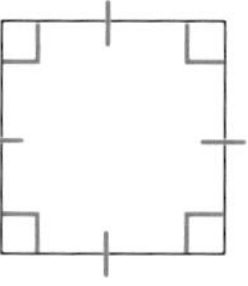

Exploration Exercise

1. Measure each angle in the rectangle to the right.
2. Add the four angles together. What is the sum?
3. Draw a quadrilateral other than a rectangle.
4. Measure each angle in the quadrilateral.
5. Add the four angles together. What is the sum?

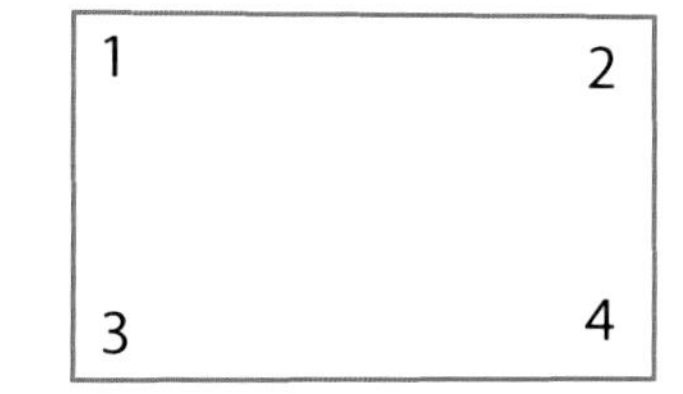

The sum of the measures of all angles in any quadrilateral is 360°.

Try THESE

Classify each quadrilateral.

1. 2.

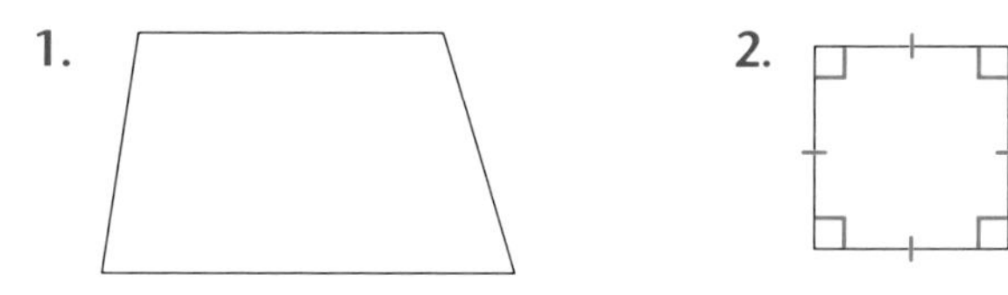

3.

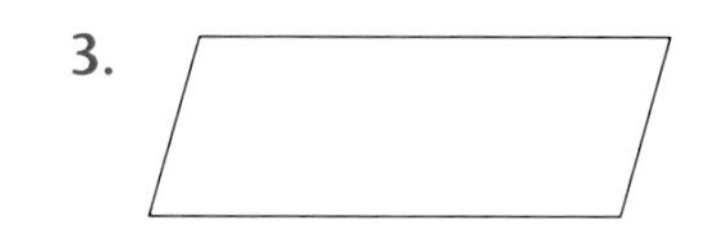

Exercises

State if the following statements are *true* or *false*.

1. A square is a rectangle.
2. All parallelograms are squares.
3. A rectangle is a quadrilateral.
4. A rhombus is a rectangle.

Draw an example of each quadrilateral.

5. rectangle
6. parallelogram
7. rhombus
8. trapezoid

Find the value of x for quadrilaterals with the given angle measures.

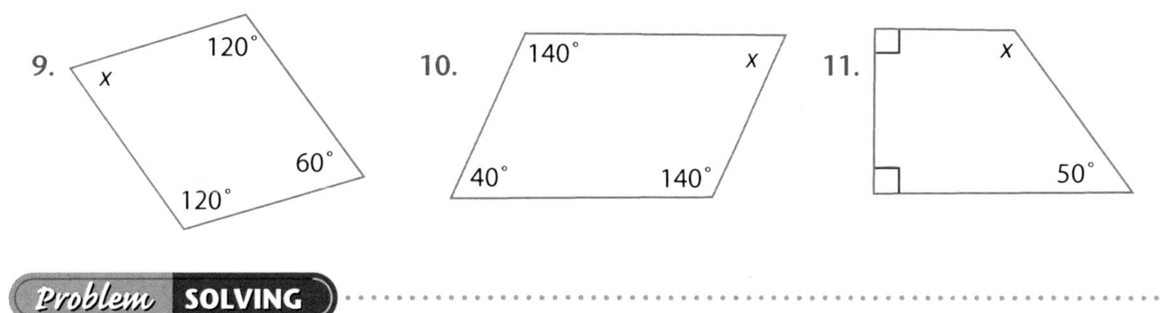

Problem SOLVING

12. Describe the similarities and differences between a rectangle and a square.
13. A quadrilateral has angles that measure 65°, 65°, and 115°. Find the measure of the fourth angle.
14. In some quadrilaterals all of the angles are equal. What is the measure of each angle in an equiangular quadrilateral?

MIXED REVIEW

Round to the underlined place-value position.

15. $6\underline{7}.203$
16. $4.\underline{3}5$
17. $8.2\underline{1}4$
18. $\underline{0}.982$
19. $8.\underline{9}99$
20. $10.0\underline{1}5$
21. $7\underline{3}.5$
22. $12.\underline{6}7$
23. $60\underline{2}.7$
24. $1.\underline{0}08$
25. $4\underline{8}2$
26. $6{,}\underline{2}67$
27. $\underline{2}49$
28. $\underline{1}{,}396$
29. $\underline{5}8$

10.7 Polygons

Objective: to name polygons

Street signs have many shapes. These signs are shaped like **polygons**. A polygon is a closed figure in a plane. It is made up of line segments that meet but do *not* cross. The figures below are *not* polygons.

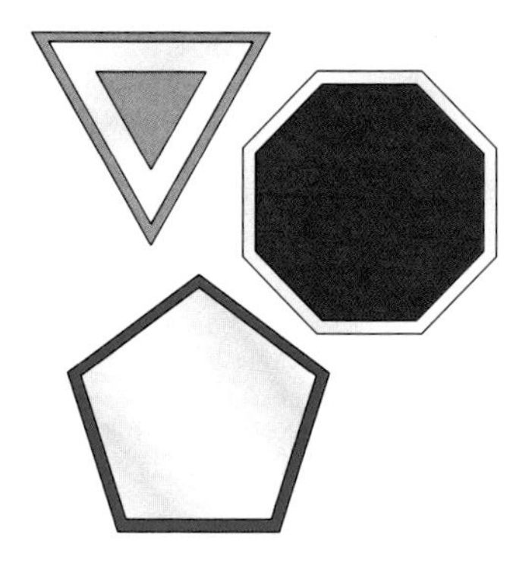

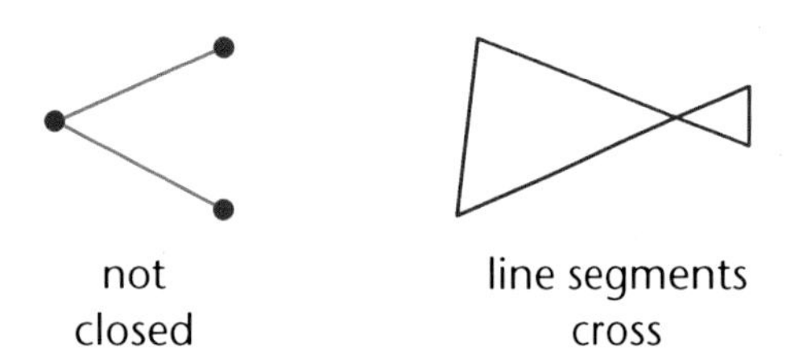

The line segments that form a polygon are called **sides.** The points where the sides meet are called **vertices** (plural of vertex).

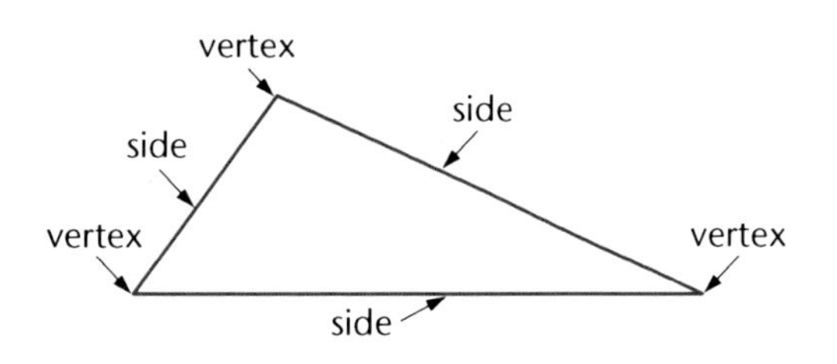

The name of a polygon is determined by the number of sides or angles. *Poly-* means "many" and *–gon* means "angle."

Polygon	Sides	Vertices
Triangle	3	3
Quadrilateral	4	4
Pentagon	5	5
Hexagon	6	6
Heptagon	7	7
Octagon	8	8

Polygons that have sides congruent and all angles congruent are called **regular polygons**. A regular triangle (equilateral triangle) and a regular quadrilateral (square) are shown.

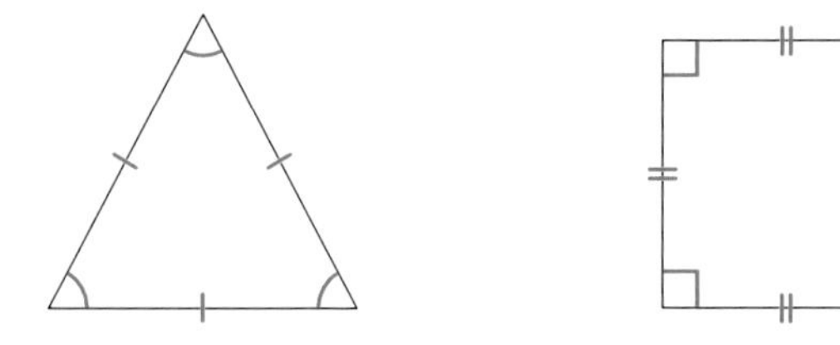

In a polygon with more than three sides you can draw diagonals. Diagonals are segments connecting two vertices that are not next to each other.

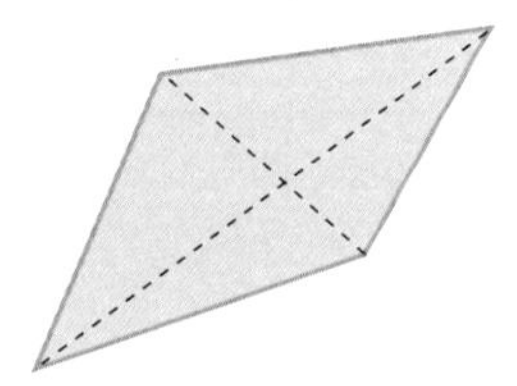

A quadrilateral has two diagonals.

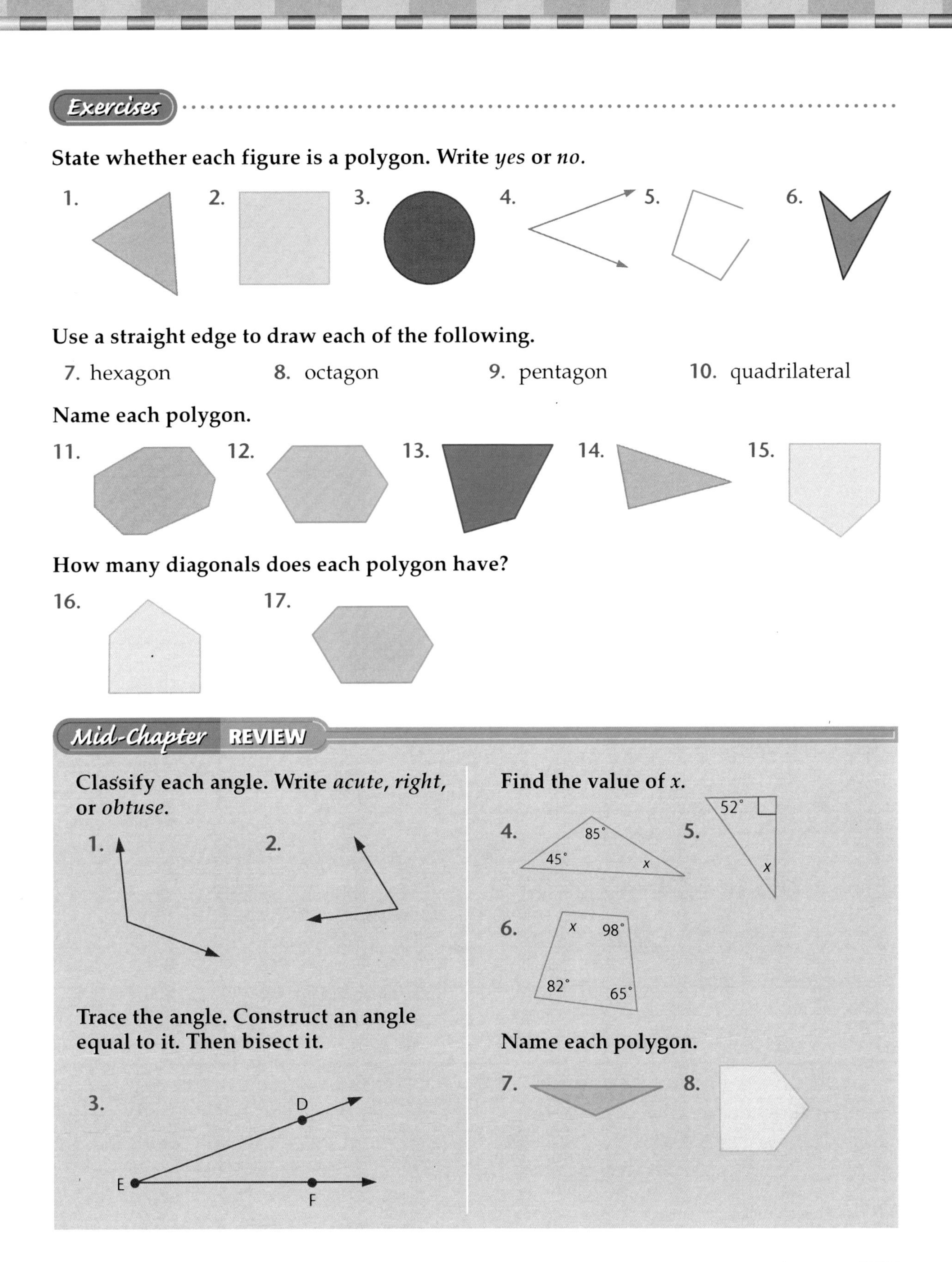

Exercises

State whether each figure is a polygon. Write *yes* or *no*.

1. 2. 3. 4. 5. 6.

Use a straight edge to draw each of the following.

7. hexagon
8. octagon
9. pentagon
10. quadrilateral

Name each polygon.

11. 12. 13. 14. 15.

How many diagonals does each polygon have?

16. 17.

Mid-Chapter REVIEW

Classify each angle. Write *acute, right,* or *obtuse*.

1. 2.

Trace the angle. Construct an angle equal to it. Then bisect it.

3.

Find the value of x.

4. 5. 6.

Name each polygon.

7. 8.

10.8 Problem-Solving Strategy: Make a Venn Diagram

Objective: to solve problems using Venn diagrams

The Smithsonian Institution is the largest museum complex in the world. Located in Washington, DC, this center houses numerous museums and research centers that collectively contribute to the nation's identity. Because of its diverse features, the people who work at the Smithsonian Institution must classify its exhibits. To *classify* means to arrange in groups that have something in common.

In mathematics, a **Venn diagram** can be used to show how objects are classified. A Venn diagram consists of two or more circles enclosed in a rectangle.

Examples

Classify the following objects into groups by looking at their attributes. What characteristics do they have in common? Then draw a Venn diagram.

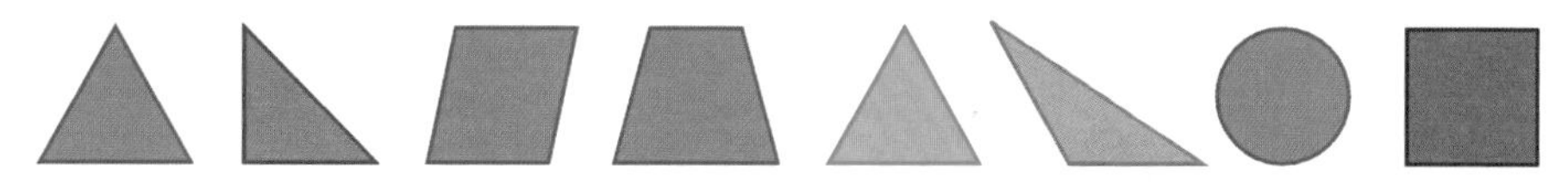

There are two obvious groups. One group has blue figures; the other group has triangles.

Draw two overlapping circles using a compass and label them as shown. Then place each figure in the correct circle.

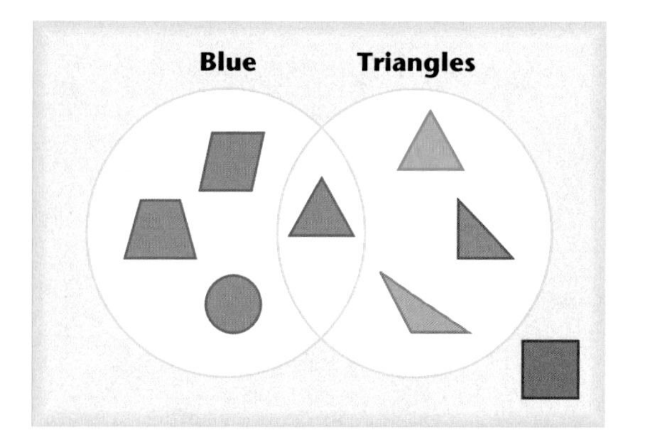

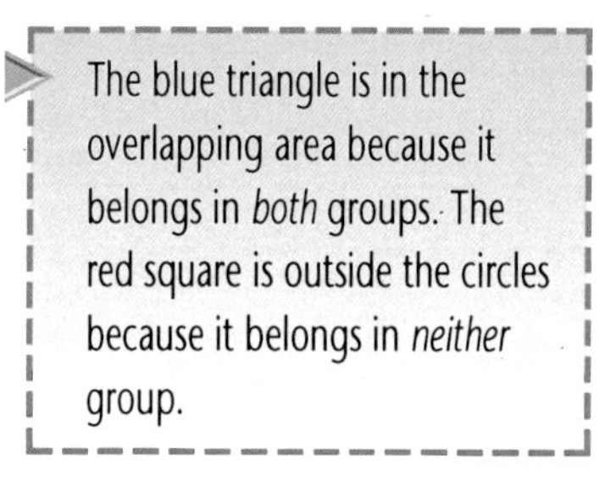

Try THESE

Name the common characteristic of the figures in each Venn diagram.

1. 2.

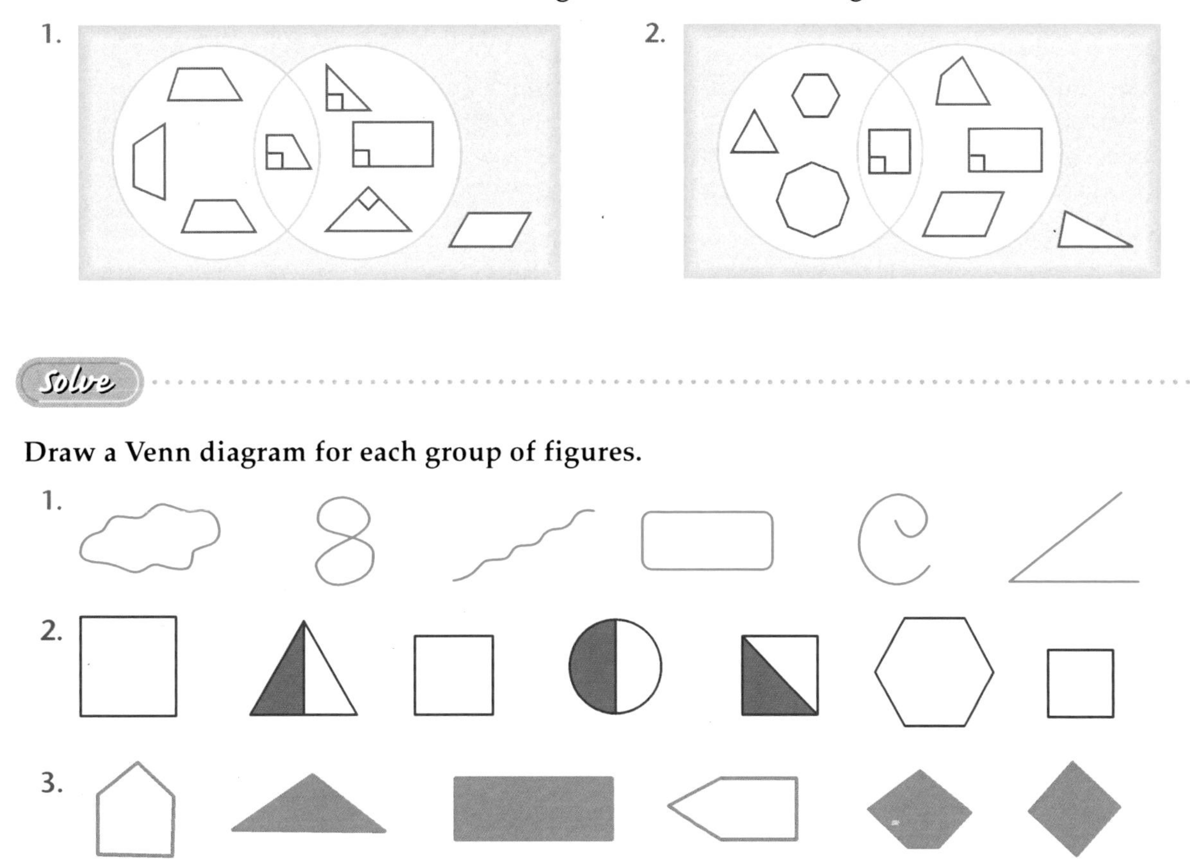

Solve

Draw a Venn diagram for each group of figures.

1.

2.

3.

Use the Venn diagram.

4. What type of figure is both a rectangle and a rhombus?
5. Why is the trapezoid not within a circle?

★6. How could you include parallelograms in the Venn diagram?

Quadrilaterals

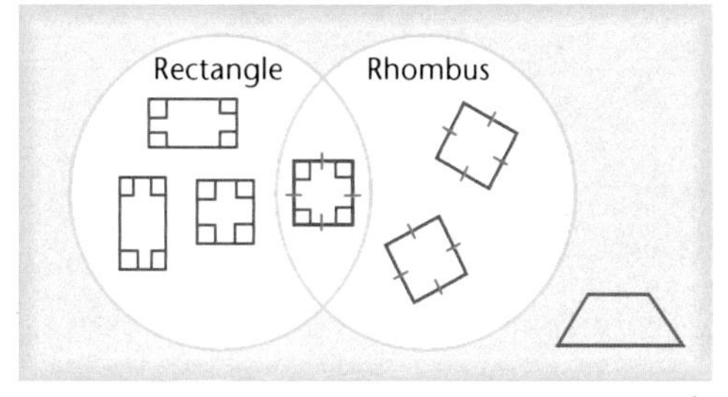

10.9 Congruent and Similar Figures

Objective: to distinguish between congruent and similar figures

If you visit two major league baseball parks, you may notice that the baseball diamonds have the same shape and size. Figures that have the *same shape and size* are called **congruent figures**. Congruent figures have parts that match. These parts are called **corresponding parts**.

The Little League baseball diamond is the *same shape* as the major league diamond, but *not the same size.* Figures that have the same shape are called **similar figures**.

Congruent figures match or fit on each other exactly. Triangles ABC and DEF are congruent. Check this by matching or tracing.

Example

△MNO and △RST are congruent. List the corresponding sides and angles.

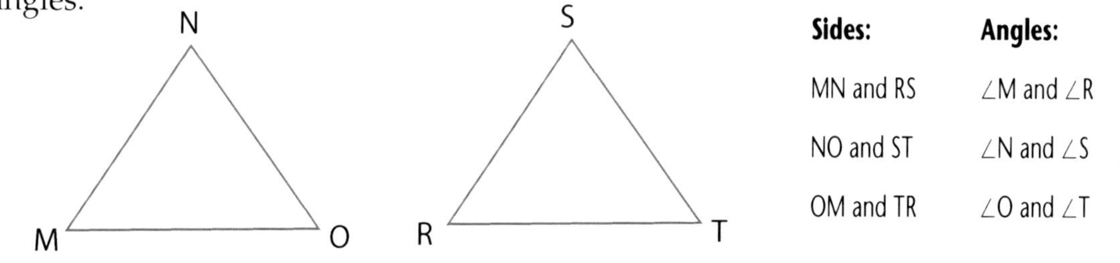

Sides:	Angles:
MN and RS	∠M and ∠R
NO and ST	∠N and ∠S
OM and TR	∠O and ∠T

Try THESE

Use tracings to draw figures congruent to the following.

1. 2. 3. 4.

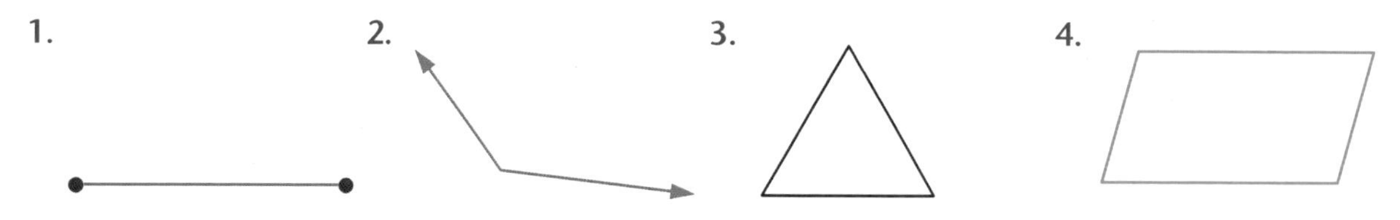

Exercises

State whether each set of figures is *congruent, similar,* or *neither.* Explain your answer.

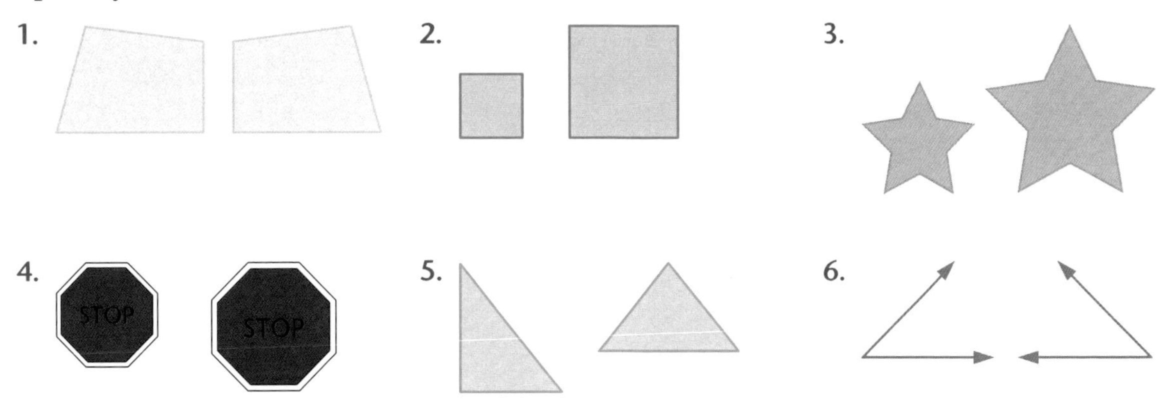

7. Explain the difference between congruent and similar figures.

△ABC and △XYZ are congruent.

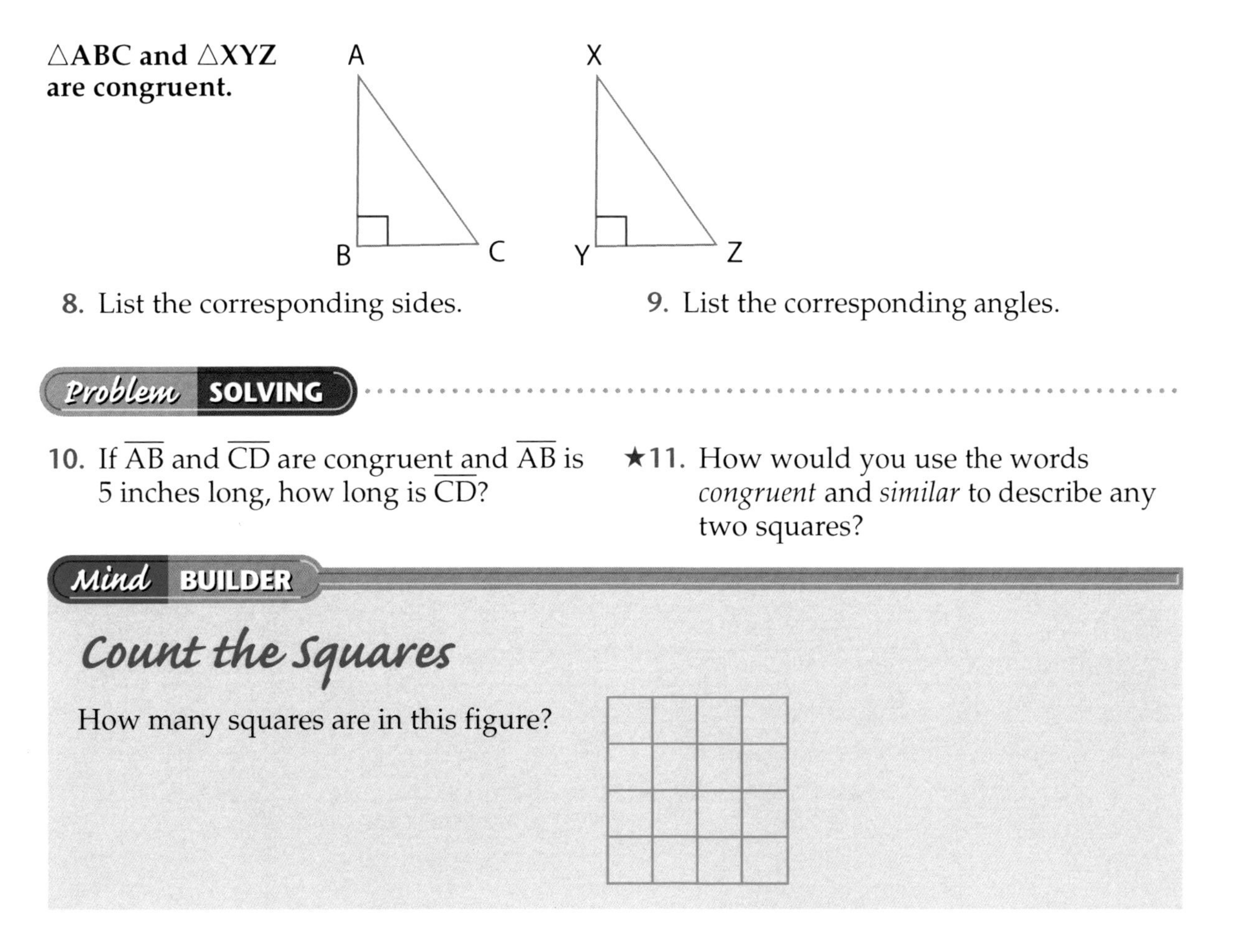

8. List the corresponding sides.

9. List the corresponding angles.

Problem SOLVING

10. If $\overline{AB}$ and $\overline{CD}$ are congruent and $\overline{AB}$ is 5 inches long, how long is $\overline{CD}$?

★11. How would you use the words *congruent* and *similar* to describe any two squares?

Mind BUILDER

Count the Squares

How many squares are in this figure?

10.10 Line Symmetry

Objective: to find lines of symmetry

Everywhere you look you can find symmetry. It is seen in nature, architecture, and art. The butterfly to the right has symmetry.

Exploration Exercise

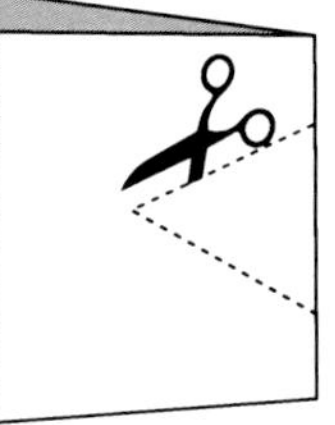

1. Fold a piece of paper in half.
2. Draw a triangle on one side of the fold line.
3. Cut out the triangle, but do not cut the fold.
4. What do you notice about the shape you cut out?

There are two triangles that are formed. They are mirror images of each other.

The figure above, as well as the butterfly, has **line symmetry**. A figure has line symmetry when a line can be drawn through the figure such that the two halves match exactly. The line that separates the figure into two halves that match exactly when the figure is folded is the **line of symmetry**.

Examples

For each figure, is the dashed line a line of symmetry? Explain.

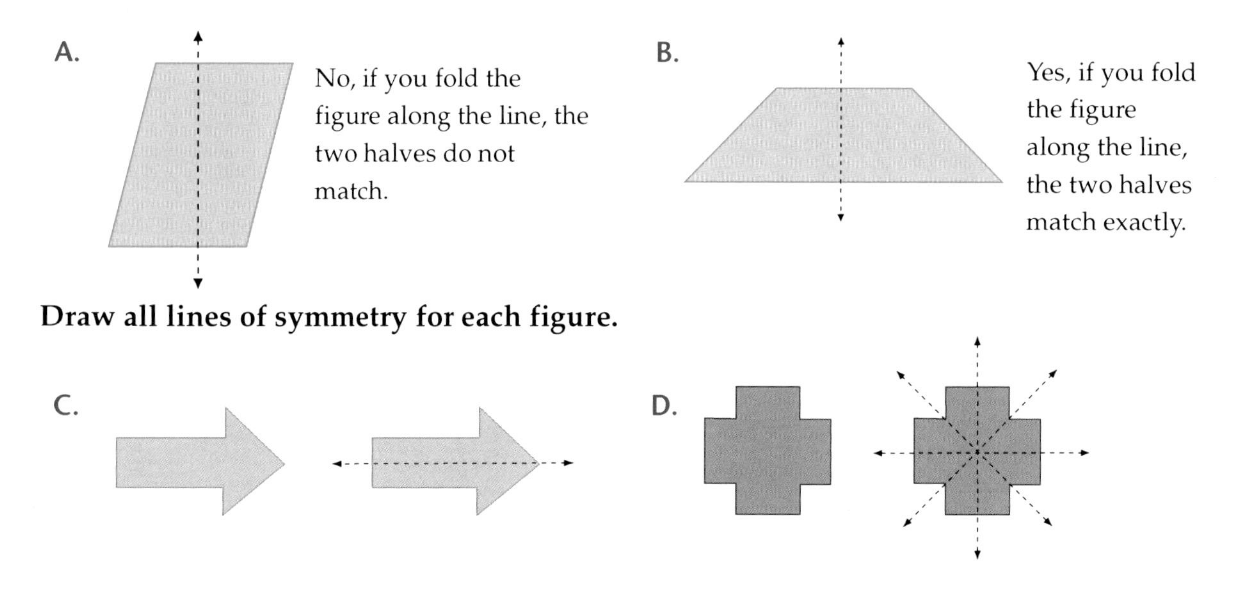

A. No, if you fold the figure along the line, the two halves do not match.

B. Yes, if you fold the figure along the line, the two halves match exactly.

Draw all lines of symmetry for each figure.

C.

D.

Try THESE

Is the dashed line a line of symmetry? Write *yes* or *no*.

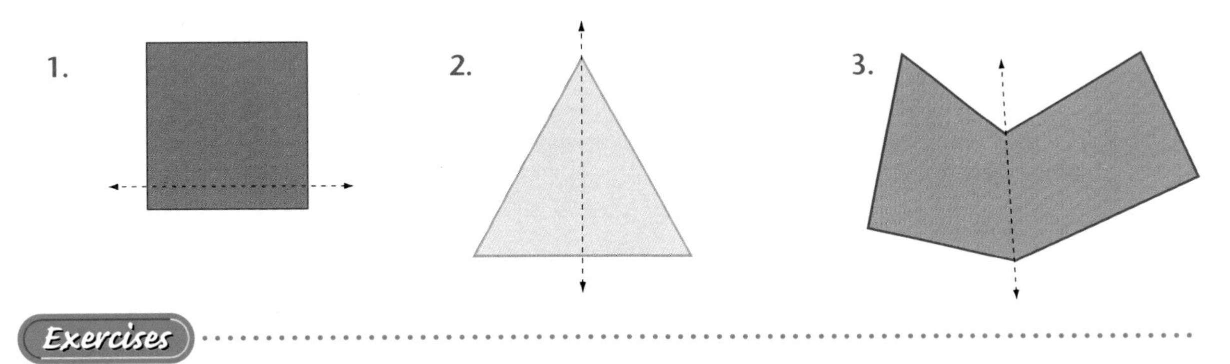

Exercises

How many lines of symmetry does each figure have? Draw the lines of symmetry.

1. 2. 3. 4.

Does each figure have line symmetry? Write *yes* or *no*.

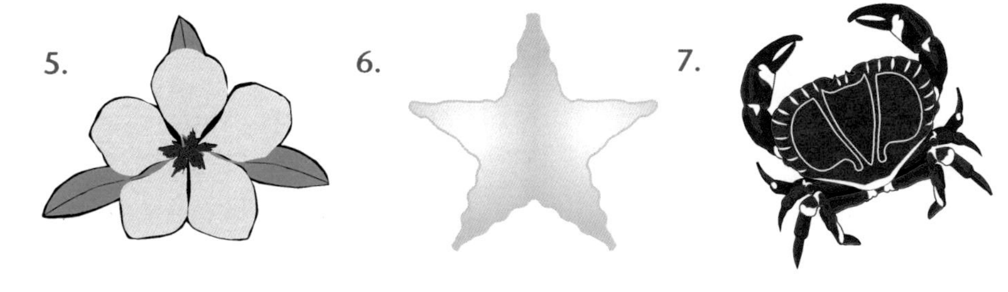

Problem SOLVING

8. How many lines of symmetry does a circle have? Explain.

9. How many lines of symmetry does a triangle have?

10. How many lines of symmetry does a square have?

11. Write all the letters of the alphabet as capital letters. How many letters have a vertical line of symmetry?

Chapter 10 Review

Language and Concepts

Write the letter of the word or phrase that best matches each description.

1. a unit used to measure angles
2. two rays with the same endpoint
3. figures having the same size and shape
4. an angle that measures more than 90° but less than 180°
5. two lines that intersect or cross to form right angles
6. lines in a plane that never cross
7. a parallelogram with all angles congruent
8. a triangle with no congruent sides
9. the endpoint where two rays meet to form an angle
10. the line that separates a figure into two halves that match exactly when the figure is folded

a. acute angle
b. angle
c. congruent figures
d. degree
e. line of symmetry
f. isosceles triangle
g. obtuse
h. parallel lines
i. perpendicular lines
j. rectangle
k. scalene triangle
l. similar figures
m. vertex

Skills and Problem Solving

Classify each angle. Write *acute, right,* or *obtuse.* (Section 10.2)

11. 12. 13. 14.

Measure each angle and then bisect. (Section 10.4)

15. 16.

Use symbols to name the following. (Section 10.1)

17. A B
18. K P
19. F G
20. C D / E F

Find the complement and supplement of each angle. (Section 10.3)

21. 65°
22. 38°
23. 42°
24. 79°

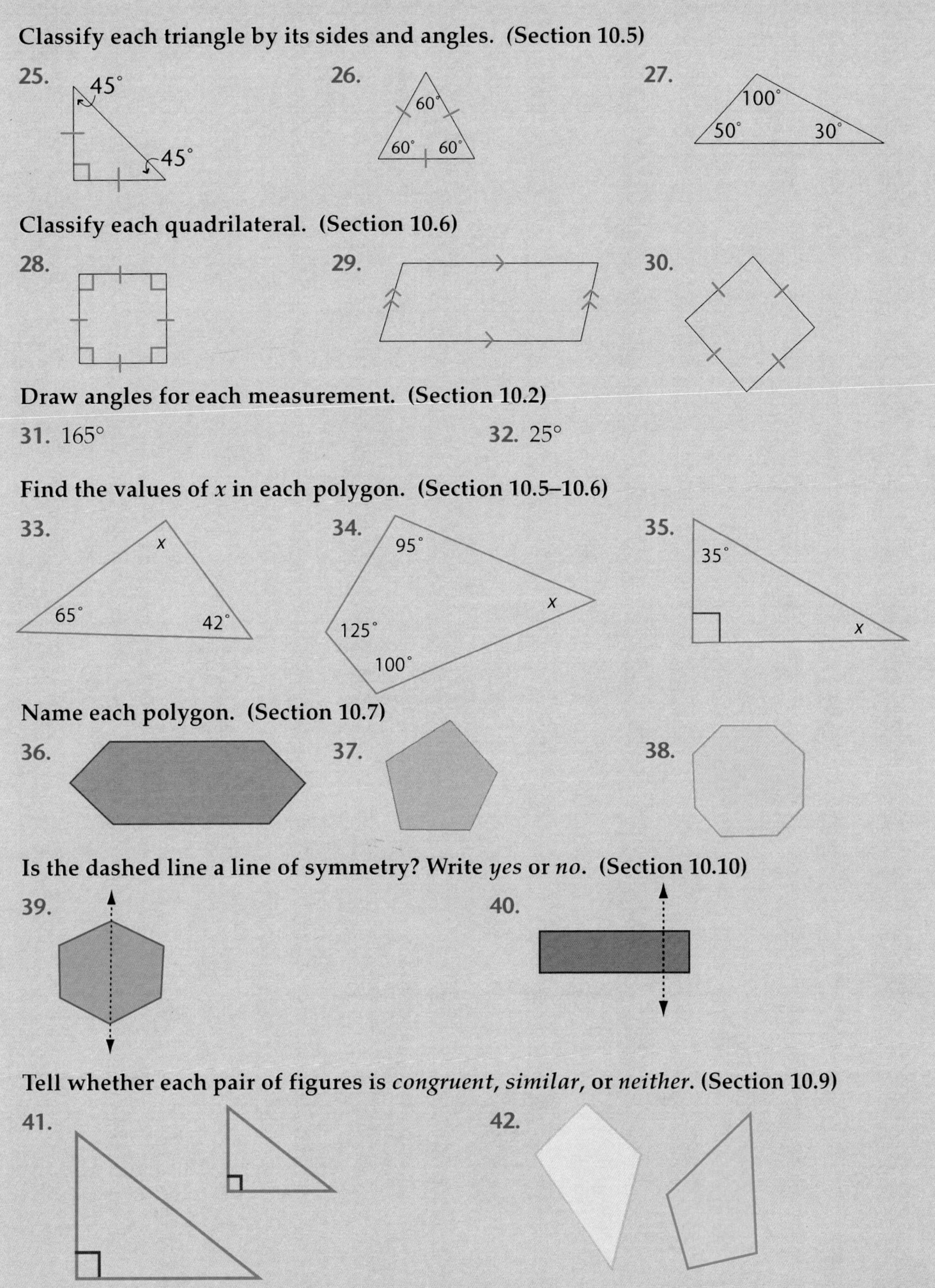

Classify each triangle by its sides and angles. (Section 10.5)

25.

26.

27.

Classify each quadrilateral. (Section 10.6)

28.

29.

30.

Draw angles for each measurement. (Section 10.2)

31. 165°

32. 25°

Find the values of x in each polygon. (Section 10.5–10.6)

33.

34.

35.

Name each polygon. (Section 10.7)

36.

37.

38.

Is the dashed line a line of symmetry? Write *yes* or *no*. (Section 10.10)

39.

40.

Tell whether each pair of figures is *congruent*, *similar*, or *neither*. (Section 10.9)

41.

42.

Chapter 10 Test

Measure each angle. Then classify each angle as *acute, right,* or *obtuse.*

1.

2.

Find the complement and supplement of each angle.

3. 73°

4. 41°

State whether each set of figures is *congruent*, *similar*, or *neither*. Explain your answer.

5.

6.

7.

Classify each triangle by its sides and angles. Then find the value of x.

8.

9.

10.

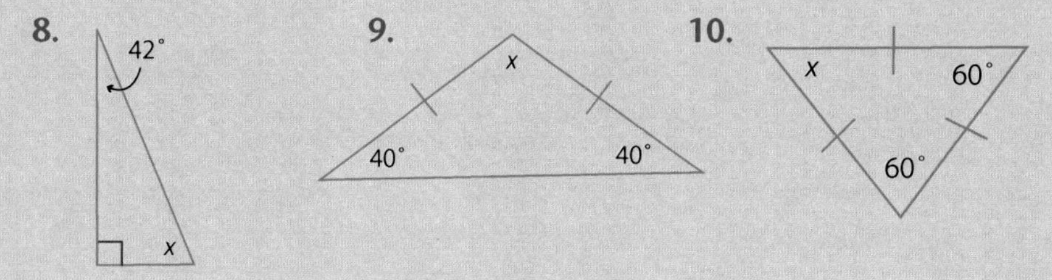

Name each polygon. Then draw all lines of symmetry.

11.

12.

13.

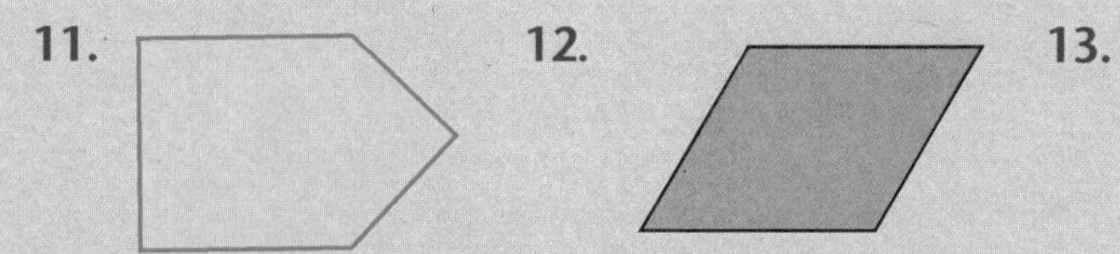

Classify the following objects into groups. Draw a Venn diagram.

14.

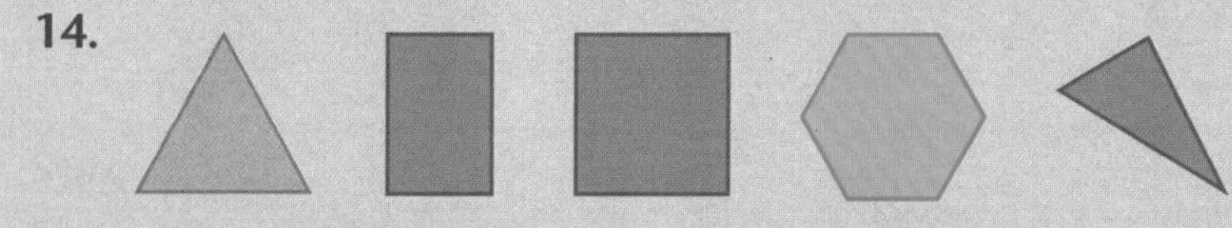

Geometric Constructions

A compass and a straightedge can be used to make designs. They can also be used to construct a line segment congruent to a given line segment and to do other constructions.

Use a compass and a ruler to make this design.

Construct a line segment congruent to $\overline{AB}$.

1. Draw a segment longer than $\overline{AB}$. Label it CD. Open your compass to match $\overline{AB}$.

2. Place the compass at point C. Draw an arc as shown in black. Label the point of intersection E.

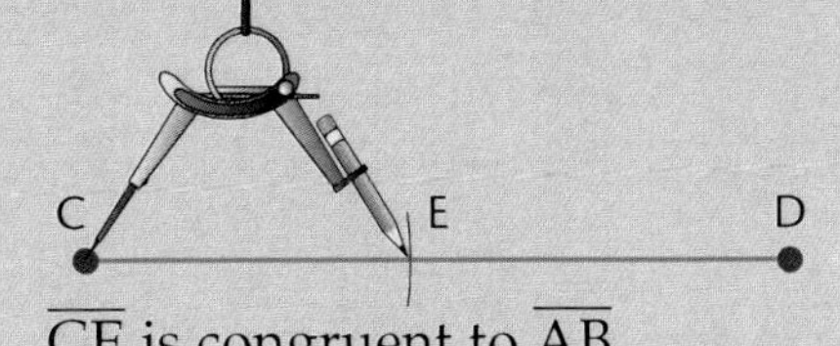

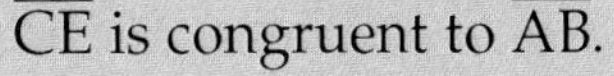

$\overline{CE}$ is congruent to $\overline{AB}$.

Complete the following to bisect a line segment.
(Bisect means to separate a figure into two congruent parts.)

1. Draw a line segment. Label it $\overline{AB}$.
2. Open your compass to more than half the length of $\overline{AB}$. With the compass point at A, draw a long arc.
3. Use the same compass setting. Draw another arc with the compass point at B. Be sure the arcs intersect.
4. Draw a line through the intersection points of the arcs. The point where this line intersects $\overline{AB}$ bisects $\overline{AB}$. This point is called the midpoint of $\overline{AB}$.

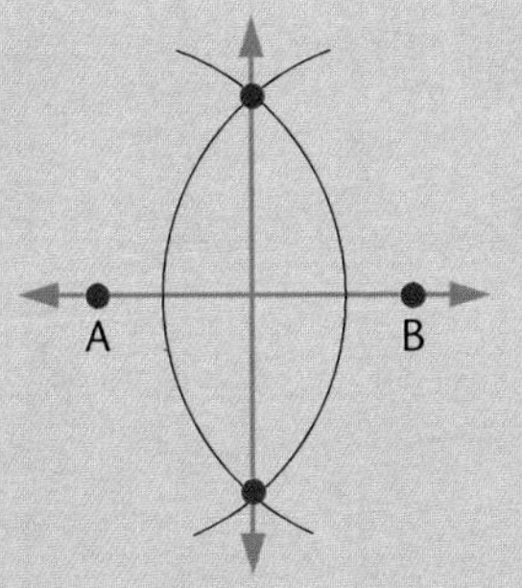

Experiment by making each construction. Use a compass and a straightedge.

5. Construct a regular hexagon in a circle; see first construction above.
6. Construct a regular octagon in a circle.
7. Trace the line segments shown at the right. Construct a triangle with sides congruent to the segments you traced.

Cumulative Test

1. Which number makes this equation true?
   406 × ■ = 406
   a. 0
   b. 1
   c. 406
   d. none of the above

2. Multiply. $\begin{array}{r} 2.94 \\ \times\ 4.7 \\ \hline \end{array}$
   a. 12.888
   b. 13.808
   c. 13.828
   d. none of the above

3. Which angle is obtuse?
   a. ∠JOL
   b. ∠JOM
   c. ∠KOL
   d. ∠KOM

4. Find 14,100 divided by 68.
   a. 27
   b. 27 R24
   c. 207
   d. 207 R24

5. A right angle measures how many degrees?
   a. 0°
   b. 45°
   c. 90°
   d. 180°

6. Subtract.
   8.36 − 4.57 = ____
   a. 2.36
   b. 2.57
   c. 3.79
   d. none of the above

7. Which pair of lines represents skew lines?

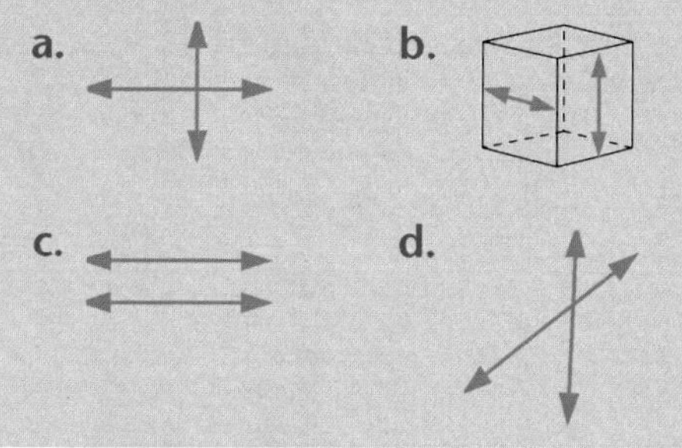

8. Which figure is a hexagon?

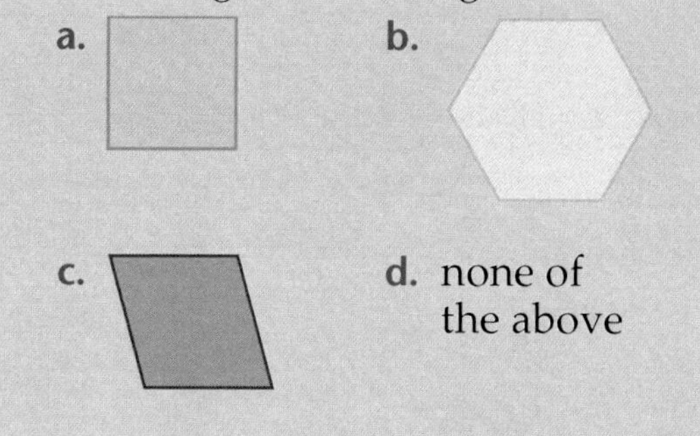

   d. none of the above

9. A post office clerk hands a customer 10 stamps costing $1.31. There are some 25¢ stamps, 18¢ stamps, and 4¢ stamps. How many of each stamp is there?
   a. 2-25¢, 4-18¢, 2-4¢ stamps
   b. 3-25¢, 2-18¢, 5-4¢ stamps
   c. 4-25¢, 1-18¢, 3-4¢ stamps
   d. none of the above

10. The fifth graders scored an average of 81 on a test while the sixth graders scored an average of 84. Which of the following is true about the average score of the fifth *and* sixth graders?
    a. The average is 84.
    b. The average is less than 84.
    c. The average is greater than 84.
    d. It is not possible to tell.

CHAPTER 11

Geometric Measurement

Faith Follansbee
Massachusetts

11.1 Perimeter and Area

Objective: to find the perimeter and area of figures

Needlepoint is a craft that uses a needle, yarn, and a grid similar to the one shown at the right. This grid is shaped like a rectangle 3 cm by 6 cm.

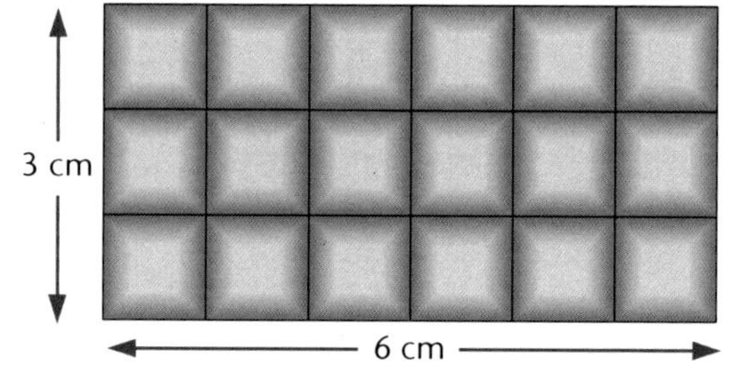

Area is the number of square units needed to completely cover a region. A **square** centimeter (cm^2) is a unit of area.

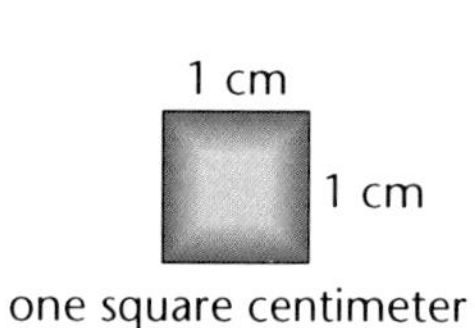

one square centimeter
1 cm^2

You can tell by counting that it takes 18 one-centimeter squares to cover the pattern. Its area is 18 cm^2.

The area measure (A) of any rectangle is found by multiplying the measures of the length (ℓ) and the width (w).

Other units of area include square meters (m^2), square feet (ft^2), and square inches ($in.^2$).

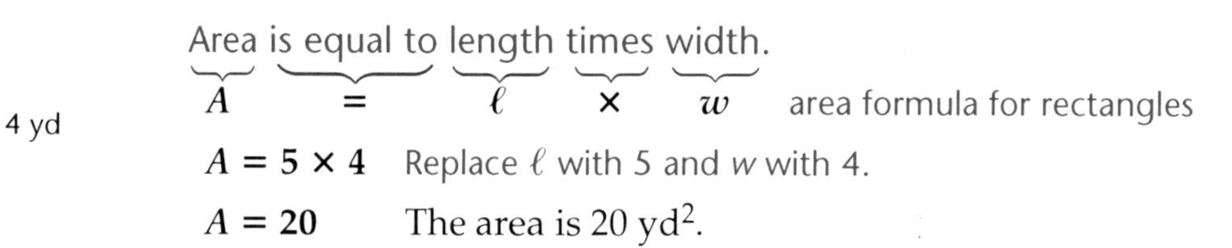

Area is equal to length times width.

$A = \ell \times w$ area formula for rectangles

$A = 5 \times 4$ Replace ℓ with 5 and w with 4.

$A = 20$ The area is 20 yd^2.

You can also find the **perimeter** of a figure. The distance around a polygon is called the perimeter. You can add up the sides of a polygon to find its perimeter. To find the perimeter of a rectangle add its length, width, length, and width.

Perimeter and Area of Rectangles

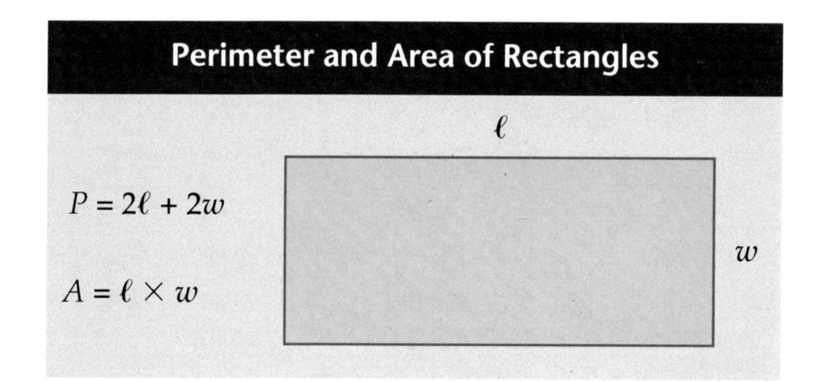

$P = 2\ell + 2w$

$A = \ell \times w$

You can also estimate the area of figures, including irregular figures, by using grid paper. The bottom of your foot is an example of an irregular figure. You can use it to do an exploration exercise.

Exploration Exercise

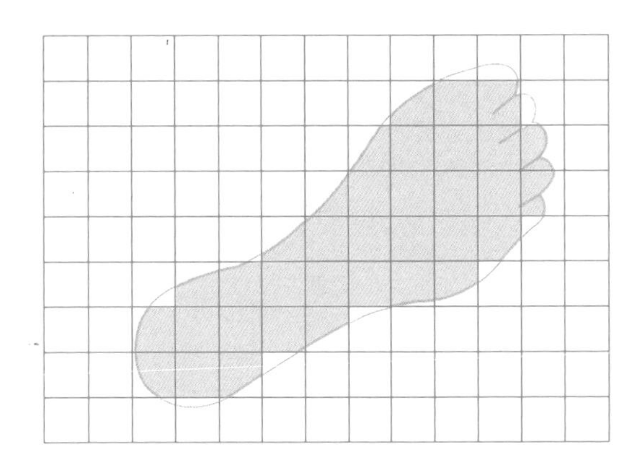

1. Lay the grid paper on the floor. Take off one of your shoes. Place your foot on the grid paper and trace around it with your pencil.
2. Shade the squares that were completely covered by your foot.
3. Count the squares that you shaded on your grid paper. Count a pair of almost full and almost empty squares as one. Count half shaded squares as a half.
4. What is the area of the bottom of your foot? (Remember this is an estimate, so your answer will include the word *about*.)
5. Compare the area of your foot to the area of your teacher's or friend's foot.

Examples

A. Find the area of a rectangle if ℓ is 58 feet and w is 79 feet.

$P = 2\ell + 2w$

$P = 2(58) + 2(79)$

$P = 116 + 158$

$P = 274$ ft

B. What is special about a square? The length is equal to the width. So, we have a special formula— $A = s^2$ (s means side).

2.5 mm

2.5 mm

$A = s^2$

$A = 2.5 \times 2.5$

$A = 6.25\text{ mm}^2$

Try THESE

Find the area of each rectangle. As a check, count the unit squares.

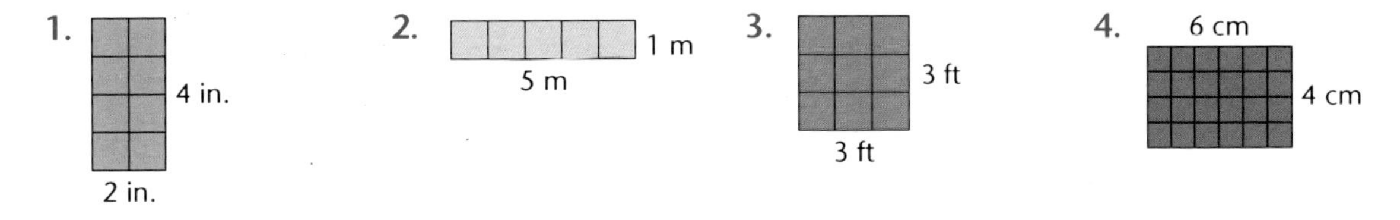

Exercises

Estimate the area of each irregular figure. Each square represents 1 cm.

1. 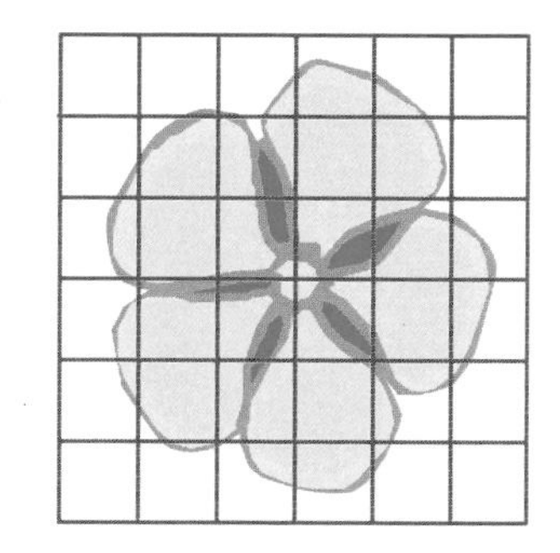

2.

Find the area and perimeter of each rectangle.

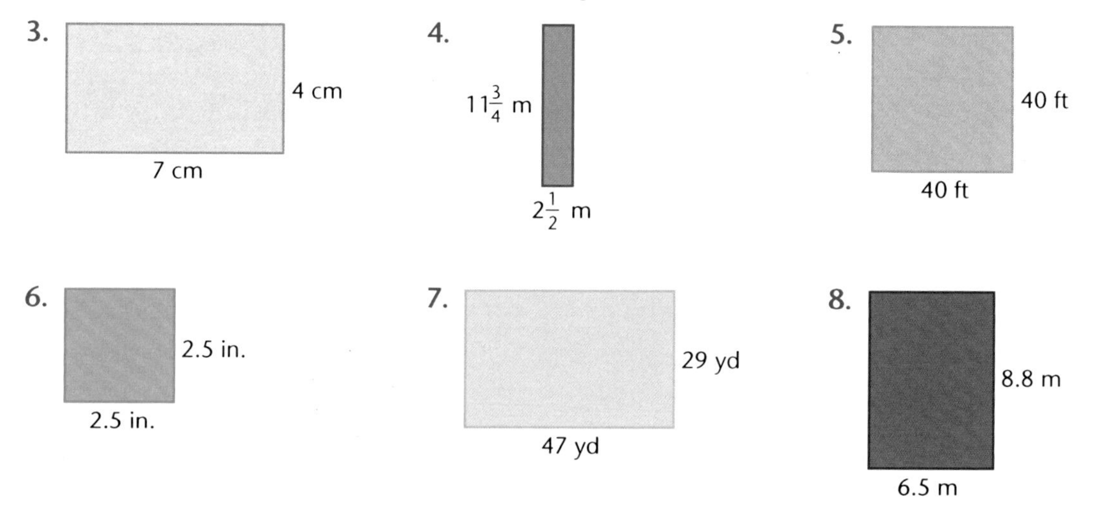

9. length, 8 yd
   width, 5 yd
10. length, 8.3 m
    width, 7 m
11. each side, $18\frac{2}{5}$ mm
12. Find the area of a square if each side is 25 inches.

★13. How many square centimeters are in a square meter?

Problem SOLVING

14. A rectangle is 10 feet wide. Suppose the length is twice the width. What is the area of the rectangle?

15. Can you write a formula for finding the area of irregular figures? Why or why not?

16. Make up your own irregular figure on grid paper. The figure should be closed, cannot intersect itself, and should be made up of a curved line. Estimate the area.

★17. What is the area of the square in square feet and square inches?

$\frac{1}{3}$ ft

★18. Draw all rectangles that have an area of 36 square units using whole numbers only. Complete the chart to help you. Which has the greatest perimeter?

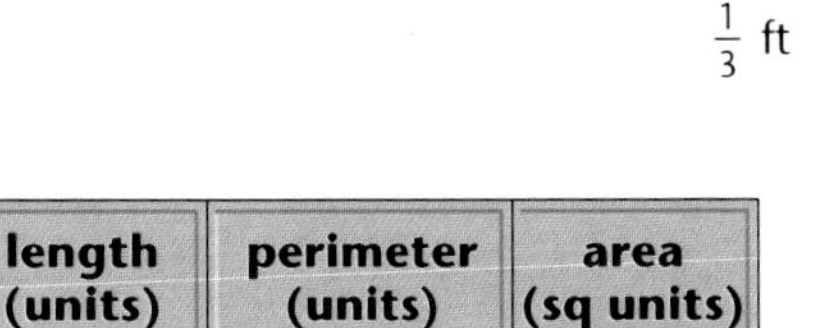

width (units)	length (units)	perimeter (units)	area (sq units)
1	36	74	36

Test PREP

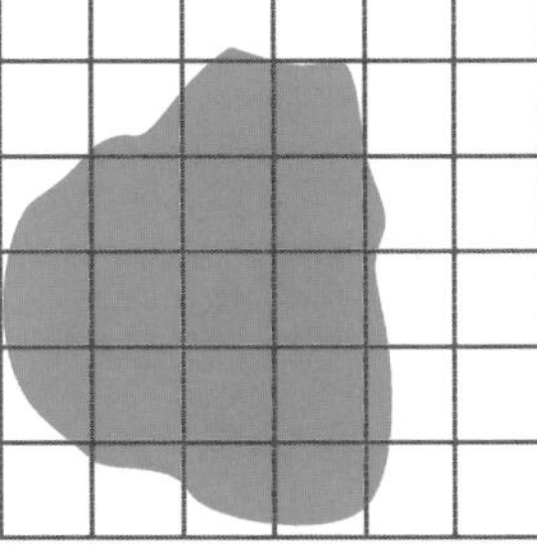

19. Each square represents 10 square meters. Which is the best estimate for the area of the park?

a. 5.5 m^2 b. 15 m^2

c. 150 m^2 d. 1,500 m^2

20. Annie would like a swimming pool that has an area of 20 square feet. If the pool is 4 feet long, how wide should the pool be?

a. 8 ft b. 16 ft c. 5 ft d. 10 ft

21. A square has an area of 36 square inches. What is the perimeter of the square?

a. 6 in. b. 24 in. c. 12 in. d. 36 in.

Mind BUILDER

Mental Images

See how many triangles you can find in this figure.

11.2 Area and Perimeter of a Parallelogram

Objective: to find the area and perimeter of a parallelogram

Exploration Exercise

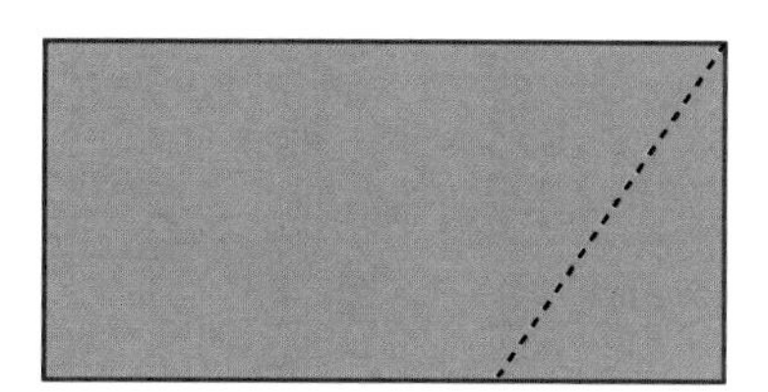

1. Draw a rectangle on grid paper.
2. Cut out a triangle from one side and move it to the other side. You will now form a parallelogram.
3. How are a rectangle and a parallelogram related?
4. How is the length of a rectangle related to the base of a parallelogram?
5. How is the width of a rectangle related to the height of a parallelogram?
6. How do you find the area of a parallelogram?

The exploration shows that the area of a parallelogram is related to a rectangle. A parallelogram has base and height instead of length and width. The height is perpendicular to the base.

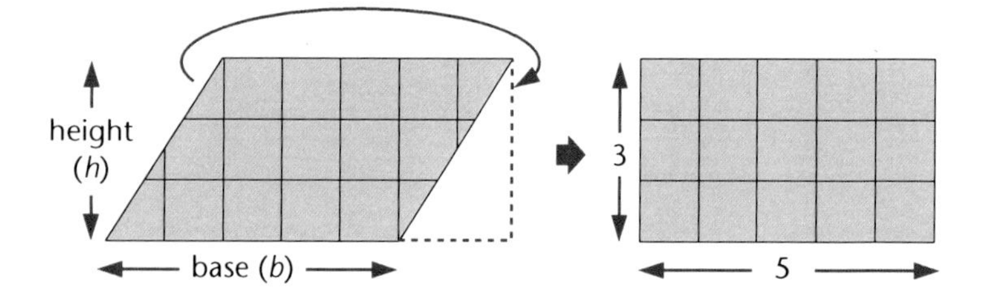

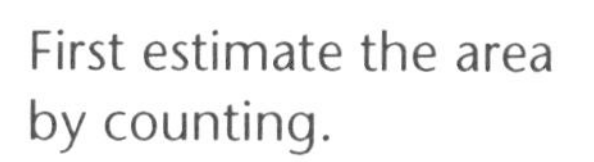

Use this formula to find the area of a parallelogram.

Area is equal to base times height.

$$A = b \cdot h$$

$A = 5 \times 3$ Replace b with 5 and h with 3.

$A = 15$ The area is 15 square units, the same area as the rectangle.

Examples

A. Find the area of the parallelogram.

B. Find the area of the parallelogram.

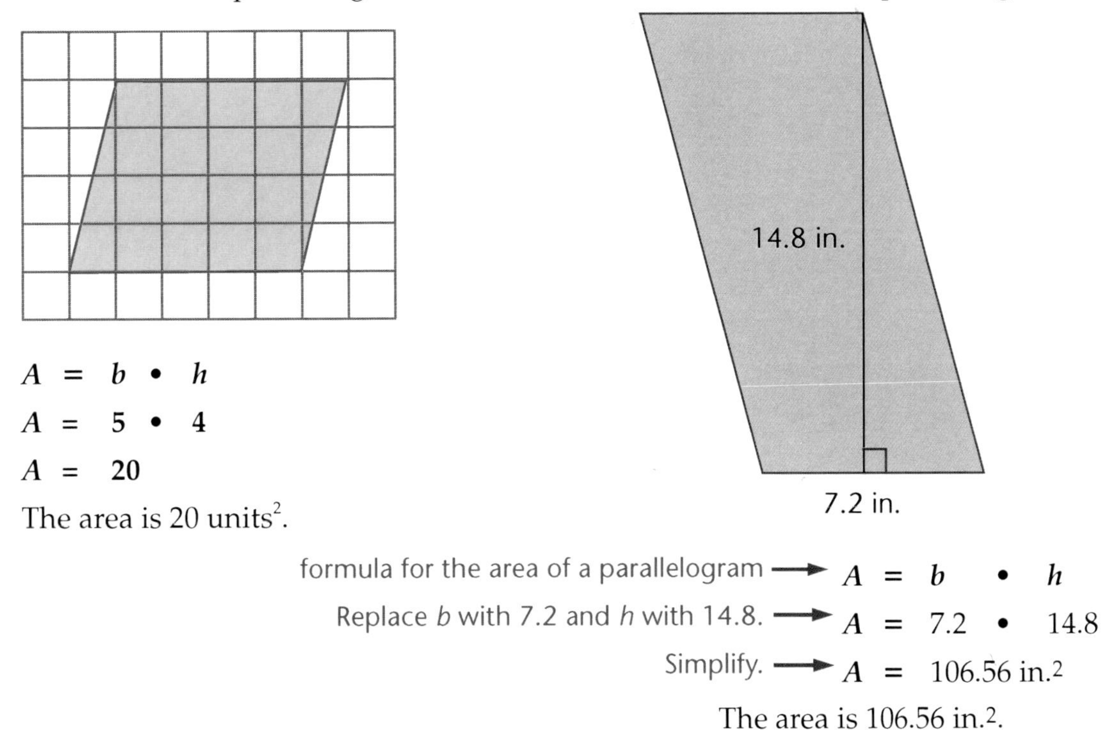

$A = b \bullet h$

$A = 5 \bullet 4$

$A = 20$

The area is 20 units^2.

formula for the area of a parallelogram → $A = b \bullet h$

Replace b with 7.2 and h with 14.8. → $A = 7.2 \bullet 14.8$

Simplify. → $A = 106.56 \text{ in.}^2$

The area is 106.56 in.^{2}.

The perimeter is the sum of the measures of the outside of a figure. To find the perimeter of a parallelogram, add the sides of the parallelogram. You do not use the height. *Remember:* the opposite sides of a parallelogram are equal.

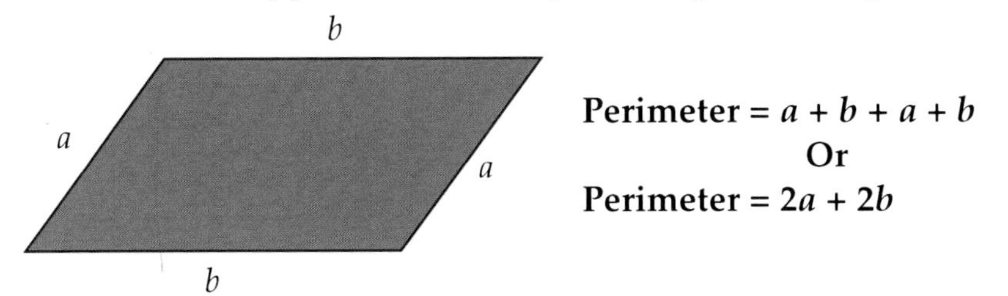

Perimeter = $a + b + a + b$

Or

Perimeter = $2a + 2b$

C. Find the perimeter of the parallelogram.

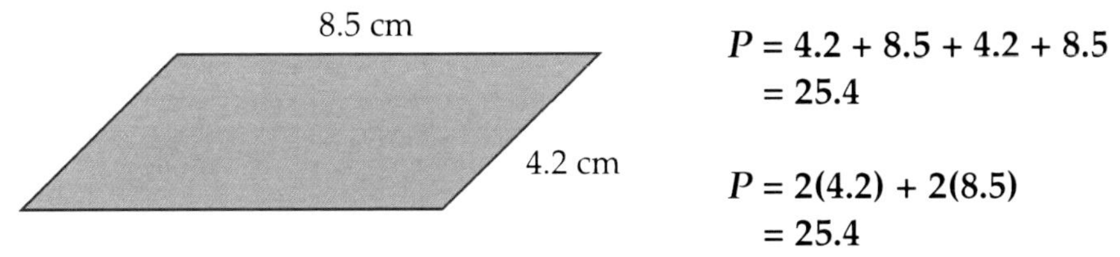

$P = 4.2 + 8.5 + 4.2 + 8.5$
$= 25.4$

$P = 2(4.2) + 2(8.5)$
$= 25.4$

The perimeter of the parallelogram is 25.4 cm.

Try THESE

Find the area of each parallelogram.

Exercises

Find the area and perimeter of each parallelogram.

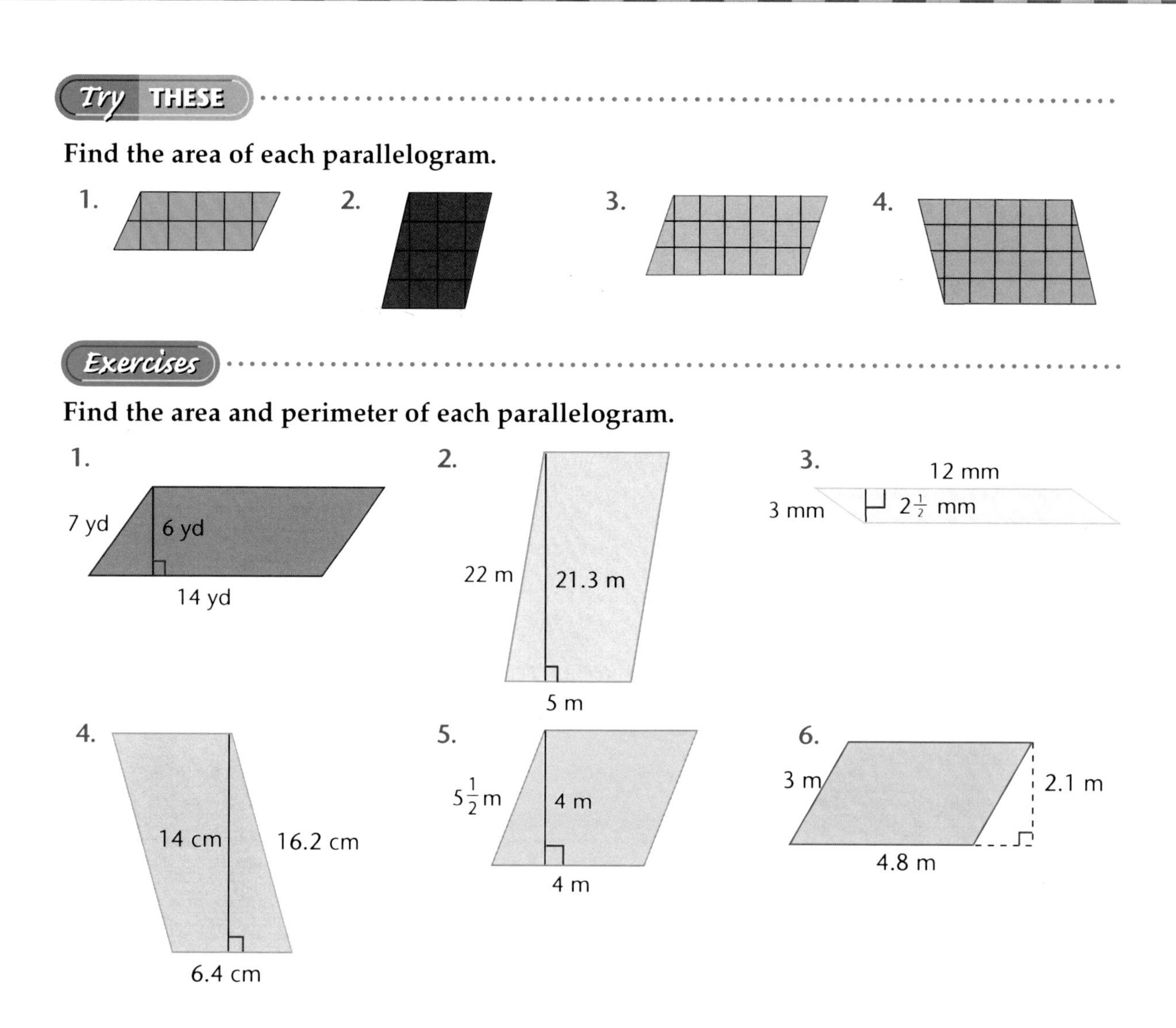

7. What is the area of a parallelogram with a base of 8.4 m and a height of 12.3 m?

8. What is the area of a parallelogram with a base of $5\frac{3}{4}$ in. and a width of $9\frac{1}{5}$ in.?

Problem SOLVING

9. A living room rug measures 13 feet by 24.5 feet. How much area will the rug cover?
10. The perimeter of a parallelogram is 40 in. If two sides each measure 8 in., what are the lengths of the other two sides?
11. How many 25-cm^2 tiles are needed to cover the floor of a room 4 meters long and 3 meters wide?

Constructed RESPONSE

12. What happens to the area of a parallelogram when you double the height? What happens when you double both the height and the base? Give examples in your explanation.

13. What happens to the perimeter of a parallelogram when you double the length of each side? Give an example in your explanation.

14. The rug to the right is a parallelogram.

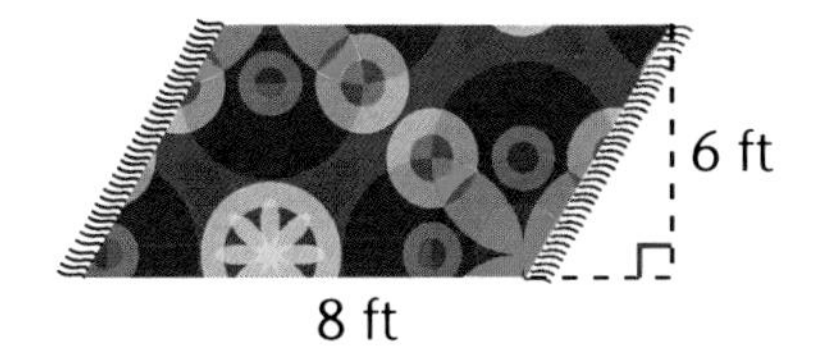

   a. What is the area of the rug?
   b. The floor that the rug will lie on is a rectangle that measures 8 ft by 10 ft. What is the area of the floor?
   c. What part of the floor will not be covered by the rug?
   d. Explain your answer to part (c).

Test PREP

15. A parallelogram has a height of 3 cm and a base of 5.5 cm. What is the area?

   a. 16.5 cm^2 b. 17 cm^2 c. 2.5 cm^2 d. 49.5 cm^2

16. The area of a parallelogram is 54 ft^2. The height is 9 ft. What is the base of the parallelogram?

   a. 18 ft   b. 12 ft   c. 36 ft   d. 6 ft

11.3 Area of a Triangle

Objective: to find the area of a triangle

A windmill can be made up of many triangles. You can find the area of the windmill by finding the area of each triangle in the windmill.

To find the area of a *triangle,* use two congruent triangles (same size and shape). Each triangle has the same area.

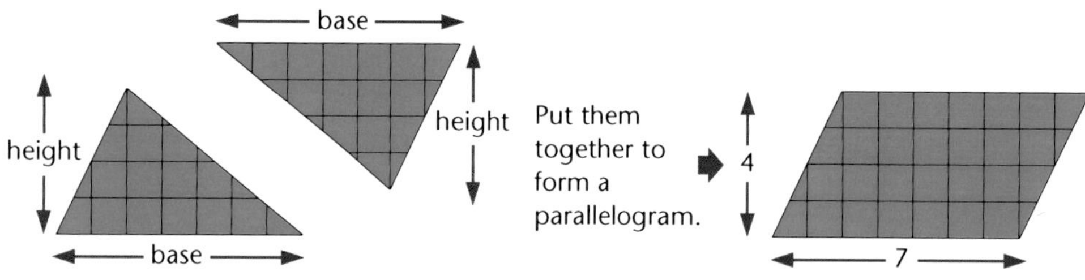

The area of each triangle is $\frac{1}{2}$ the area of the parallelogram. Do you see why? Use this formula to find the area of a triangle.

Area is equal to one-half times base times height.

$$A = \frac{1}{2} \bullet b \bullet h$$

$A = \frac{1}{2} \bullet 7 \bullet 4$ Replace b with 7 and h with 4.

$A = \frac{1}{2} \bullet \frac{28}{1}$

$A = 14$ The area is 14 square units, one-half the area of the parallelogram.

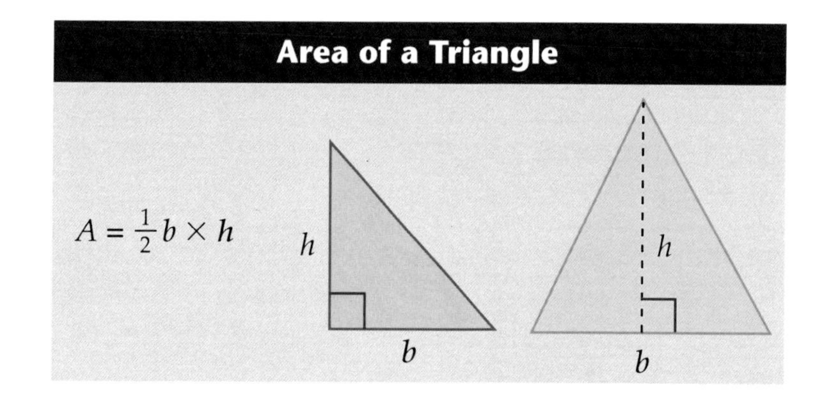

Examples

Find the area of each triangle.

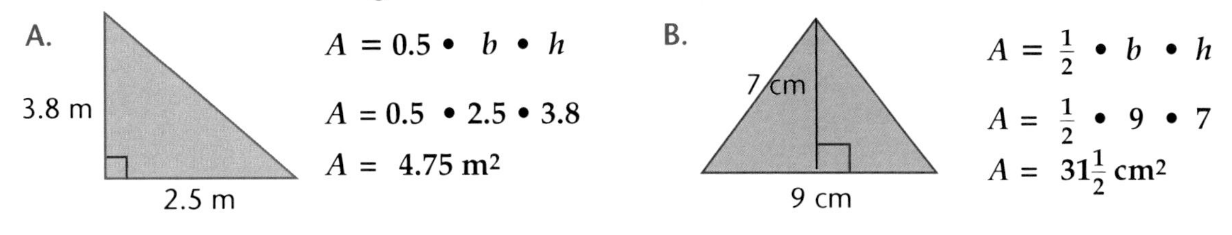

A. $A = 0.5 \bullet b \bullet h$

$A = 0.5 \bullet 2.5 \bullet 3.8$

$A = 4.75 \text{ m}^2$

B. $A = \frac{1}{2} \bullet b \bullet h$

$A = \frac{1}{2} \bullet 9 \bullet 7$

$A = 31\frac{1}{2} \text{ cm}^2$

C. Find the area of the complex figure. The figure is made up of a rectangle and a triangle. Find the area of each figure, and then add.

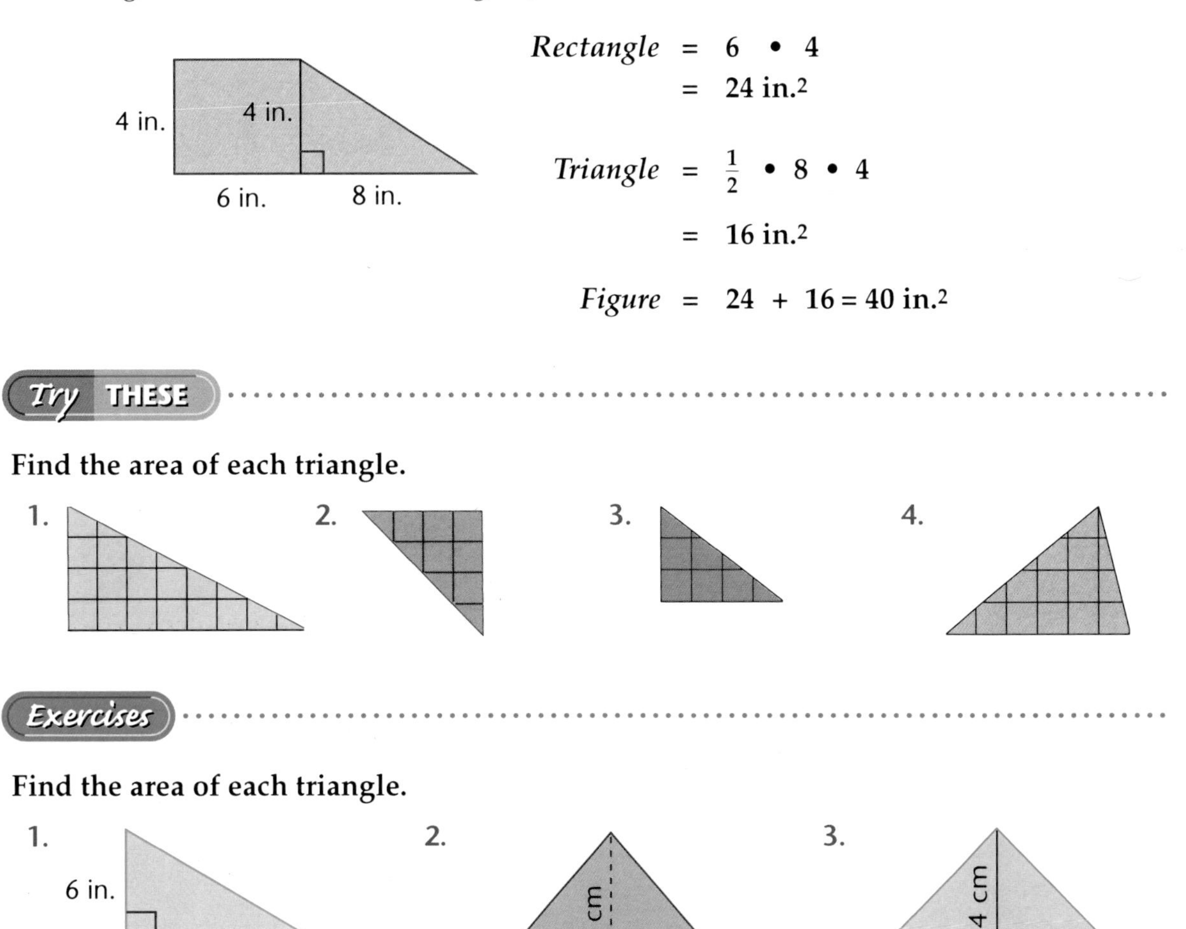

Rectangle $= 6 \bullet 4$

$= 24 \text{ in.}^2$

Triangle $= \frac{1}{2} \bullet 8 \bullet 4$

$= 16 \text{ in.}^2$

Figure $= 24 + 16 = 40 \text{ in.}^2$

Try THESE

Find the area of each triangle.

1. 2. 3. 4.

Exercises

Find the area of each triangle.

1. 6 in., 15 in.

2. 28 cm, 18 cm

3. 4 cm, 8.2 cm

Find the area of each triangle.

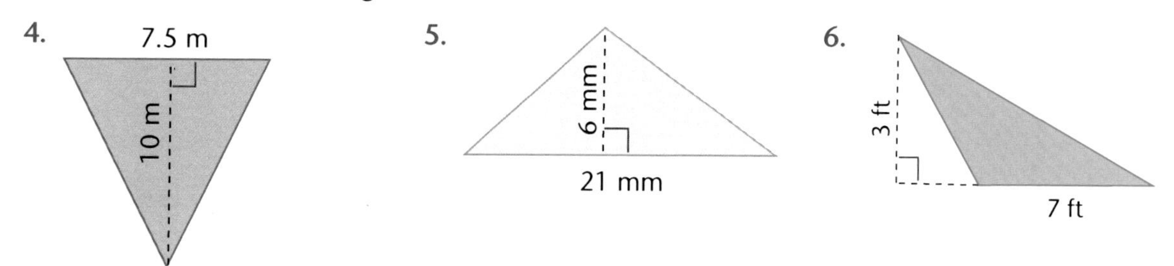

Find the area of each complex figure.

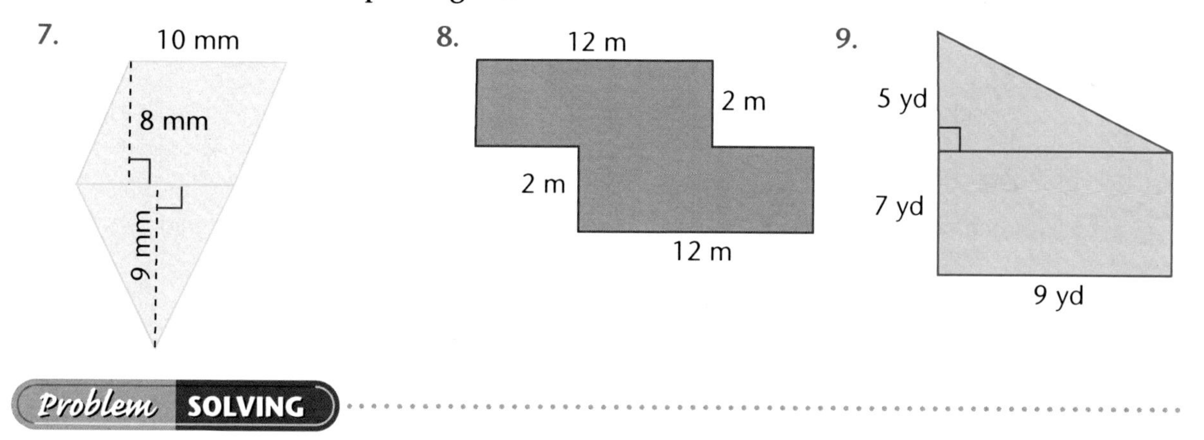

Problem SOLVING

10. A triangular magnet has a base of 2.5 inches and a height of 6 inches. What is the area of the magnet?

11. Yasmine and Juan found the area of the triangle below. Who completed the problem correctly? Explain.

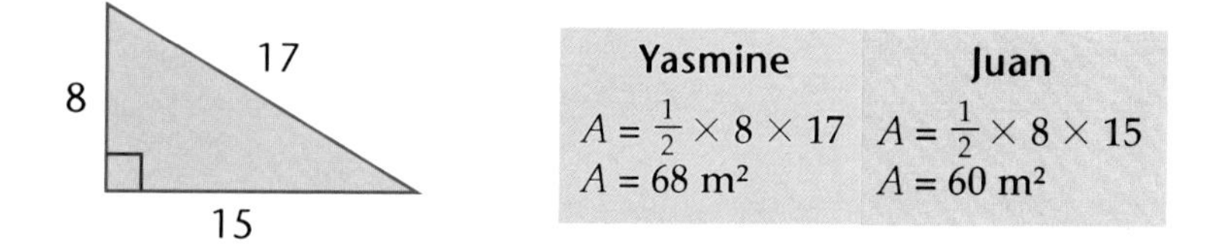

Yasmine	Juan
$A = \frac{1}{2} \times 8 \times 17$	$A = \frac{1}{2} \times 8 \times 15$
$A = 68 \text{ m}^2$	$A = 60 \text{ m}^2$

12. Karen is making a sail for her sailboat to the right. She needs to find the area of the sails so that she knows how much fabric she will need. Find the area of the sails.

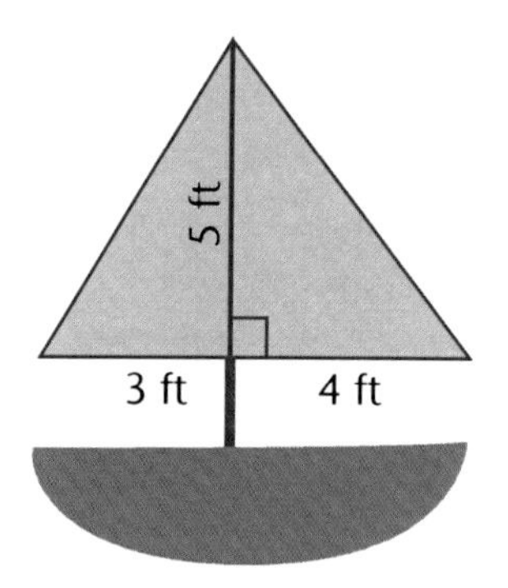

Cumulative Review

Compare using $<$, $>$, or $=$.

1. 0.8 ● 0.801
2. 465 ● 456
3. 4 ● 4.00
4. 0.011 ● 0.010
5. $\frac{6}{18}$ ● $\frac{4}{12}$
6. $\frac{7}{8}$ ● $\frac{9}{10}$

Compute.

7. $\begin{array}{r} 41 \\ +\ 59 \\ \hline \end{array}$
8. $\begin{array}{r} 1{,}074 \\ 965 \\ +\ 8{,}409 \\ \hline \end{array}$
9. $\begin{array}{r} 9{,}614 \\ -\ 1{,}088 \\ \hline \end{array}$
10. $\begin{array}{r} 0.22 \\ \times\ \ 0.5 \\ \hline \end{array}$
11. $9\overline{)747}$
12. 86,950 − 42,080
13. \$95 − \$18.09
14. 6.5 + 0.66
15. 16 × 100
16. 80 ÷ 10
17. 52.8 ÷ 0.6

Compute. Write each answer in simplest form.

18. $\frac{1}{5} + \frac{1}{5}$
19. $\frac{5}{6} - \frac{1}{6}$
20. $\frac{3}{4} - \frac{1}{8}$
21. $\frac{1}{12} + \frac{1}{6}$
22. $\frac{1}{2} \times 6$
23. $\frac{8}{9} \times \frac{3}{4}$
24. $5 \div \frac{1}{5}$
25. $\frac{2}{3} \div \frac{2}{3}$

Translate each statement into an equation. Then solve each equation.

26. x increased by 23 is 52.9.
27. Two less than z is 28.1.

Write each ratio as a fraction in simplest form.

28. eight to ten
29. one to two
30. seven to three
31. fourteen to four
32. six to twenty
33. thirty to nine

Solve.

34. Lori paid \$16, \$1.98, and \$0.59 for various quilting supplies. How much did she pay altogether?
35. Gary scored 88, 94, and 76 on three English tests. What was his average score?
36. What is the area of the parallelogram?

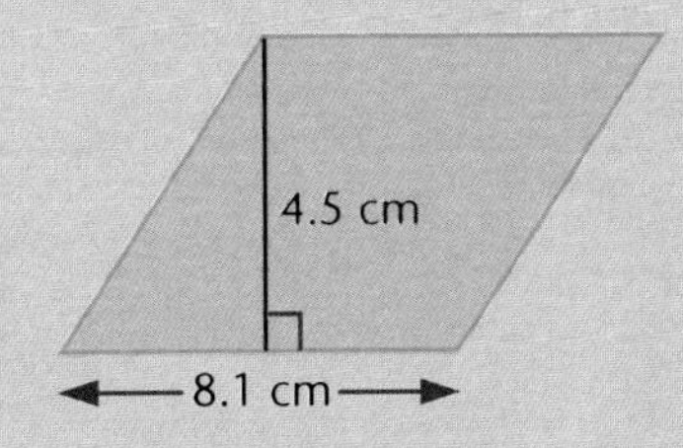

37. What is the area of the square?

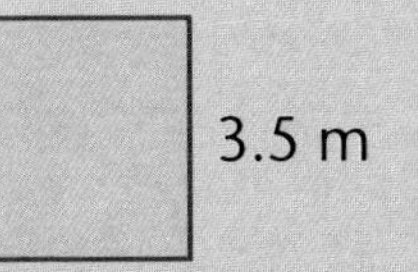

11.4 Area of a Trapezoid

Objective: to find the area of a trapezoid

A trapezoid is a quadrilateral with exactly one pair of opposite sides parallel. A trapezoid can be separated into two triangles with the same height. You can find the area of a trapezoid by using the triangle area formula.

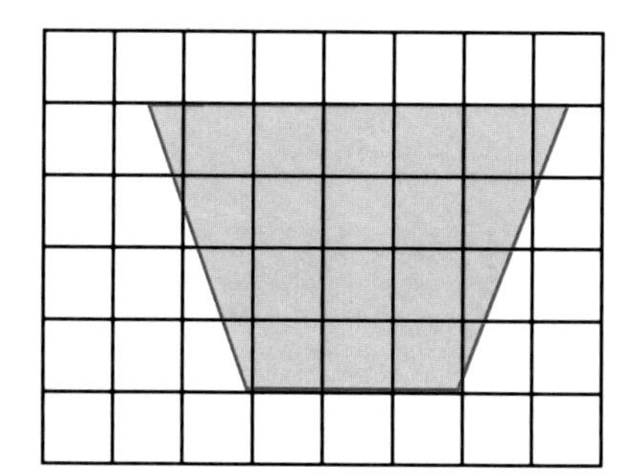

Exploration Exercise

1. Draw a trapezoid on grid paper.
2. Estimate the area by counting the number of squares covered by the trapezoid.
3. Draw a diagonal separating the trapezoid into two triangles.
4. Label the height and base for each triangle.
5. Find the area of each triangle.

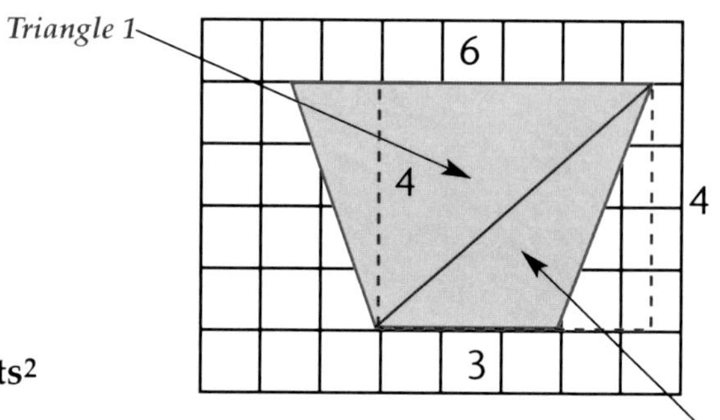

Triangle 1		**Triangle 2**
$A_1 = \frac{1}{2} \bullet b_1 \bullet h$	+	$A_2 = \frac{1}{2} \bullet b_2 \bullet h$
$A_1 = \frac{1}{2} \bullet 6 \bullet 4$	+	$A_2 = \frac{1}{2} \bullet 3 \bullet 4$
$A_1 = \textbf{12 units}^2$	+	$A_2 = \textbf{6 units}^2 = \textbf{18 units}^2$

You can add the areas of the triangles to find the area of the trapezoid or you can combine the formulas to make one formula by using the distributive property.

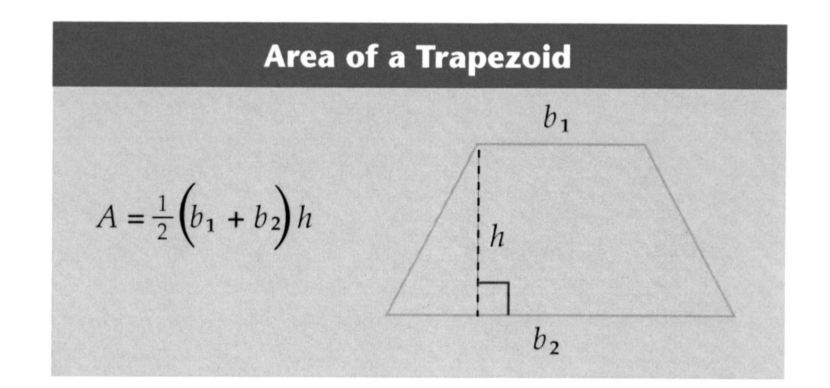

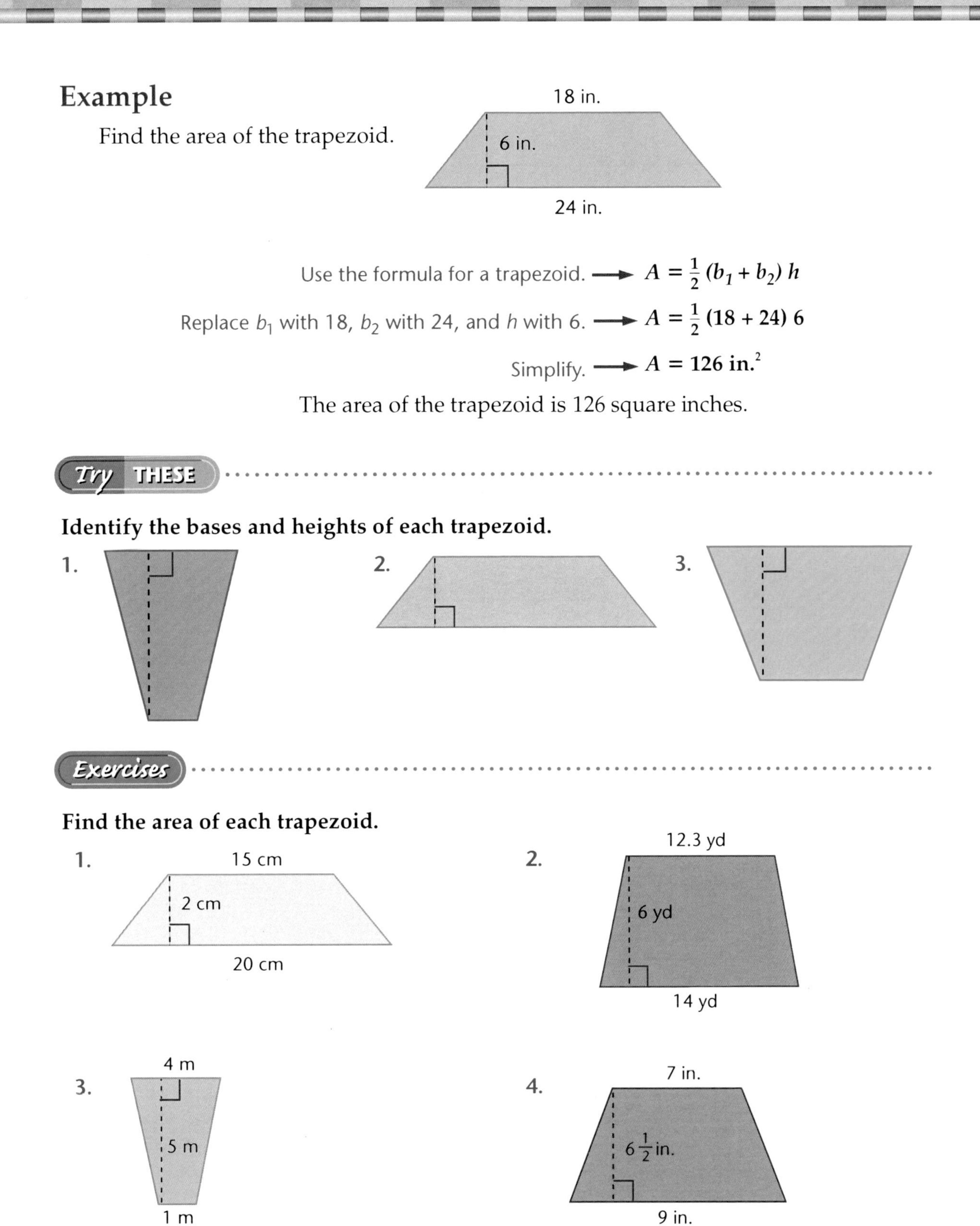

Example

Find the area of the trapezoid.

Use the formula for a trapezoid. ⟶ $A = \frac{1}{2}(b_1 + b_2)h$

Replace b_1 with 18, b_2 with 24, and h with 6. ⟶ $A = \frac{1}{2}(18 + 24)\,6$

Simplify. ⟶ $A = 126 \text{ in.}^2$

The area of the trapezoid is 126 square inches.

Try THESE

Identify the bases and heights of each trapezoid.

1.

2.

3.

Exercises

Find the area of each trapezoid.

1.

2.

3.

4.

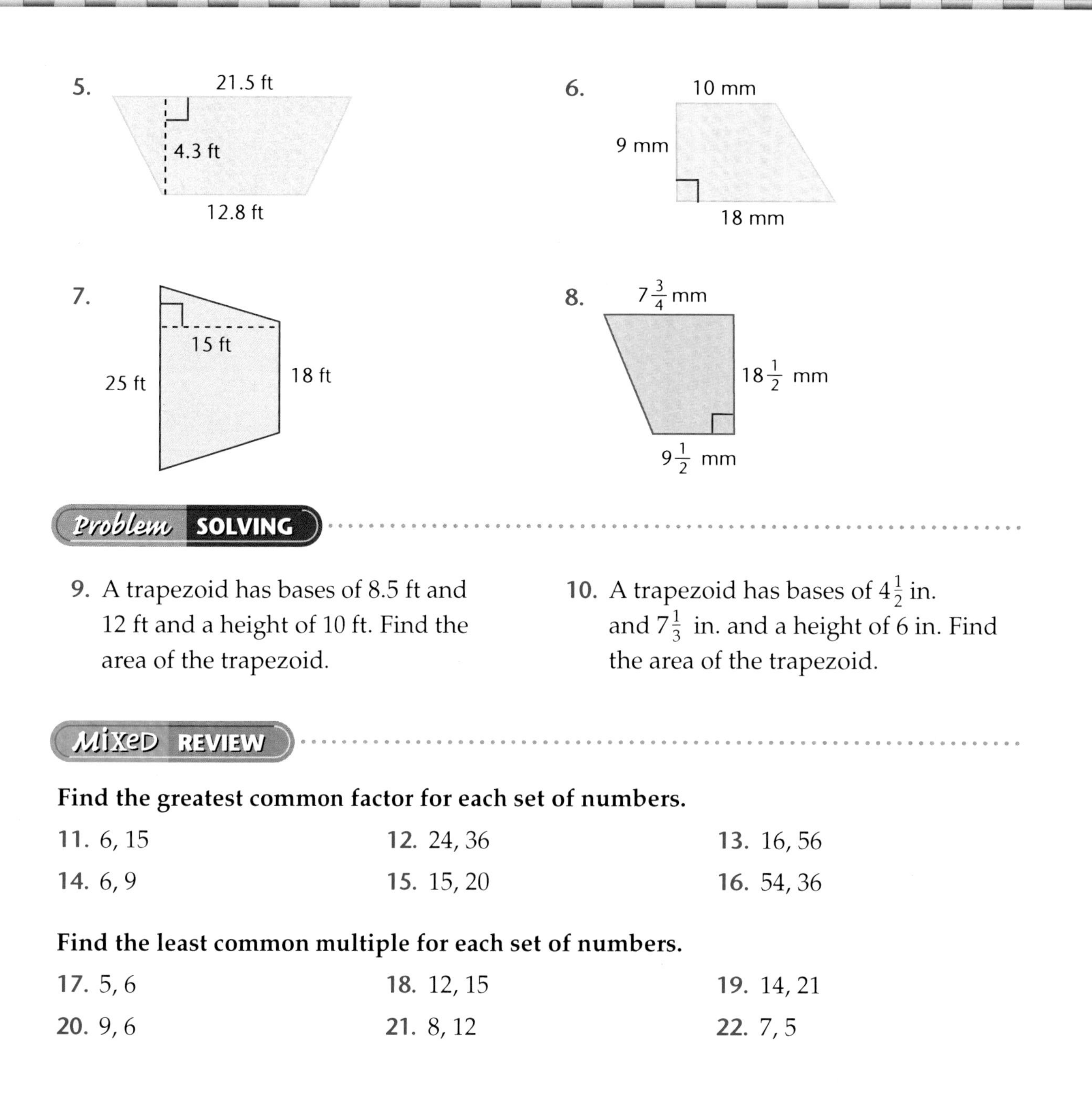

Problem SOLVING

9. A trapezoid has bases of 8.5 ft and 12 ft and a height of 10 ft. Find the area of the trapezoid.

10. A trapezoid has bases of $4\frac{1}{2}$ in. and $7\frac{1}{3}$ in. and a height of 6 in. Find the area of the trapezoid.

MixeD REVIEW

Find the greatest common factor for each set of numbers.

11. 6, 15
12. 24, 36
13. 16, 56
14. 6, 9
15. 15, 20
16. 54, 36

Find the least common multiple for each set of numbers.

17. 5, 6
18. 12, 15
19. 14, 21
20. 9, 6
21. 8, 12
22. 7, 5

To Catch a Thief

Detective Kolombo was assigned to catch a burglar who robbed one home in four different cities on the same night. While at the first crime scene in Boomtown, Detective Kolombo found a map that the burglar had dropped. The map showed all the routes that the burglar would take. The map was planned so that the burglar would not be going over any route twice. Detective Kolombo knows by looking at the map that the burglar does not live in the same city as his first robbery. Can you help Kolombo figure out in what city the burglar lives? Copy the map on your paper to figure out the burglar's path. The city where your path ends is the city in which the burglar lives.

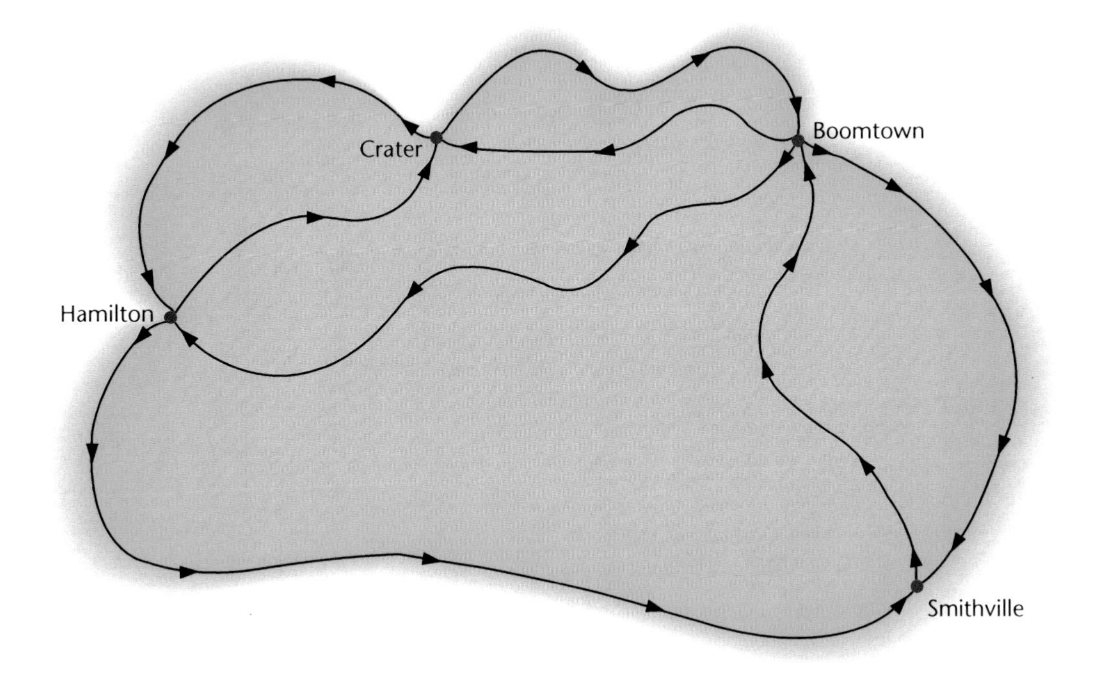

Extension

The map above is traceable. That means you must use all routes but never the same one twice. You must trace the routes without lifting your pencil. Are the figures below traceable?

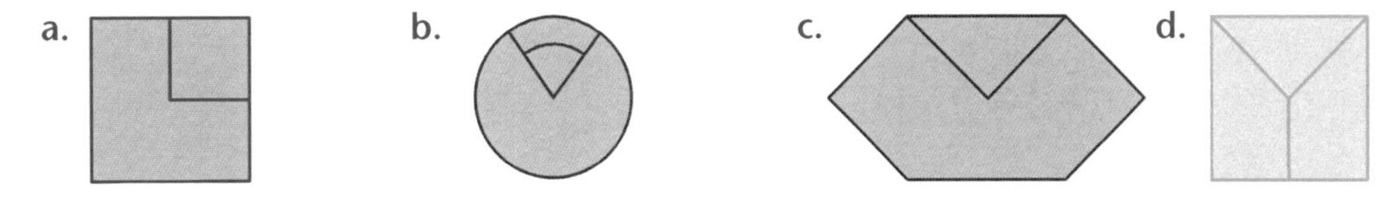

11.5 Circumference of a Circle

Objective: to find the circumference of circles

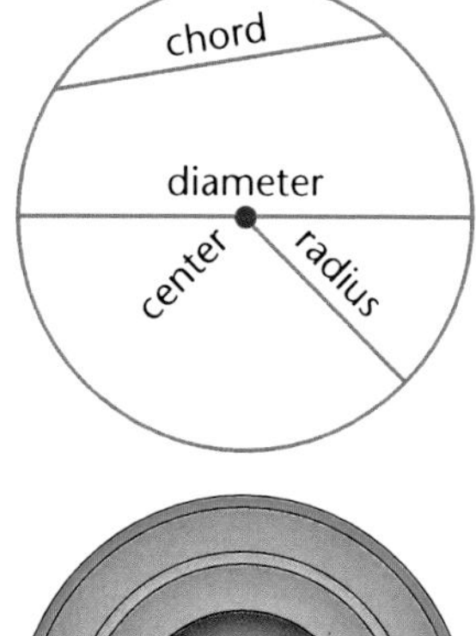

A **circle** is a curved figure in a plane. All points on the circle are the same distance from a given point in the plane called the **center**. A **radius** (r) is a line segment that has one endpoint at the center and the other endpoint on the circle. A **diameter** (d) is a line segment that passes through the center with both endpoints on the circle. A **chord** is a segment that has both endpoints on the circle. The **circumference** (C) is the distance around the circle.

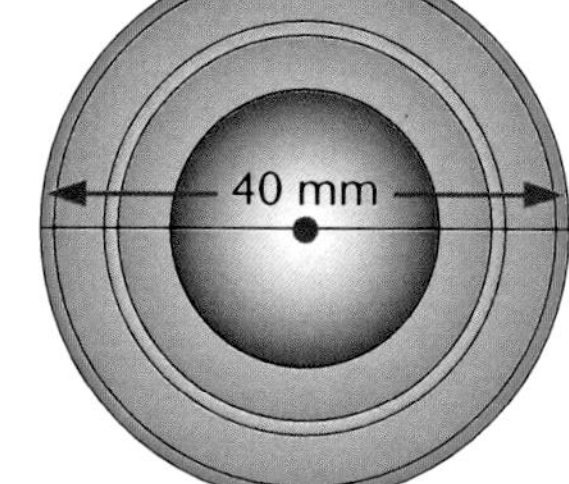

Exploration Exercise

Use a metric ruler, string, and a circular object to determine how the circumference is related to the diameter.

1. Measure the length of the diameter.
2. Wrap a string around the circle once so it matches the circumference.
3. Measure the length of the string. (126 mm in this example.)
4. Divide the length of the string (circumference) by the diameter. (126 ÷ 40)

126 mm

For any circle, this division results in the same number. The Greek letter π (*pi*) stands for this number.

$$\pi = \frac{C}{d} \quad \frac{\text{measure of circumference}}{\text{measure of diameter}}$$

$\pi \approx 3.14$ or $\frac{22}{7}$

$\approx$ means "is approximately equal to."

Note that $\frac{126}{40} = 3.15$. The difference is due to measurement error.

> You can use the Associative Property to think of the formula as $C = \pi \times 2 \times r$ or $C = \pi \times d$.

Examples

Use these formulas to find the circumference.

A.

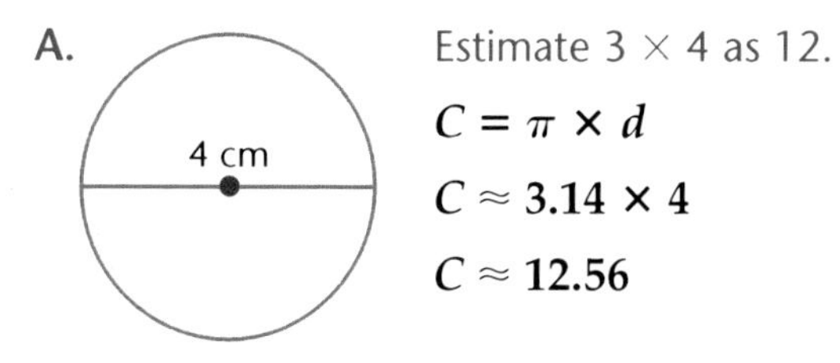

Estimate 3 × 4 as 12.

$C = \pi \times d$

$C \approx 3.14 \times 4$

$C \approx 12.56$

The circumference is about 12.56 cm.

B.

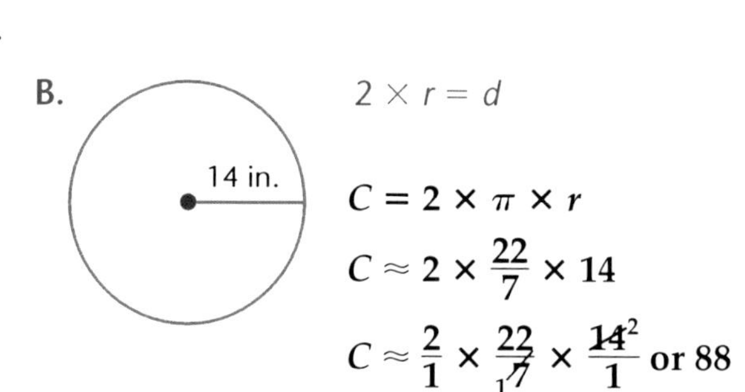

$2 \times r = d$

$C = 2 \times \pi \times r$

$C \approx 2 \times \frac{22}{7} \times 14$

$C \approx \frac{2}{1} \times \frac{22}{\not{7}_1} \times \frac{\not{14}^2}{1}$ or 88

The circumference is about 88 in.

Try THESE

Match. Use the circle on the right.

1. A — a. radius
2. $\overline{BC}$ — b. circumference
3. $\overline{AD}$ — c. chord
4. E — d. center
5. F — e. circle
6. GH — f. diameter

Exercises

Find the circumference of each circle. Use 3.14 for π in exercises 1–4. Use $\frac{22}{7}$ for π in exercises 5–8.

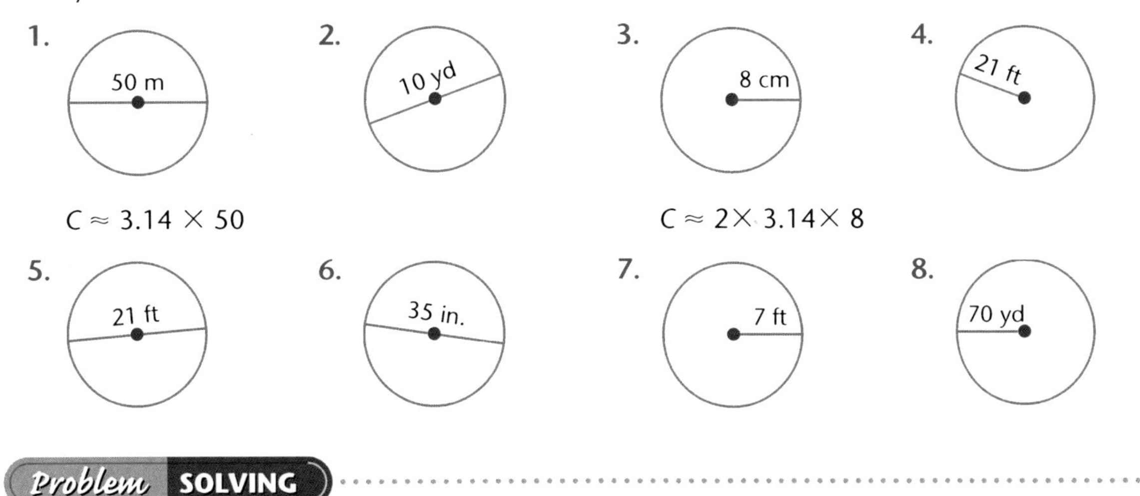

Problem SOLVING

★9. The radius of a circle is doubled. How does the circumference change?

★10. Kate's bicycle tire has a diameter of 2 feet. How far will Kate's bicycle travel when the tire has made 4 revolutions?

Mind BUILDER

Circumference

Use the formula $C = \pi d$ to compute the circumference of each of these planets.

Planet	Radius	Circumference
Mars	2,106 miles	
Jupiter	43,441 miles	
Saturn	36,184 miles	
Earth	3,959 miles	

Objective: to find the area of a circle

You know you can estimate the area of this circle by counting the shaded squares.

16	squares all shaded
8	squares almost all shaded
+ 4	8 halves partly shaded = 4 ones
28	

You can also find the area of circles using a formula. The circular design below can be cut apart and rearranged into what looks like a parallelogram. Remember, the formula for the area of a parallelogram is $A = b \times h$.

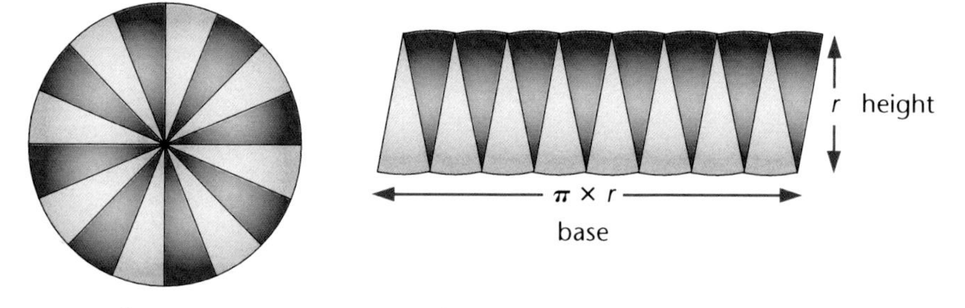

base → $\frac{1}{2}$ the circumference or $\frac{1}{2}(2 \times \pi \times r)$ or $\pi \times r$.

height → the radius or r. So, $A = \pi \times r \times r$ or $\pi \times r^2$.

Example

Find the area of the circle using $A = \pi \times r \times r$. Use 3.14 for π.

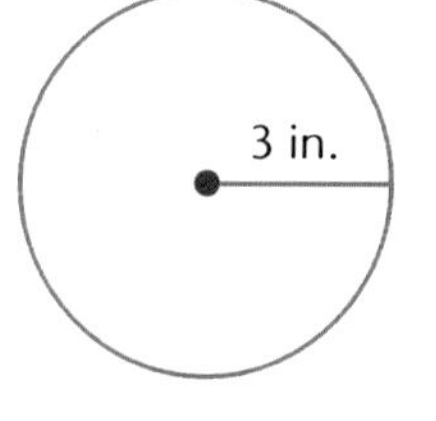

Area is equal to pi times radius times radius. An estimate is $3 \times 3 \times 3$ or 27.

$A \quad = \quad \pi \quad \times \quad r \quad \times \quad r$

$A \approx$ **3.14 × 3 × 3** Replace r with 3. ≈ means "is about."

$A \approx$ **28.26** The area is about 28.26 in.^{2}.

Try THESE

Estimate the area of each circle.

1.

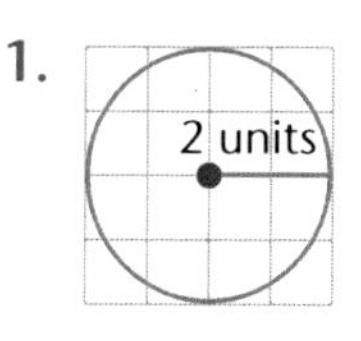

2.

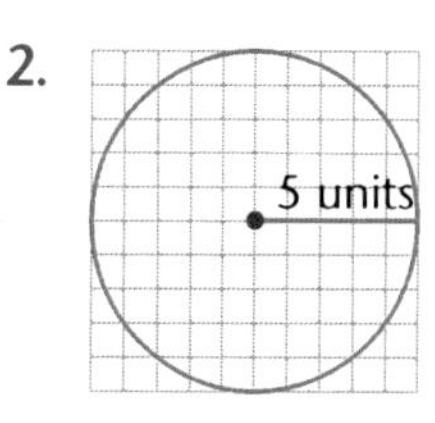

3. 4.

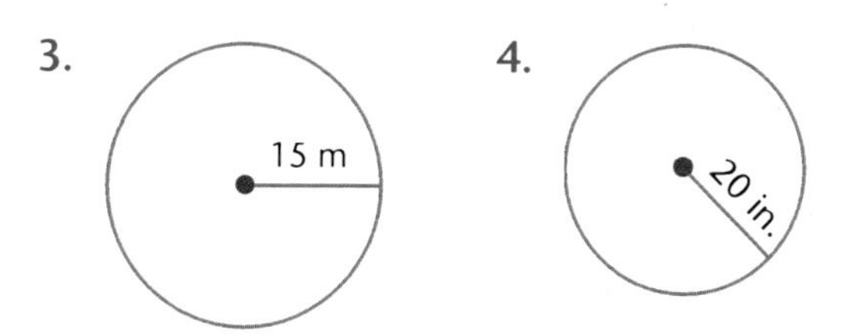

Exercises

Find the area of each circle. Use 3.14 for π.

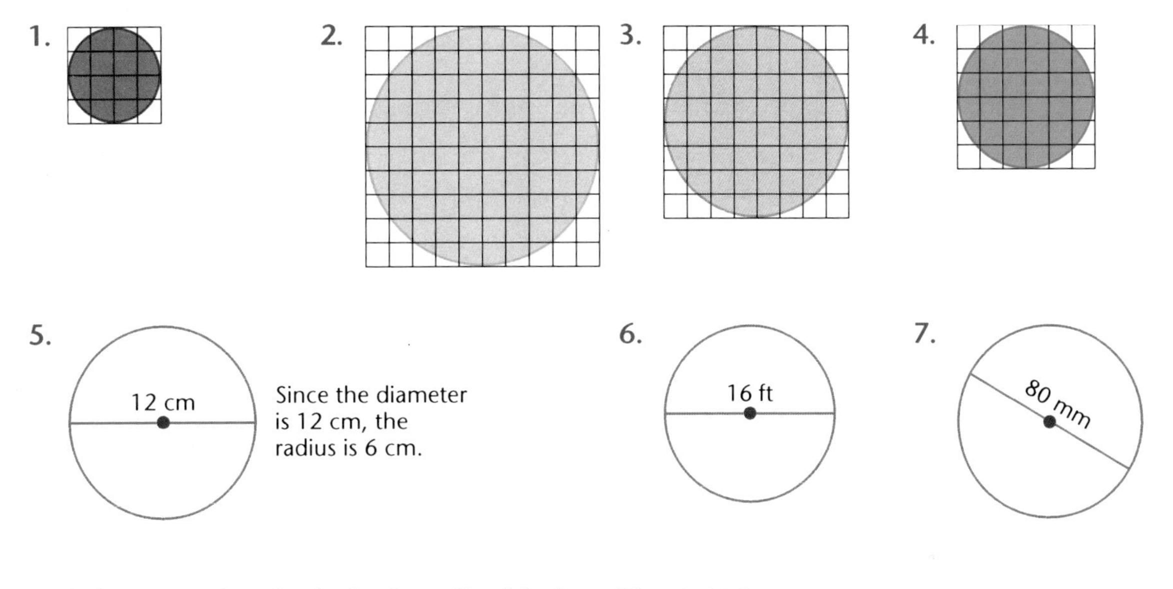

Find the area of each circle described below. Use 3.14 for π.

8. radius, 1 mi
9. diameter, 24 m
10. diameter, 80 in.
11. radius, 10 yd
12. radius, 18 mm
13. diameter, 100 cm

Problem SOLVING

★14. Find the area of the shaded part of the square. Use 3.14 for π.

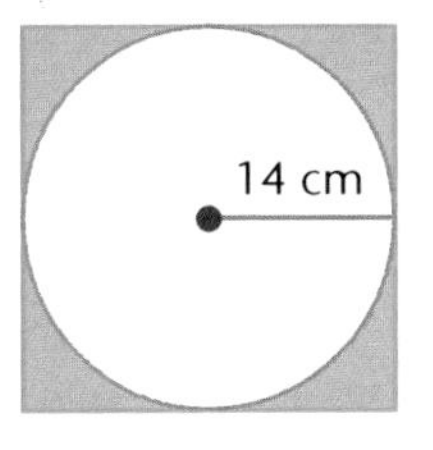

15. What happens to the area of a circle when the radius is doubled?

★16. Find the area of the shaded part of the circle. Use 3.14 for π.

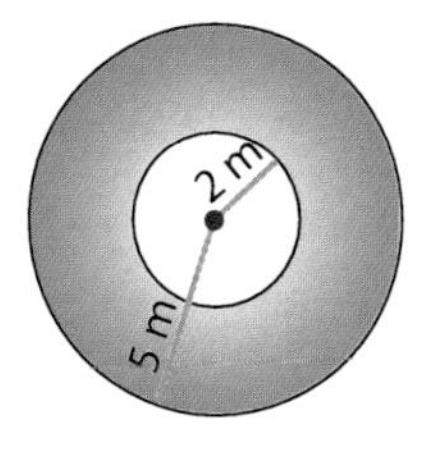

MIXED REVIEW

Rename each fraction or decimal as a percent.

17. $\frac{18}{100}$
18. $\frac{9}{50}$
19. $\frac{2}{25}$
20. 0.55
21. 0.7

11.7 Problem-Solving Application: Using Perimeter and Area

Objective: to find the perimeter and area of figures

G. E. Bergstrom, an architect, designed the Pentagon that is located near Washington, DC. He designed it so that each side would be 906.5 feet long. What is the perimeter or distance around this regular polygon?

You need to find the perimeter of a regular pentagon (5 sides of equal length). The length of each side is given.

You can add or multiply to find the perimeter.

An estimate is 900×5 or 4,500.

Add. **906.5 + 906.5 + 906.5 + 906.5 + 906.5 = 4,532.5** Why is 906.5 used as an addend 5 times?

Multiply. **906.5 × 5 = 4,532.5** Why is 5 a factor?

The perimeter of the polygon is 4,532.5 feet.

Compared to the estimate, the answer is reasonable.

The distance around a polygon is called perimeter. What is the distance around a circle called?

4. CHECK

Since the result is the same whether you add or multiply, the solution seems reasonable.

Try THESE

Circle the correct answer.

1. Which formula is *not* a formula for the perimeter of a rectangle?

   a. $P = \ell + \ell + w + w$

   b. $P = 2 + (\ell + w)$

   c. $P = 2 \times \ell + 2 \times w$

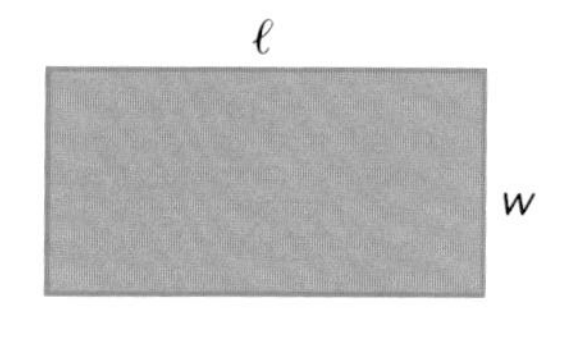

2. Which formula is *not* a formula for the perimeter (P) of a square?

   a. $P = 4 \times s$

   b. $P = 4 + s$

   c. $P = s + s + s + s$

Solve

1. Find the perimeter and area of the figure below.

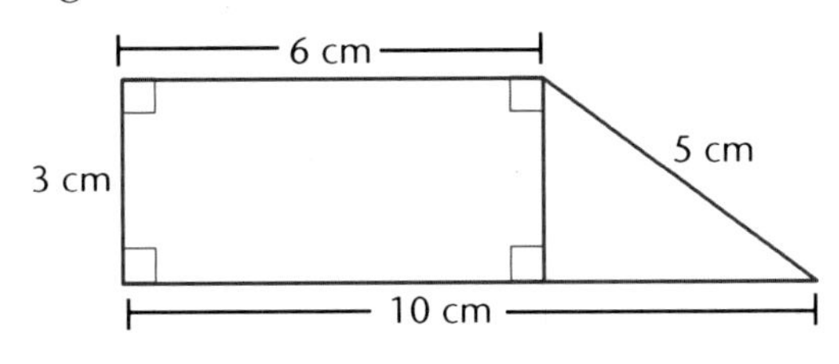

2. Find the area of the red part of the rectangle. (The figure inside is a square.)

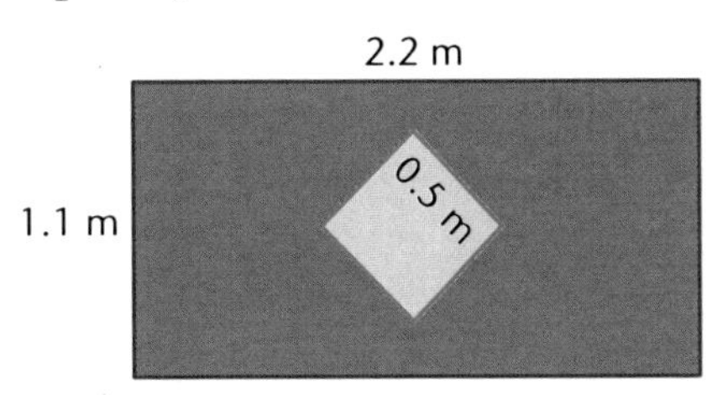

3. A piece of wood for a chest is shaped like a pentagon. It is made up of a square and a triangle. Find the area of the piece of wood.

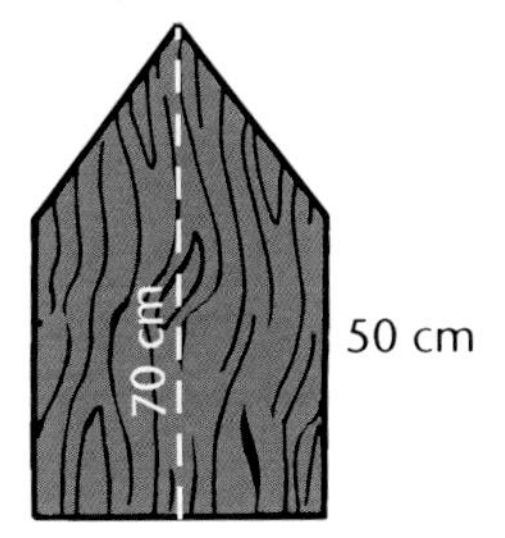

4. A horse has a circular walking arena that has a radius of 12 feet. If the horse walks around the circle once, how far does it walk? What is the area of the circle?

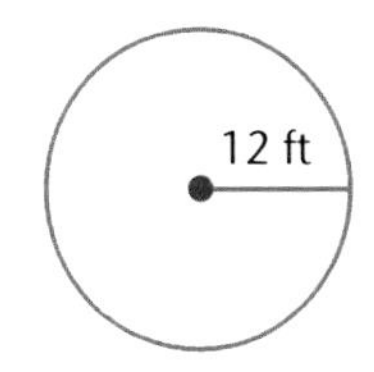

5. One bag of fertilizer will cover 600 square feet. Darnell's lawn is 50 ft by 75 ft. How many bags of fertilizer should Darnell buy to fertilize his lawn?

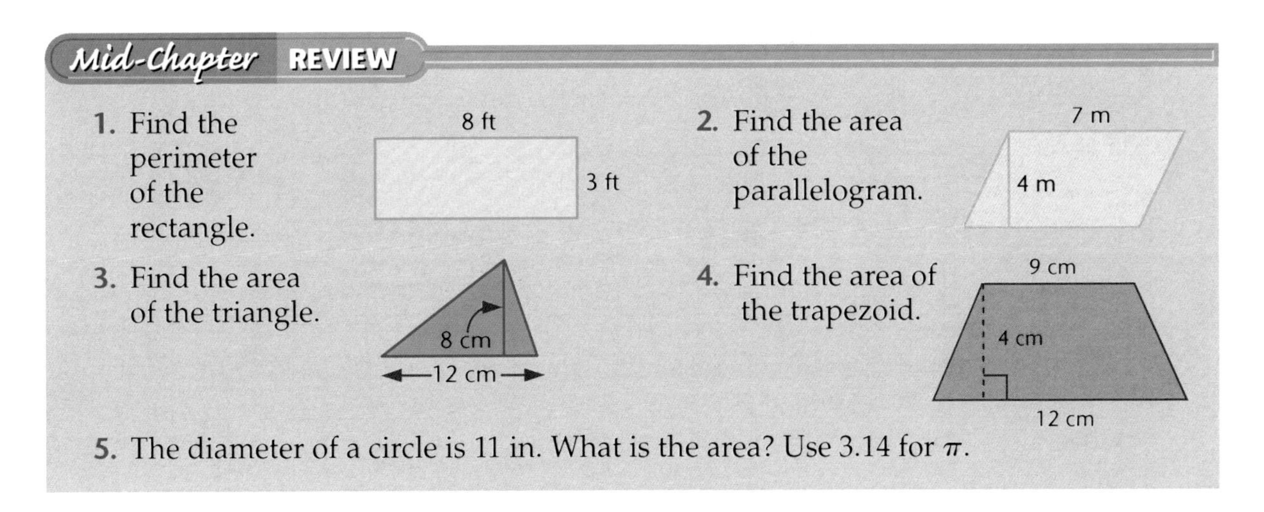

Mid-Chapter REVIEW

1. Find the perimeter of the rectangle.

2. Find the area of the parallelogram.

3. Find the area of the triangle.

4. Find the area of the trapezoid.

5. The diameter of a circle is 11 in. What is the area? Use 3.14 for π.

11.8 Three-Dimensional Figures

Objective: to identify three-dimensional figures

Many shapes that we see every day are three-dimensional.
A **three-dimensional figure** is a figure that has three dimensions: length, width, and height. It does not lie in a plane.
Three-dimensional figures have special parts as shown on the figure to the right.

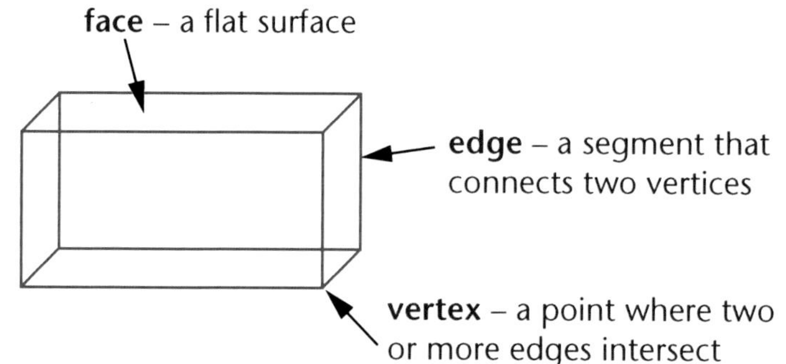

There are many types of three-dimensional figures.

Prism	• has two parallel congruent faces called **bases** • is named by the shape of its bases Triangle □ Triangular Prism Rectangle □ Rectangular Prism Pentagon □ Pentagonal Prism Hexagon □ Hexagonal Prism Heptagon □ Heptagonal Prism Octagon □ Octagonal Prism	**Rectangular Prism**
Pyramid	• has one polygon for a base • is named by the shape of its base • all other faces are triangles	**Rectangular Pyramid**
Cone	• has one circular base and one vertex • has no edges	
Cylinder	• has two parallel congruent circular bases • has no vertices or edges	
Sphere	• all points are the same distance from a given point called the center • has no faces, bases, or edges	

Examples

A. Name the prism.

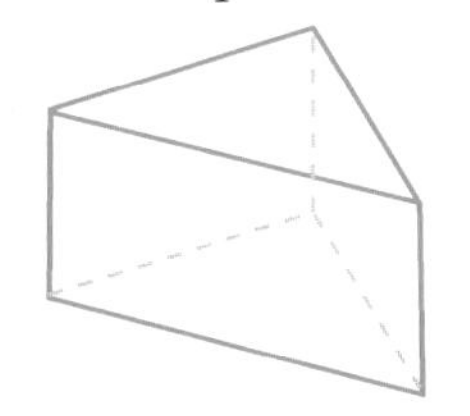

The bases are triangles, so it is a triangular prism.

B. Name the prism.

The bases are pentagons, so it is a pentagonal prism.

Name each prism.

1. 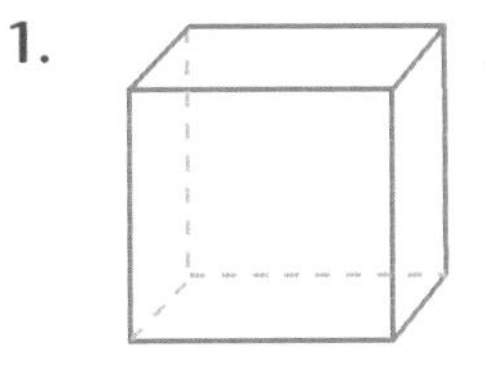

2.

3.

Exercises

Name each figure.

1. 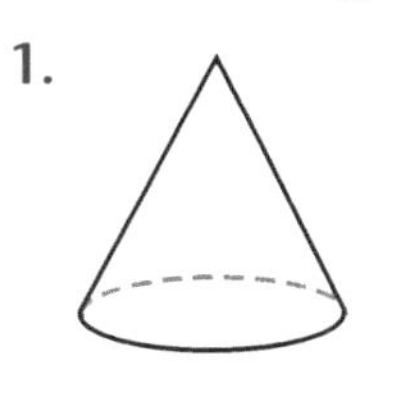

2. 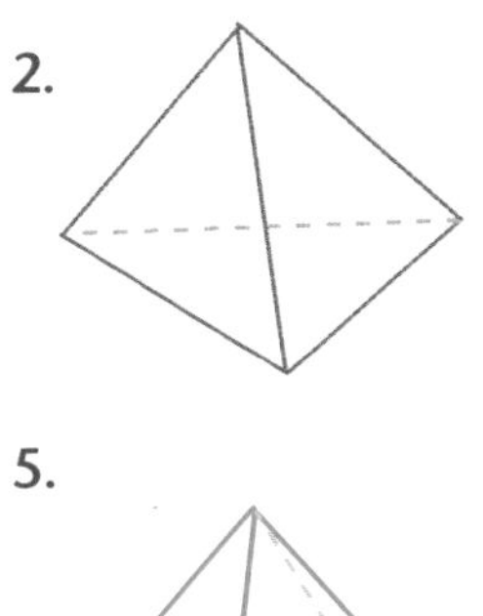

3. 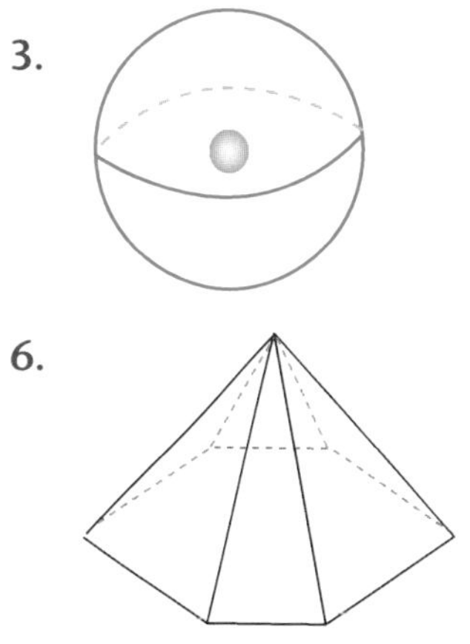

4. 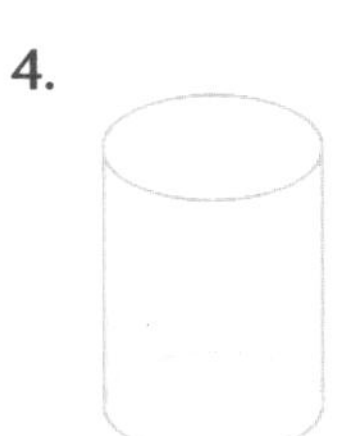

5. 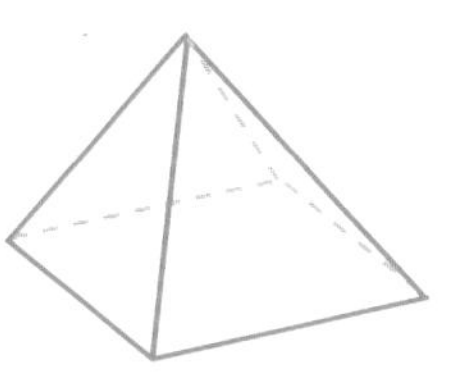

6.

Sketch each three-dimensional figure.

7. triangular prism

8. cone

9. rectangular pyramid

11.9 Surface Area of a Rectangular Prism

Objective: to find the surface area of rectangular prisms

Exploration Exercise

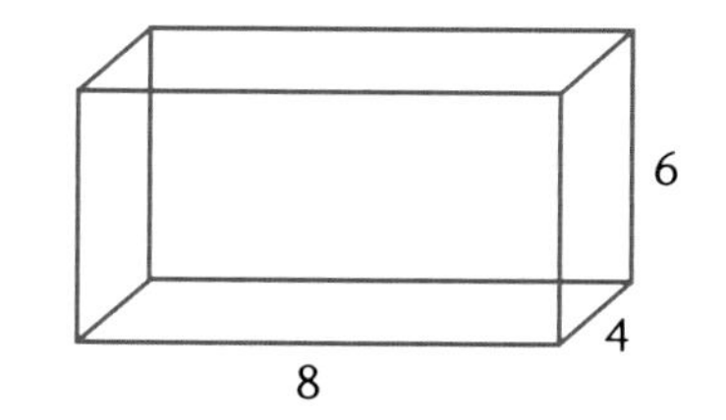

In this exploration you will use a **net**, a two-dimensional figure that can be folded into a three-dimensional figure.

1. Use grid paper to draw and cut out the net of the rectangular prism to the right.
2. Label the faces of the net as *top, bottom, front, back,* and *sides.*
3. Find the area of each face.
4. What do you notice about the area of the top and bottom?
5. What do you notice about the area of the front and back?
6. What do you notice about the area of the sides?
7. Add the areas.

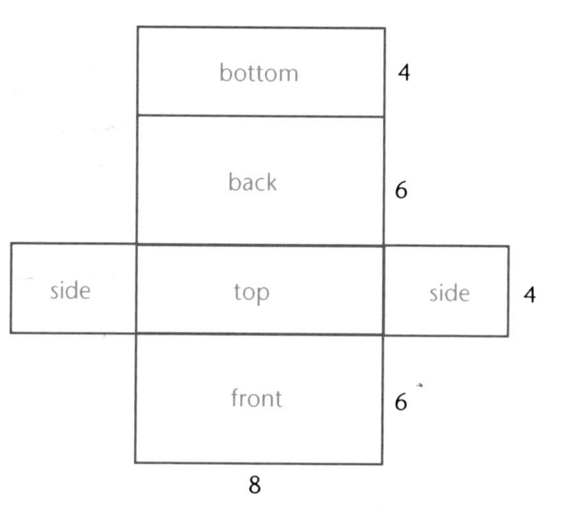

Front	$8 \bullet 6 = 48$ units^2	**Back**	$8 \bullet 6 = 48$ units^2
Top	$8 \bullet 4 = 32$ units^2	**Bottom**	$8 \bullet 4 = 32$ units^2
Side	$6 \bullet 4 = 24$ units^2	**Side**	$6 \bullet 4 = 24$ units^2

Sum of the faces = 48 + 32 + 48 +32 + 24 +24 = 208 units^2

By adding the areas of the faces and the bases together you get the **surface area** of the prism. The surface area is the sum of the area of all of the faces. You can use a net to find the surface area of a prism or you can use a formula.

Surface Area of a Rectangular Prism

The surface area of a rectangular prism with length ℓ, width w, and height h is the sum of the area of its faces.

$$S = 2\ell w + 2\ell h + 2wh$$

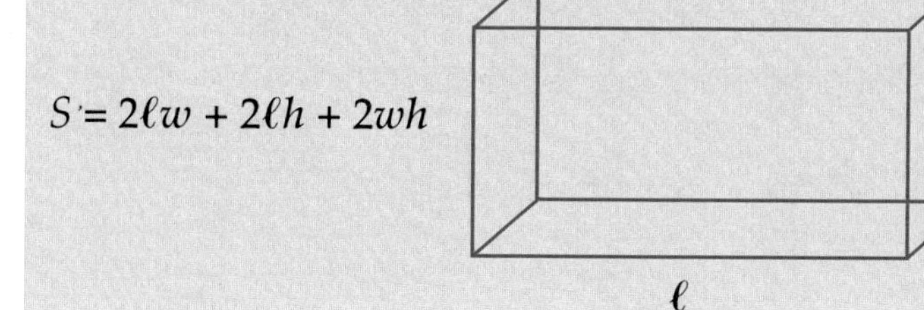

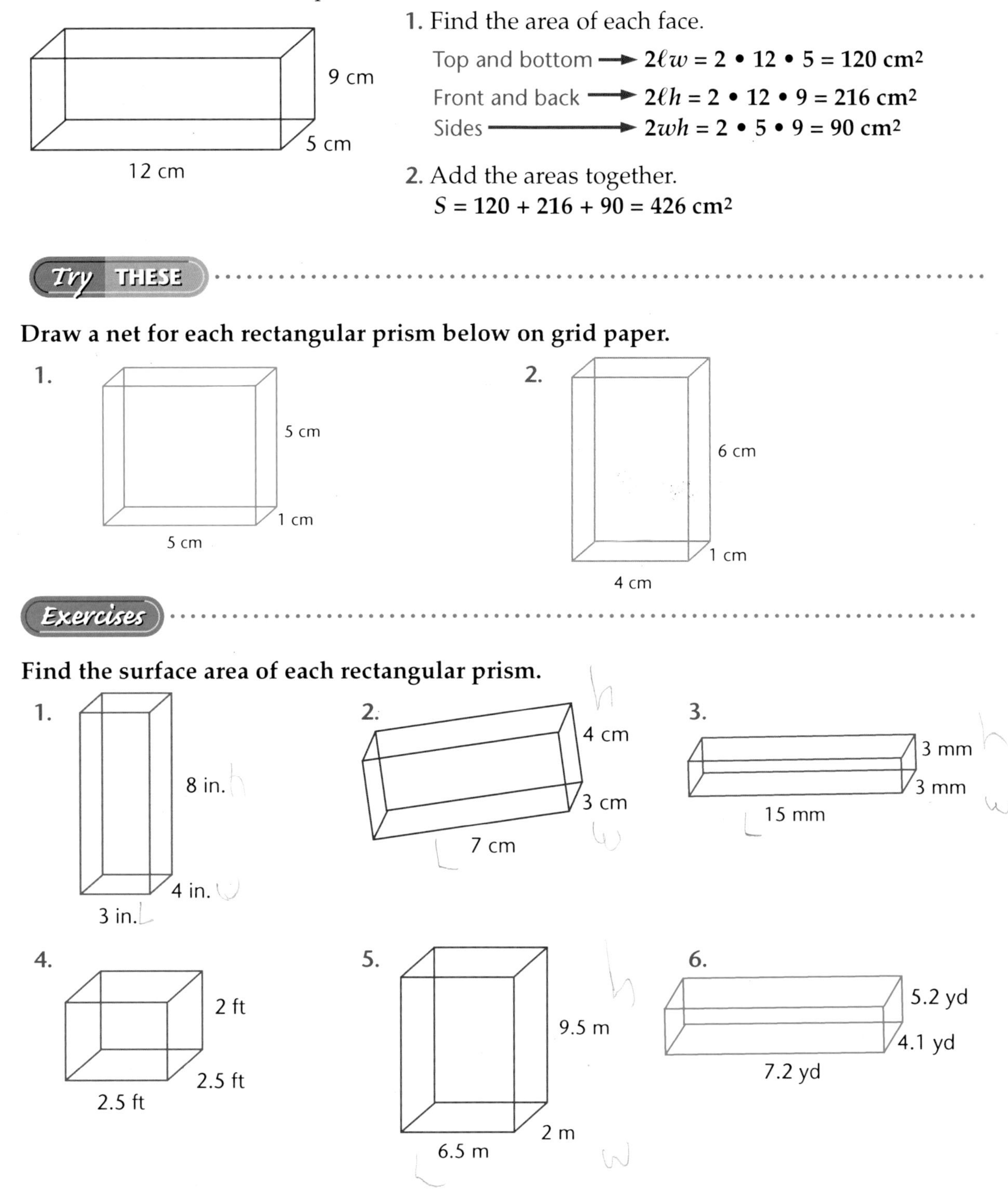

Example

Find the surface area of the prism.

1. Find the area of each face.

   Top and bottom → $2\ell w = 2 \bullet 12 \bullet 5 = 120\ \text{cm}^2$

   Front and back → $2\ell h = 2 \bullet 12 \bullet 9 = 216\ \text{cm}^2$

   Sides → $2wh = 2 \bullet 5 \bullet 9 = 90\ \text{cm}^2$

2. Add the areas together.

   $S = 120 + 216 + 90 = 426\ \text{cm}^2$

Try THESE

Draw a net for each rectangular prism below on grid paper.

1.

2.

Exercises

Find the surface area of each rectangular prism.

1.

2.

3.

4.

5.

6.

Problem SOLVING

7. Name the six faces on the rectangular prism to the right. Are any faces parallel? If yes, which ones?

Constructed RESPONSE

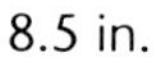

8. Look at the cereal box to the right.
   a. Estimate the surface area of the box.
   b. Find the surface area of the box.
   c. How does the surface area change if each dimension is doubled? Explain.
9. What is the surface area of the cube below?

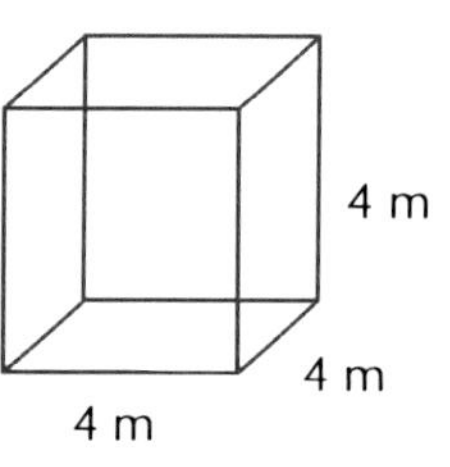

★ 10. How can you simplify the surface area formula $S = 2lw + 2lh + 2wh$ when finding the surface area for a cube?

Mixed REVIEW

Complete.

11. $\begin{array}{r} 41.38 \\ +\ 2.59 \\ \hline \end{array}$

12. $\begin{array}{r} 121.35 \\ -\ 54.81 \\ \hline \end{array}$

13. $\begin{array}{r} 15.3 \\ +\ 0.9 \\ \hline \end{array}$

14. $\begin{array}{r} 48.3 \\ \times\ 2.9 \\ \hline \end{array}$

15. $16.4\overline{)45.92}$

16. $\begin{array}{r} 14.003 \\ \times\ 1.08 \\ \hline \end{array}$

Double Negatives

Lianna, a photographer's assistant, was developing a roll of very similar photos that the photographer had shot that day. However, in the middle of her developing, the phone rang. While she chatted on the phone, she lost track of which prints were made from which negatives. Can you help Lianna match the negatives with their prints?

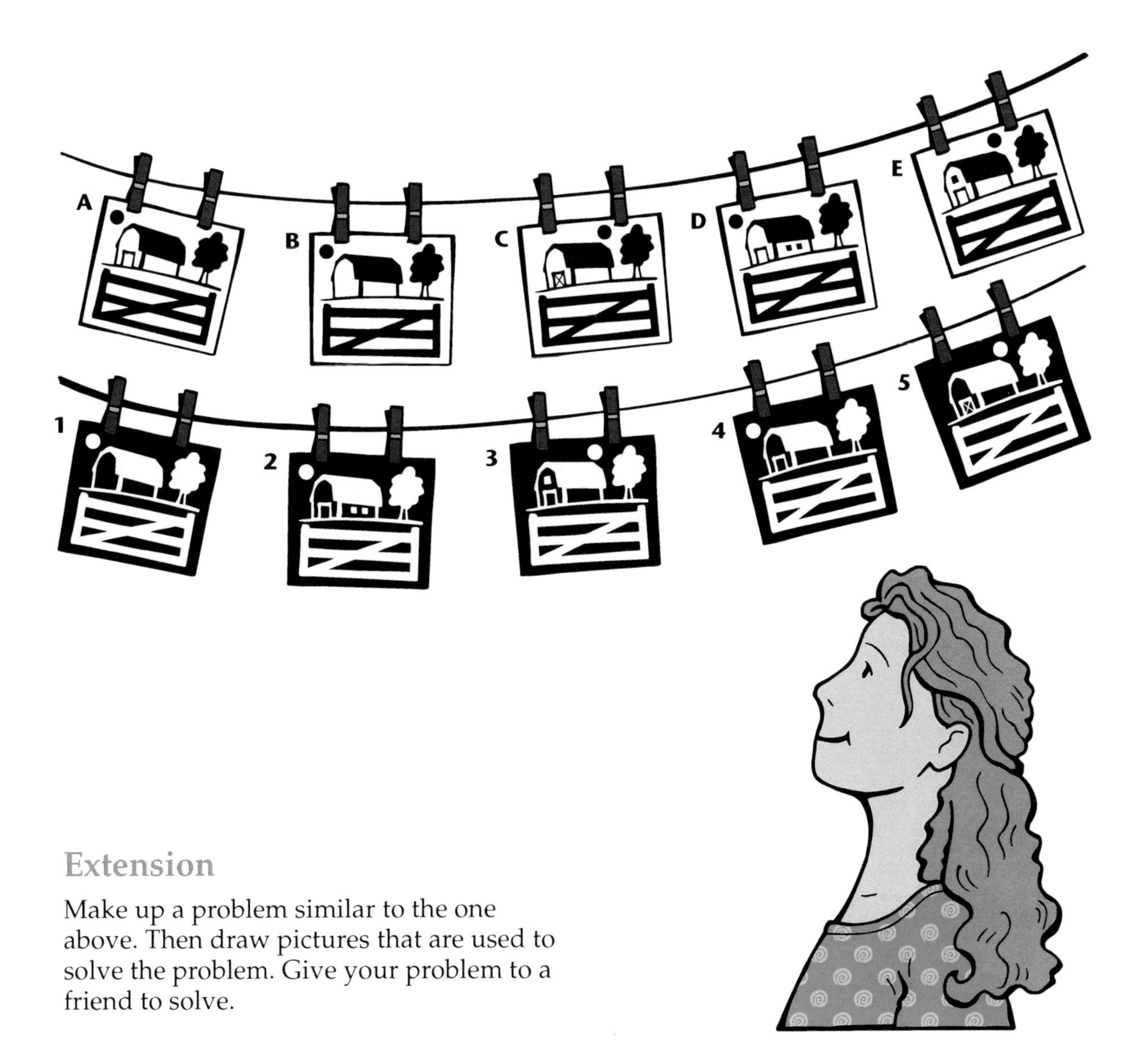

Extension

Make up a problem similar to the one above. Then draw pictures that are used to solve the problem. Give your problem to a friend to solve.

11.10 Volume of Rectangular Prisms

Objective: to find the volume of rectangular prisms

Volume is the amount of space that a solid figure contains. A **cubic** centimeter (cm^3) is a unit of volume. Other units of volume include cubic meters (m^3), cubic feet (ft^3), and cubic inches ($in.^3$).

Find the volume of the rectangular prism as follows.

- The top layer contains 6×3 or 18 one-inch cubes.
- There are 4 layers (4 in.).
- The volume is 18×4 or 72 cubic inches.

More Examples

A. The volume measure (V) of any rectangular prism can be found using this formula.

Volume is equal to length times width times height.

$$V = \ell \times w \times h$$

$V = 6 \times 3 \times 4$ Replace ℓ with 6, w with 3, and h with 4.

$V = 18 \times 4$

$V = 72$ The volume is 72 $in.^3$.

B. Find the volume of a cube that is 13 cm on a side.

$V = \ell \times w \times h$

$V = 13 \times 13 \times 13$

$V = 2{,}197$ The volume is 2,197 cm^3.

Try THESE

Count the unit cubes to find the volume of each rectangular prism.

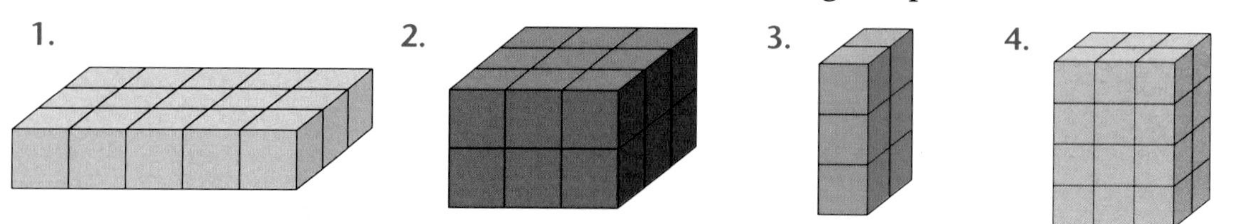

Exercises

Find the volume of each rectangular prism.

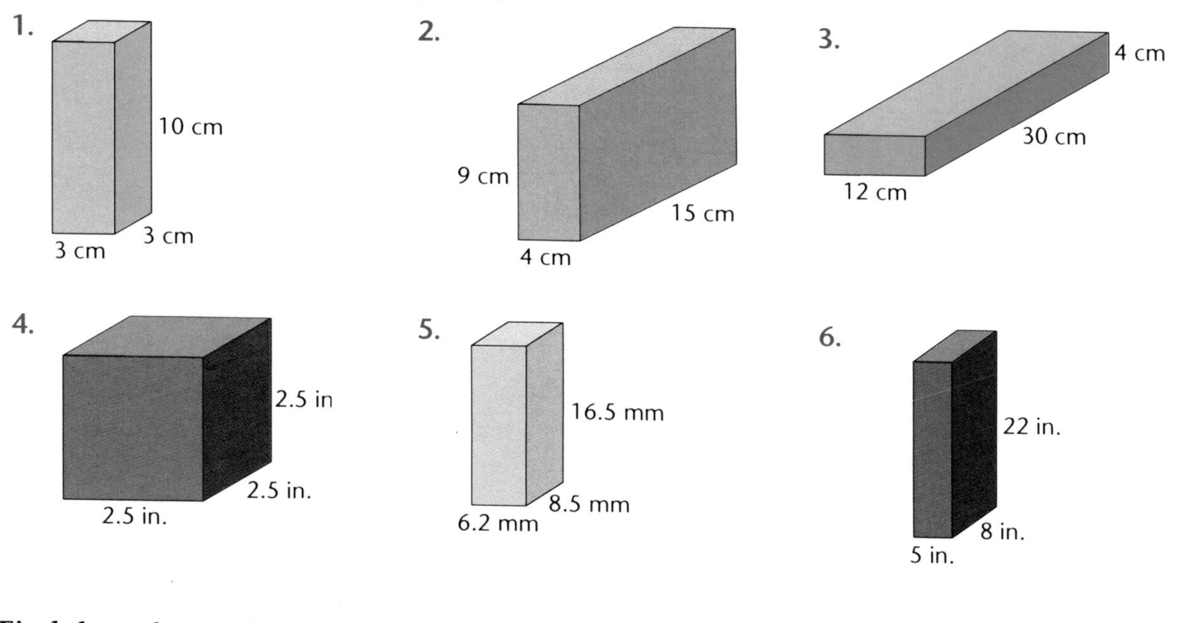

Find the volume of each figure.

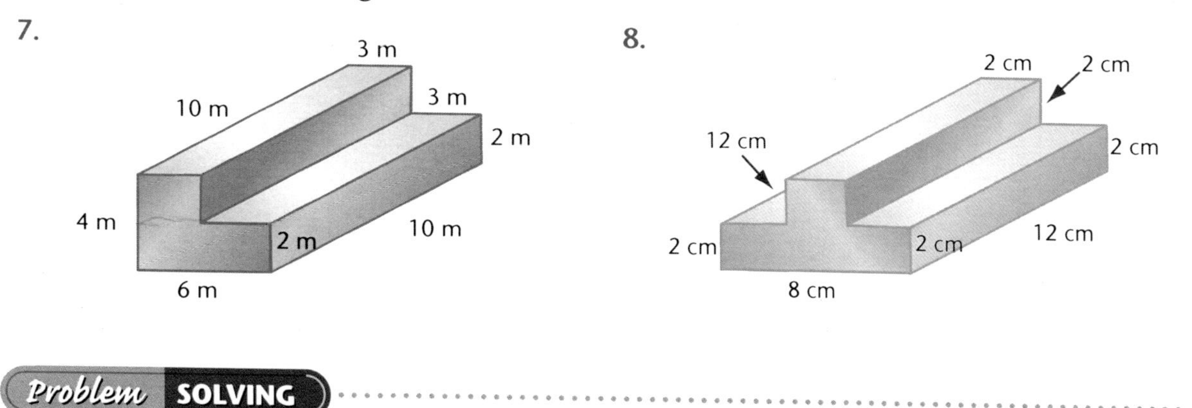

Problem SOLVING

9. Mr. Simpson laid a concrete driveway that was 30 feet long, 10.5 feet wide, and 0.6 feet thick. How many cubic feet of concrete were used?

10. The interior of a microwave oven is 1.3 feet long, 1.3 feet wide, and 0.8 feet high. What is the volume of the cooking space?

Chapter 11 Review

Language and Concepts

Choose a word or formula from the list at the right that best completes each sentence.

$A = \frac{1}{2} \times b \times h$
$A = \pi \times r \times r$
area
circle
circumference
diameter
pi
radius
triangle
volume

1. The number of square units needed to completely cover a region is the _____.
2. The area of a _____ is $\frac{1}{2}$ the area of a parallelogram with the same base and height.
3. A line segment that has one endpoint at the center of a circle and the other endpoint on the circle is called the _____.
4. The distance around a circle is called the _____.
5. The Greek letter _____ represents 3.14.
6. To find the area of a circle, use the formula _____.
7. The amount of space that a solid figure contains is called the _____.

Skills and Problem Solving

Find the area of each figure. As a check, count the unit squares. (Sections 11.1–11.3)

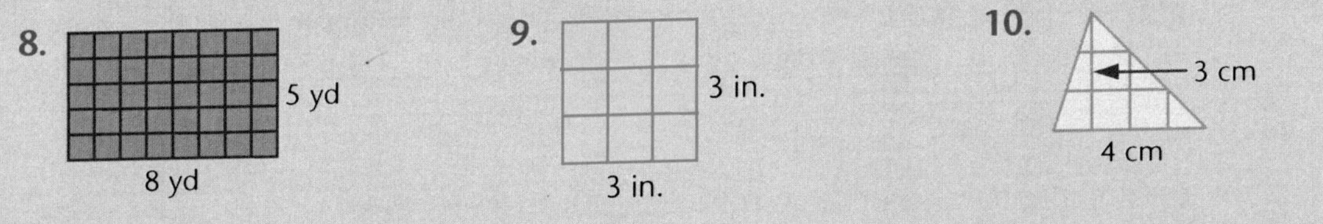

Find the area of each figure. (Sections 11.1–11.4)

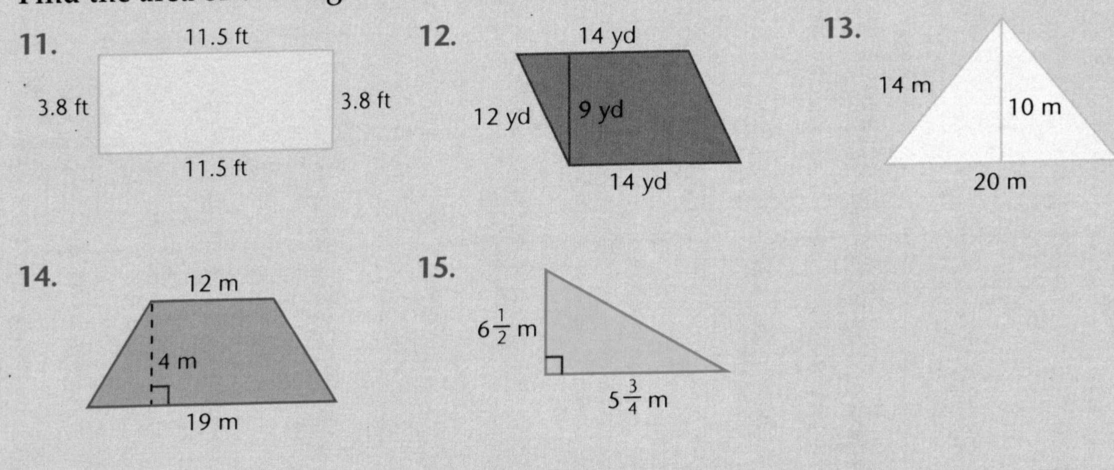

Find the circumference and area of each circle. Use 3.14 for π. (Sections 11.5–11.6)

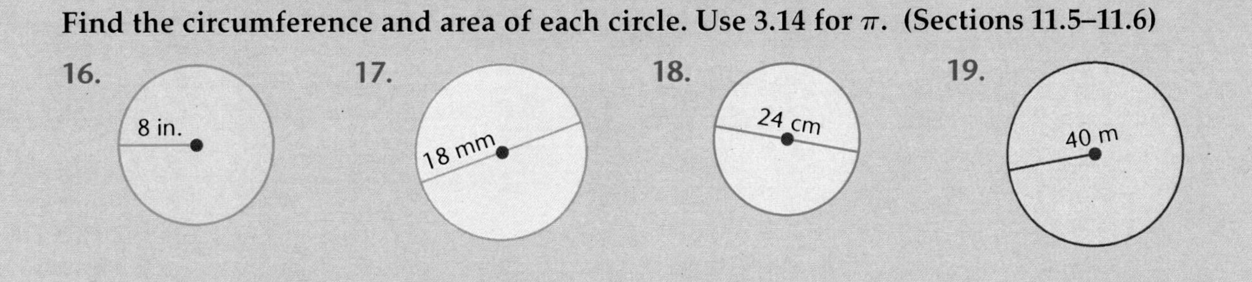

Find the surface area and volume of each rectangular prism. (Sections 11.9–11.10)

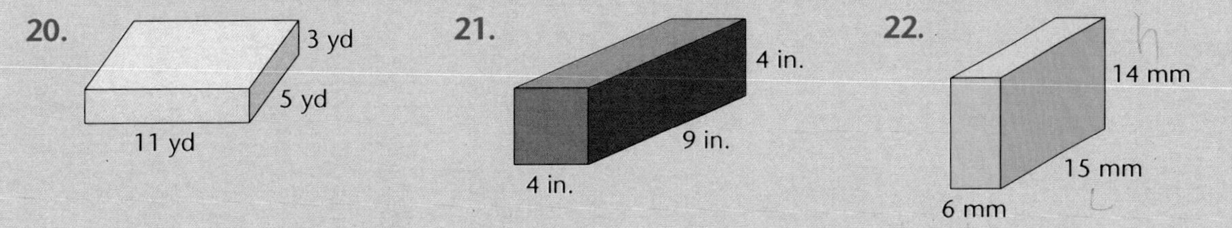

Solve. (Sections 11.2–11.4)

23. Find the area of the blue part of the square.

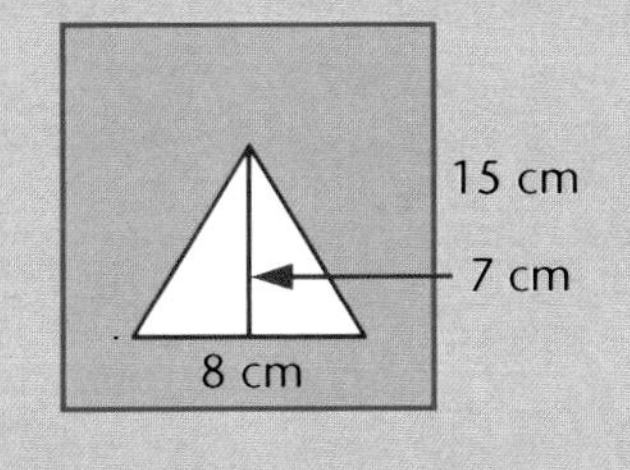

24. The Gonzalas family is buying carpet for their family room and hallway. What is the area of the floor they want to cover?

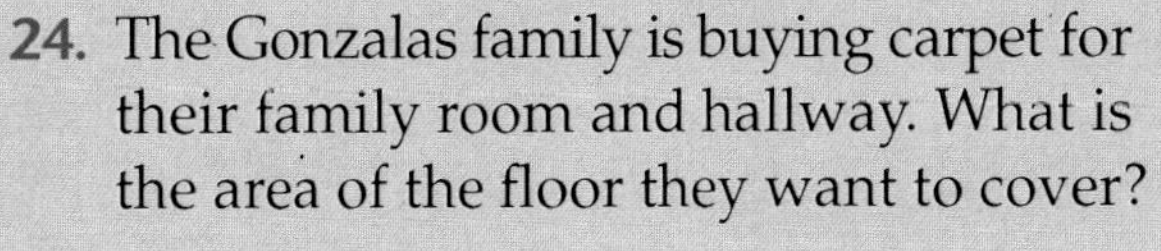

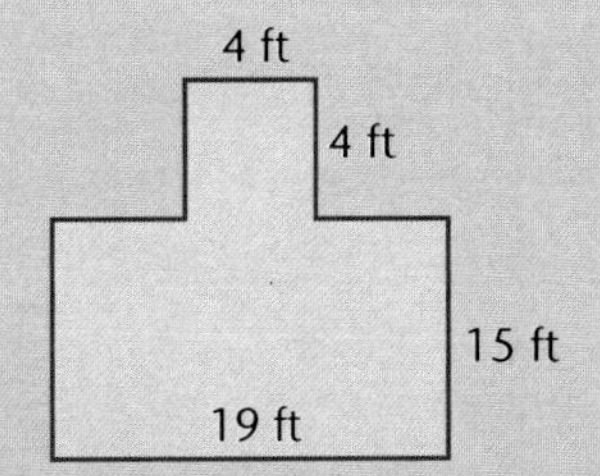

In Your Own Words

25. Explain why square units are used to name the surface area of rectangular solids and why cubic units are used to name the volume of rectangular solids.

Chapter 11 Test

Find the area of each figure.

1. 6 mm; 12 mm

2. 18 yd; 26 yd

3. 12 ft; 24 ft

Find the area and circumference of each circle. Use 3.14 for π.

4. **5.** **6.**

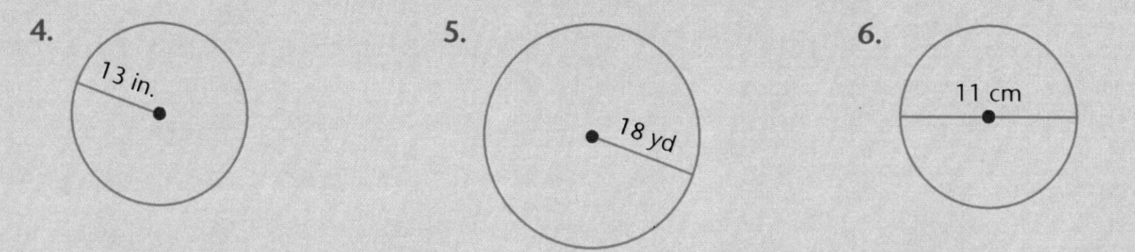

Find the surface area and volume of each figure.

7.

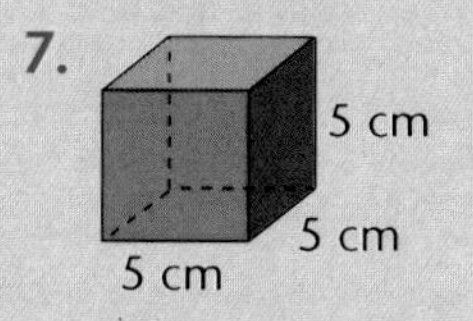

8.

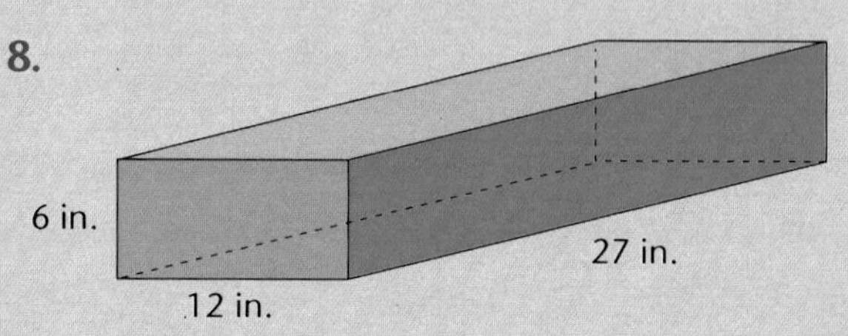

Solve.

9. Find the area of the figure below.

10. The perimeter of a square is 96 centimeters. What is the length of each side?

11. A jewelry box is damaged on one side. To refinish the damaged side, its area must be found. What is the area of the side shown?

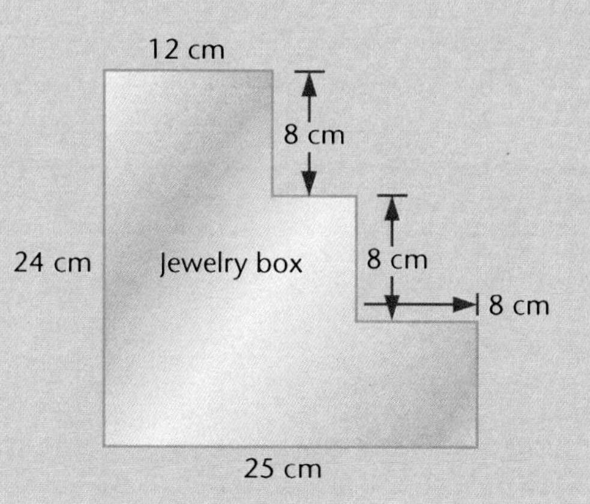

Change of Pace

Pick's Theorem

Use the *geoboard* at the right to find the area of figures A, B, C, and D.

- To find area, count the units of area.

Nails touching a rubber band are called *border nails*. Nails inside but not touching the rubber bands are called *interior nails*.

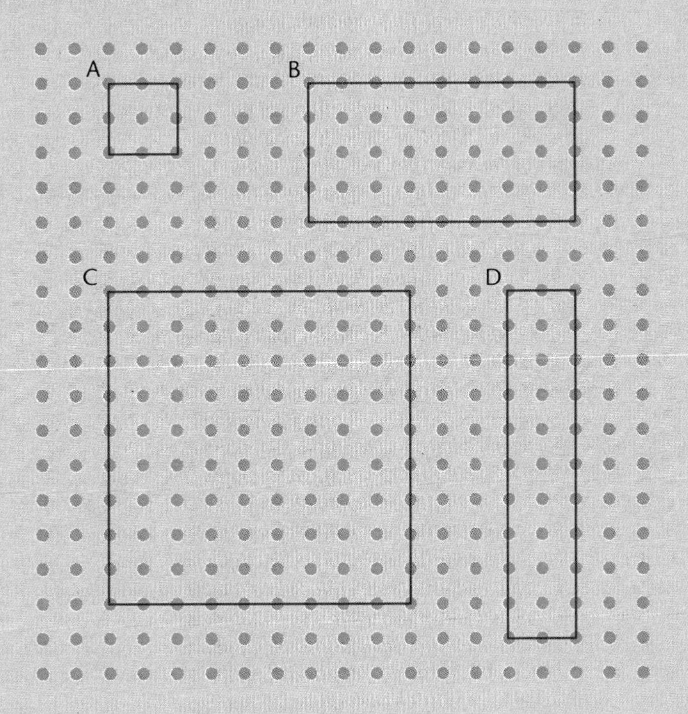

Copy and complete the table for figures B, C, and D.

Figure	Border Nails	Interior Nails	Area
A	8	1	4

Put the information on figures E and F in your table.

Pick's Theorem gives this formula for finding area.

$$A = \frac{B}{2} + I - 1$$

A is area.
B is the number of border nails.
I is the number of interior nails.

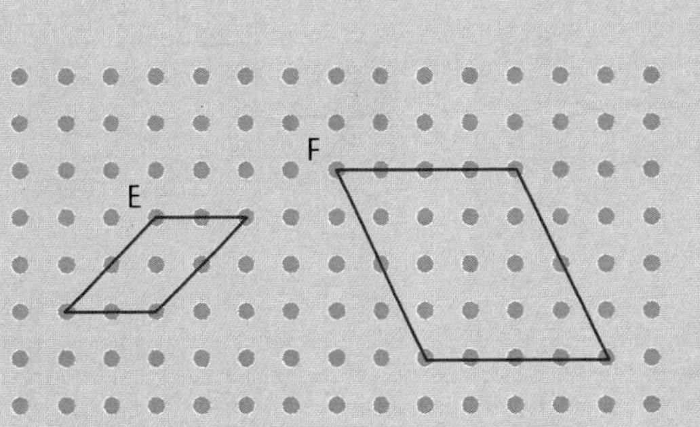

Show that this formula works for figures A, B, C, D, E, and F.

Collect Data.

Make five other figures using dot paper. Use Pick's Theorem to find the area of each figure.

Cumulative Test

1. ⟷ ⟷

These lines are ____.

a. perpendicular
b. intersecting
c. parallel
d. none of the above

2. What is the area of a rectangle 2 cm by 4 cm?

a. 8 cm
b. 8 cm^2
c. 16 cm
d. 16 cm^2

3. What is the perimeter of this figure?

a. 8 cm
b. 8.1 cm
c. 27 cm
d. none of the above

0.8 cm
2 cm
4 cm
1.3 cm

4. Solve.

$\frac{x}{15} = \frac{18}{25}$

a. 8
b. 10
c. 11
d. none of the above

5. What is the surface area of this figure?

9 in.
1 in.
4 in.

a. 14 $in.^2$
b. 98 $in.^2$
c. 72 $in.^2$
d. 36 $in.^2$

6. Delwyn has $14.96. Veda has $37.07. How much more money does Veda have than Delwyn?

a. $14.96
b. $22.11
c. $23.89
d. $37.07

7. Mr. Loy wants to know how many calculators to buy for his class of 30. Each calculator will be shared by 2 students. Which equation represents this situation?

a. $c \times 2 = 30$
b. $c \div 2 = 30$
c. $c \times 30 = 2$
d. $c + 2 = 30$

8. Kyron ordered a 12-inch round pizza from Tony's Pizza.

The distance across the pizza, 12 inches, represents the ____.

a. circle
b. circumference
c. diameter
d. radius

9. Laurelle wants to paint a small cube to use as a table. The cube is 15 in. by 15 in. by 15 in. How many cans of paint should she buy if each can covers 150 $in.^2$?

a. 3
b. 6
c. 9
d. 12

10. For more than three years, Sue has doubled the number of books she reads each year. If she read 32 books this year, how many books did she read 3 years ago?

a. 2
b. 4
c. 10
d. none of the above

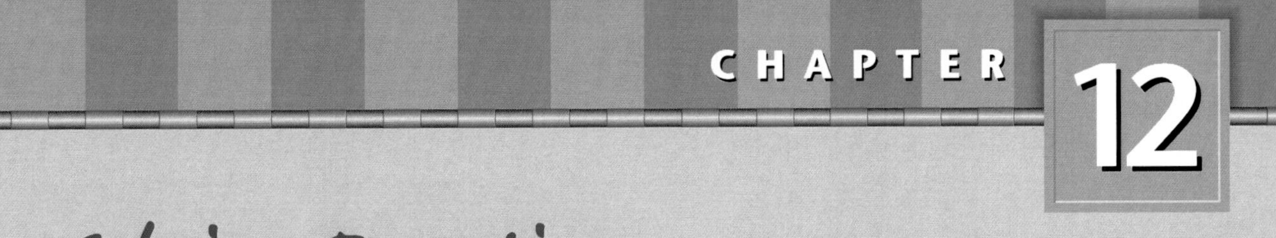

CHAPTER 12

Solving Equations

Harry Larimer
Florida

12.1 Properties

Objective: to identify and use the Commutative Property, Associative Property, and Identity Property

Sam's bakery sells 32 loaves of rye bread each week. How many loaves of rye bread will Sam sell in 3 weeks?

To solve this problem you can multiply 3 • 32 or you can multiply 32 • 3. The order in which you multiply the two factors does not change the product. This is an example of the Commutative Property of Multiplication. This is one special property that you use as you solve problems.

Properties of Addition and Multiplication

Property	Definition	Example
Commutative Property	Changing the order of the addends or factors does not change the sum or product.	$4 + 2 = 2 + 4$ $4 \cdot 2 = 2 \cdot 4$
Associative Property	The way addends or factors are grouped does not change the sum or product.	$2 + (5 + 6) = (2 + 5) + 6$ $2 \cdot (5 \cdot 6) = (2 \cdot 5) \cdot 6$
Additive Identity	Adding zero to a number results in that number.	$5 + 0 = 5$
Multiplicative Identity	Multiplying a number by 1 results in that number.	$7 \cdot 1 = 7$

Examples

Identify the property shown by the equation.

A. $6 + 8 = 8 + 6$

This is an example of the Commutative Property of Addition because the order of the addends changes.

B. $(7 \cdot 3) \cdot 8 = 7 \cdot (3 \cdot 8)$

This is an example of the Associative Property of Multiplication because the grouping of the factors changes.

You can use these properties to help you compute mentally.

C. Find $15 + 7 + 5$ mentally.

You can easily add 15 and 5, so you should change the order.

$\mathbf{15 + 7 + 5}$	
$\mathbf{= 15 + 5 + 7}$	Commutative Property of Addition
$\mathbf{= (15 + 5) + 7}$	Group the numbers. Associative Property of Addition
$\mathbf{= 20 + 7}$	Add mentally.
$\mathbf{= 27}$	

By rearranging the numbers, it is easier to compute the numbers mentally.

Try THESE

Replace each ■ with a number.

1. $4 \bullet 0 = ■$
2. $1 \bullet ■ = 12$
3. $(9 \bullet 7) \bullet 2 = 9 \bullet (■ \bullet 2)$
4. $14 + ■ = 6 + 14$
5. $16 \bullet 8 = 8 \bullet ■$
6. $137 \bullet ■ = 137$

Exercises

Identify the property shown by each equation.

1. $14 \bullet 6 = 6 \bullet 14$
2. $7 + (5 + 8) = (7 + 5) + 8$
3. $3 + 0 = 3$
4. $6 + 19 = 19 + 6$
5. $(8 \bullet 6) \bullet 5 = 8 \bullet (6 \bullet 5)$
6. $(14 + 5) + 0 = (14 + 5)$
7. $12 \bullet 1 = 12$
8. $8 \bullet (10 \bullet 15) = (8 \bullet 10) \bullet 15$

Replace each ■ with a number.

9. $■ \bullet 19 = 19 \bullet 2$
10. $■ \bullet 24 = 24$
11. $(14 + 3) + 8 = 14 + (3 + ■)$
12. $12 + 15 = ■ + 12$
13. $18 \bullet ■ = 18$
14. $16 \bullet 8 = 8 \bullet ■$
15. $182 + ■ = 182$
16. $137 + ■ = 137$
17. $(42 \bullet 3) \bullet 8 = ■ \bullet (3 \bullet 8)$
18. $(15 + 6) + ■ = (15 + 6)$

Find each sum or product mentally.

19. $35 + 9 + 5$
20. $2 \bullet 23 \bullet 10$
21. $8 \bullet 3 \bullet 5$
22. $4 \bullet 3 \bullet 25$
23. $105 + 3 + 15$
24. $15 + 19 + 20$

Problem SOLVING

25. In the sixth grade there are 30 students who like pepperoni pizza. In the seventh grade there are 12 students who like pepperoni pizza and in the eighth grade there are 25 students who like pepperoni pizza. Use mental math to compute how many students in the middle school like pepperoni pizza.

Constructed RESPONSE

26. So far you have seen that the Commutative Property and Associative Property are true for whole numbers. Are they also true for fractions? Explain and give examples.

Test PREP

27. What property can be used to find the missing number in $4 + (5 + 9) = (4 + ■) + 9$?

    a. Commutative Property of Addition
    b. Associative Property of Addition
    c. Associative Property of Multiplication
    d. Additive Property of Addition

28. What property can be used to find the missing number in $4 \bullet 7 \bullet 10 = 10 \bullet 4 \bullet ■$?

    a. Commutative Property of Addition
    b. Commutative Property of Multiplication
    c. Associative Property of Addition
    d. Associative Property of Multiplication

Mind BUILDER

Logic

Study the statements below.

All elephants are gray.
An elephant is in the zoo.

Some roses are red.
None of the roses in the park are red.

What conclusion can you make about these two statements?

1. The elephant in the zoo is gray.
2. A rose may or may not be red.

Complete the following.

3. Write an explanation of the words *all*, *some*, and *none*.
4. Write a statement that uses *all*, a statement that uses *some*, and a statement that uses *none*.

Cumulative Review

Compare using <, >, or = .

1. 261 ● 216
2. 31.85 ● 30.85
3. 42,012 ● 42,102
4. $\frac{11}{12}$ ● $\frac{10}{12}$
5. $\frac{1}{2}$ ● $\frac{9}{14}$
6. $\frac{3}{4}$ ● $\frac{15}{20}$

Estimate.

7. $\begin{array}{r} 22.3 \\ +\ 18.91 \\ \hline \end{array}$
8. $\begin{array}{r} 7{,}432 \\ -\ \ 956 \\ \hline \end{array}$
9. $\begin{array}{r} 15.46 \\ \times\ \ 12.1 \\ \hline \end{array}$
10. $22\overline{)315}$
11. $1.3\overline{)26}$

Compute.

12. $\begin{array}{r} 258 \\ +\ \ 93 \\ \hline \end{array}$
13. $\begin{array}{r} 87.19 \\ -\ 28.37 \\ \hline \end{array}$
14. $\begin{array}{r} 23.3 \\ \times\ \ 6.2 \\ \hline \end{array}$
15. $8\overline{)1{,}024}$
16. $6\overline{)28.8}$
17. $8.4 + 0.26$
18. $52{,}825 - 20{,}025$
19. 33×120

Use factor trees to write the prime factorization.

20. 21
21. 42
22. 18
23. 27

Compute. Write each answer in simplest form.

24. $\frac{5}{9} + \frac{7}{9}$
25. $\frac{3}{4} - \frac{9}{16}$
26. $\frac{5}{8} \times 4$
27. $\frac{4}{5} \div \frac{1}{10}$

Evaluate each expression.

28. $a + 6$, if $a = 15$
29. $g - 2.7$, if $g = 9.6$
30. $m \times 10$, if $m = 11$

Solve.

31. Mr. Kelly filled up his camper with 22.8 gallons of gasoline. He paid $26.22. What was the cost per gallon?

32. At the end of a 478-mile trip, the odometer reading is 15,015. What was the reading at the start?

33. Jackie drives 45 miles each hour. She drives 5 hours on Monday, 4 hours on Tuesday, and 7 hours on Friday. How far does she drive in all?

34. Paulette works part-time at a pet store. She works $2\frac{3}{4}$ hours each day. About how many hours does she work in 5 days?

12.2 Solving Addition Equations

Objective: to solve addition equations

Over the weekend Keenan made some kites for his trip to the beach. He made some on Saturday and 7 on Sunday. He made a total of 12 kites. How many did he make on Saturday?

You can write an equation for this problem.

k	$+$	7	$=$	12
number of kites made on Saturday		number of kites made on Sunday		total number of kites made

To solve an equation means to find the number that will replace the variable and make the equation true.

You can solve an equation using **inverse operations**. Inverse operations undo each other. Subtraction undoes addition. To solve an addition equation, use subtraction.

$k + 7 - 7 = 12 - 7$ To undo the addition, subtract 7 from each side.

$k + 0 = 5$

$k = 5$

Keenan made 5 kites on Saturday.

Every time you solve an equation you need to check your solution. To check the solution substitute the solution back into the equation.

$5 + 7 = 12$ ✓

When you solve an equation by subtracting the same number from each side of the equation, you are using the Subtraction Property of Equality.

Subtraction Property of Equality

If a number is subtracted from each side of an equation, the two sides remain equal.

$10 - 6 = 10 - 6$	$x + 3 - 3 = 9 - 3$
$4 = 4$	$x = 6$

Example

$$d + 12 = 28.7$$

$$d + 12 - 12 = 28.7 - 12$$ Undo the addition; subtract 12 from each side.

$$d = 16.7$$ Simplify.

$$16.7 + 12 = 28.7 \checkmark$$ Check your solution.

Try THESE

State whether the given number is a solution of the equation. Write *yes* or *no*.

1. $t + 6 = 9$
   $t = 3$
2. $b + 3 = 11$
   $b = 11$
3. $n + 18 = 25$
   $n = 6$

Exercises

Solve each equation. Check your solution.

1. $21 + d = 30$
2. $8 + f = 72$
3. $g + 15.3 = 19.7$
4. $2.6 + k = 6.24$
5. $16 = p + 9$
6. $y + 8 = 25$
7. $a + 9.3 = 15.4$
8. $14 + m = 35$
9. $45 = g + 21$
10. $2.8 + r = 15.1$
11. $p + \frac{1}{2} = 3\frac{3}{4}$
12. $t + 0.55 = 4.2$
13. $h + 2\frac{1}{4} = 5\frac{1}{2}$
14. $m + 1\frac{1}{3} = 5\frac{1}{3}$
15. $4.6 = n + 2.4$

Problem SOLVING

Write an equation and solve.

16. The sum of 7 and u is 14.
17. y plus 12 is 57.6.
18. v increased by 9 is 12.4.
19. n added to $3\frac{1}{4}$ is $5\frac{1}{2}$.
20. Some number added to 7.5 equals 18.4.
21. 21 added to some number equals 72.1.
22. You bought lunch and spent $5.25. You bought a sandwich, which cost $3.85, and a soda. How much did the soda cost?
23. Write a word problem that can be modeled by the equation $x + 3 = 8$.
24. Jamal and Keisha went running. Keisha ran 2.4 miles more than Jamal. If Keisha ran 8.5 miles, how many miles did Jamal run?

12.3 Solving Subtraction Equations

Objective: to solve subtraction equations

Josh made birdhouses for people who collect them. He sold 9 of them and had 7 left. How many birdhouses did Josh have at first?

You can write an equation for this problem.

b	$-$	9	$=$	7
number of birdhouses		number sold		number left

Addition and subtraction are inverse operations. You can solve this subtraction equation using addition.

$b - 9 + 9 = 7 + 9$ Undo the subtraction; add 9 to each side.

$b + 0 = 16$ Simplify.

$b = 16$

Josh had 16 birdhouses.

You need to check the solution.

$16 - 9 = 7$ ✓

When you solve an equation by adding the same number to each side of the equation, you are using the addition property of equality.

Addition Property of Equality

If a number is added to each side
of an equation, the two sides remain equal.

$10 + 6 = 10 + 6$ $\qquad$ $x - 5 + 5 = 9 + 5$

$16 = 16$ $\qquad$ $x = 14$

Example

A. $m - 15 = 5$

$m - 15 + 15 = 5 + 15$ Undo the subtraction. Add 15 to both sides.

$m = 20$ Simplify.

Check your solution.

$20 - 15 = 5$ ✓

Try THESE

State whether the given number is a solution of the equation. Write *yes* or *no*.

1. $s - 4 = 14$
   $s = 10$
2. $g - 8 = 7$
   $g = 15$
3. $n - 14 = 35$
   $n = 46$

Exercises

Solve each equation. Check your solution.

1. $e - 28 = 14$
2. $p - 14 = 28$
3. $a - 124 = 256$
4. $c - 12 = 19$
5. $p - 14.8 = 12.4$
6. $s - 15 = 20$
7. $x - 4\frac{1}{5} = 3\frac{1}{10}$
8. $g - 5.2 = 4.9$
9. $r - 239 = 57$
10. $5 = x - 9$
11. $1 = t - 6$
12. $n - \frac{1}{4} = 2\frac{1}{2}$
13. $y - \frac{4}{5} = \frac{1}{3}$
14. $t - \frac{7}{9} = \frac{1}{3}$
15. $a - 1.2 = 3.8$

Use addition or subtraction to solve each equation. Check your solution.

16. $b + 3 = 7.7$
17. $6 + c = 17$
18. $21 = x - 5$
19. $8.7 = d + 3.5$
20. $m - 6\frac{3}{4} = 2\frac{2}{5}$
21. $8.4 = s - 0.5$

Problem SOLVING

Write an equation and solve.

22. y decreased by 23 is 52.9.
23. v minus 11 is 4.4.
24. A number minus 7 is 15.
25. The difference of x and 9 is 27.
26. Mrs. Gonzalez baked cookies for her class. After 12 students ate a cookie, there were 6 left. How many cookies did Mrs. Gonzalez bake?
27. Can inverse operations be used to solve $27 - x = 9$? Explain your reasoning.

Constructed RESPONSE

28. Leila and Clark are solving the problem $x - 9 = 15$. Who is correct? Explain.

Leila thinks you should add 9 to each side of the equation.

Clark thinks you should subtract 9 from each side of the equation.

12.4 Solving Multiplication and Division Equations

Objective: to solve multiplication and division equations

The drawing at the right represents the equation $2x = 6$. Notice that the scale is balanced.

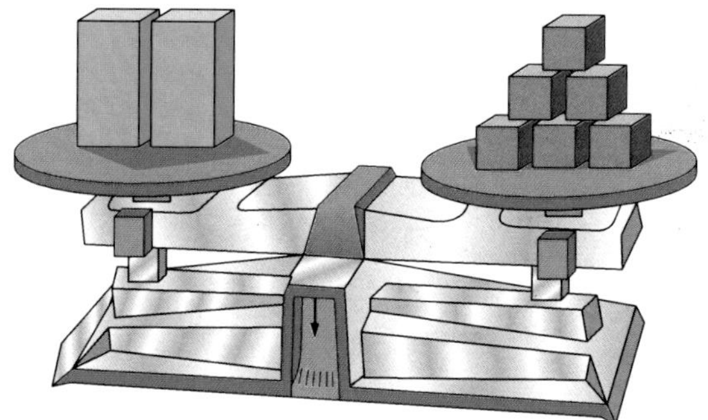

Divide the objects on each side of the balance into two equal groups.

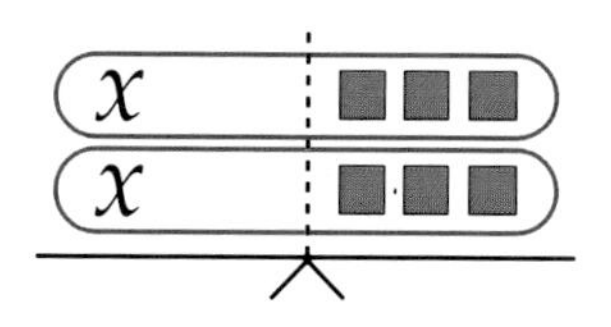

Since $2x = 6$, it follows that $x = 3$.

Just as you can subtract and add the same number to each side of an equation, you can also divide and multiply to isolate the variable to find its value.

Examples

If you divide each side of an equation by the same nonzero number, the two sides remain equal. This is called the **Division Property of Equality**.

A. Solve $7p = 63$.

$7p = 63$

$\frac{7p}{7} = \frac{63}{7}$ ➤ Divide each side by 7. The fraction bar means to divide.

$p = 9$

Check your solution.

If you multiply each side of an equation by the same number, the two sides remain equal. This is called the **Multiplication Property of Equality**.

B. Solve $\frac{x}{8} = 9$.

$\frac{x}{8} = 9$

$\frac{x}{8} \bullet 8 = 9 \bullet 8$ ➤ Multiply each side by 8.

$x = 72$

Check your solution.

Try THESE

Name the number by which you would divide each side to solve each equation.

1. $5x = 15$
2. $7m = 49$
3. $4d = 3.6$
4. $8.1 = 0.3b$

Name the number by which you would multiply each side to solve each equation.

5. $\frac{m}{9} = 7$
6. $\frac{r}{16} = 9$
7. $1.6 = \frac{m}{4}$
8. $9 = \frac{p}{6}$

Exercises

Solve each equation. Check your solution.

1. $9a = 45$
2. $42 = 6n$
3. $8p = 48$
4. $9c = 81$
5. $3z = 24$
6. $5r = 35$
7. $185 = 5c$
8. $88 = 8d$
9. $\frac{a}{7} = 6$
10. $\frac{x}{8} = 6$
11. $7 = \frac{s}{22}$
12. $\frac{w}{2} = 17$
13. $\frac{m}{15} = 4$
14. $9 = \frac{d}{41}$
15. $\frac{t}{13} = 9$
16. $\frac{r}{3} = 61$

Write an equation and solve.

17. The product of b and 11 is 66.
18. 60 times a number is 240.
19. A number multiplied by 7 is 91.
20. t divided by 7 is 212.

Problem SOLVING

21. A farmer sells 1,140 eggs and receives $200.00. How much did he receive per egg?
22. The area of a parallelogram is 48 square meters. The base is 6 meters. Write and solve an equation to find the height.
23. At 55 miles per hour, how long will it take to drive 1,430 miles?
24. Can inverse operations be used to solve $27 \div x = 9$? Explain your reasoning.

Test PREP

25. What is the solution of $4.5x = 22.5$?
    a. 4.5 b. 5 c. 100 d. 101.25
26. What is the solution of $\frac{x}{6.4} = 8$?
    a. 64 b. 48.4 c. 50.8 d. 51.2

Mid-Chapter REVIEW

Identify the property shown by each equation.

1. $3 \bullet 5 = 5 \bullet 3$
2. $6 + (5 + 7) = (6 + 5) + 7$
3. $120 + 0 = 120$
4. $3 + 8 = 8 + 3$

Solve each equation. Check your solution.

5. $k - 147 = 359$
6. $f + 81 = 108$
7. $30.1 + p = 42.5$
8. $7n = 56$
9. $\frac{s}{8} = 12$
10. $9m = 198$

12.5 Solving Two-Step Equations

Objective: to solve two-step equations

Denise bought two shirts from an online clothing store. She was charged \$5 for shipping and handling. Her total bill was \$35. How much did each shirt cost?

You can write an equation to model this situation.

$2s$	$+$	5	$=$	35
Denise bought two shirts.		The shipping was \$5.		The total cost was \$35.

This equation cannot be solved using only one operation. We need to use two operations. To solve a two-step equation, begin by undoing the addition or subtraction then undo the multiplication or division.

$2s + 5 - 5 = 35 - 5$ Subtract 5 from each side to undo the addition.

$2s = 30$ Divide each side by 2 to undo the multiplication.

$\frac{2s}{2} = \frac{30}{2}$ Simplify.

Check the solution.

$$2s + 5 = 35$$
$$2(15) + 5 = 35$$
$$35 = 35 \checkmark$$

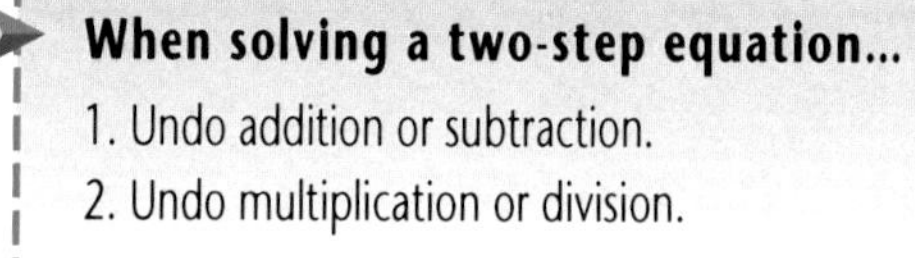

When solving a two-step equation...

1. Undo addition or subtraction.
2. Undo multiplication or division.

Each shirt that Denise bought cost \$15.

Examples

A. Solve $3x - 8 = 4$.

$3x - 8 + 8 = 4 + 8$ Add 8 to both sides.

$\frac{3x}{3} = \frac{12}{3}$ Divide both sides by 3.

$x = 4$

Check your solution.

$$3(4) - 8 = 4$$
$$4 = 4 \checkmark$$

B. Solve $\frac{x}{4} + 7 = 12$.

$\frac{x}{4} + 7 - 7 = 12 - 7$ Subtract 7 from both sides.

$4 \bullet \frac{x}{4} = 5 \bullet 4$ Multiply both sides by 4.

$x = 20$

Check your solution.

$$\frac{20}{4} + 7 = 12$$
$$12 = 12 \checkmark$$

Try THESE

State the first step and then the second step when solving each equation.

1. $5x - 9 = 24$
2. $14 = 7x - 8$
3. $\frac{x}{3} + 12 = 25$
4. $8 + 3x = 9$
5. $\frac{x}{6} - 3 = 12$
6. $10 = 2b - 8$

Exercises

Solve each equation. Check your solution.

1. $7x - 5 = 9$
2. $2b + 8 = 26$
3. $\frac{x}{4} - 7 = 3$
4. $6 = 2d - 10$
5. $\frac{f}{8} + 5 = 8$
6. $7 + 3x = 28$
7. $\frac{d}{2} - 5 = 6.5$
8. $3h + 4 = 10$
9. $4m - 6.8 = 5.6$
10. $2.2x + 3.1 = 5.3$
11. $1 = 2d - 9$
12. $3x - 7.4 = 5.5$
13. $\frac{x}{6} + 9 = 15$
14. $3 + 2g = 12$
15. $6 + \frac{h}{4} = 12$

Problem SOLVING

Write an equation and solve.

16. Three less than four times a number is 9. What is the number?

17. Fourteen is eight more than two times a number. What is the number?

18. Danielle went ice-skating. It cost $3 an hour and $5 for skate rental. If Danielle spent $11, how many hours did she ice-skate?

★19. It took Preston two more than three times the number of minutes to run a mile than his brother. If it took Preston 14 minutes, how long did it take his brother?

Constructed RESPONSE

20. Find the error in the problem to the right and explain how to correct it.

$$\frac{x}{3} + 2 = 5$$
$$\frac{x}{3} = 3$$
$$x = 1$$

MiXeD REVIEW

Compute.

21. $56.3 \bullet 4$
22. $24.6 \div 0.2$
23. $153.4 + 14.8$
24. $16.8 - 8.55$
25. $12.24 \div 3.6$
26. $28.54 - 14.8$
27. $0.002 \bullet 40.5$
28. $38.57 + 110.47$

12.6 Problem-Solving Strategy: Use an Equation

Objective: to solve word problems using equations

Allison has collected 100 leaves for her science project. She wants to display them in a notebook with four leaves on each page. Write an equation that can be used to find how many pages she needs.

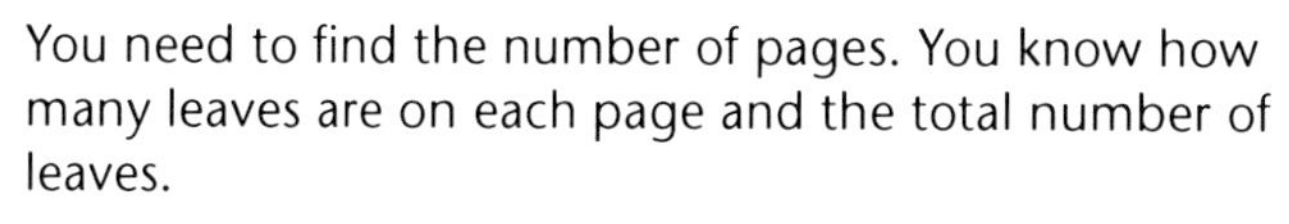

You need to find the number of pages. You know how many leaves are on each page and the total number of leaves.

4	•	p	=	100
leaves on each page		number of pages		total number of collected leaves

Solve the equation.

$$4p = 100$$

$$\frac{4p}{4} = \frac{100}{4}$$

$$p = 25$$

The solution is 25. Allison needs 25 pages.

If Allison uses 25 pages, each with 4 leaves, she will have a total of 100 leaves. The solution is correct.

Try THESE

Choose the equation that can be used to find the answer. Then solve.

1. Sara spends \$18 at the hardware store. She has \$13 left. How much money did Sara have before shopping?

   a. $18 - m = 13$ b. $13m = 18$

   c. $m - 18 = 13$ d. $m + 13 = 18$

2. A nursery plants 4 rows of trees with the same number of trees in each row. Find the number of trees in each row if 124 trees are planted.

   a. $t + 4 = 124$ b. $4t = 124$

   c. $124t = 4$ d. $\frac{4}{t} = 124$

Solve

Write an equation and solve.

1. Bill owns 66 CDs and DVDs. He owns 27 CDs. How many DVDs does Bill own?

2. Miss Mendoz ran 3.7 miles in 2 days. The first day she ran 1.8 miles. Find the miles that Miss Mendoz ran on the second day.

3. Mr. Sims distributes 30 pencils evenly among the 15 students in his class. How many pencils did each student receive?

4. Vicki Rush sells 5 times as many napkins as place mats. If she sells 105 napkins, how many place mats does she sell?

5. Jane has 73 more coins in her collection than Ned has in his. If Ned has 254 coins, how many does Jane have?

6. Four friends split a lunch bill. The total amount was $25. How much did each person pay?

7. Suppose a skateboarder travels a distance of 20 meters in 16 seconds. Find the average speed. Use the formula $d = rt$.

8. What is the average speed of a baseball if it takes 0.5 seconds for a pitcher to throw the baseball to home plate 60 feet away? Use $d = rt$.

9. Wesley biked three times as far as his sister rollerbladed. If his sister rollerbladed 1.5 miles, how far did Wesley bike?

10. The perimeter of a rectangle is 36 inches. Find its width if the length is 5 inches.

Test PREP

11. It cost $100 to join a gym plus a monthly fee. You spent $700 last year at the gym. How much was the monthly fee? Use m for monthly fee. Choose the equation that best fits this problem.

    a. $m + 100 = 700$ b. $100m = 700$ c. $12m + 100 = 700$ d. $12m = 700$

12. Three friends go to lunch and split the bill evenly. If the bill was $28.50, how much did each person pay?

    a. $9.00 b. $8.50 c. $8.00 d. $9.50

Chapter 12 Review

Language and Concepts

Choose the letter of the word to complete each statement.

1. When zero is added to a number and the same number results, it is the ________ Identity.
2. Operations that undo each other are ______ operations.
3. A ________ is used to represent some unknown quantity.
4. When you change the order in which the numbers are added, you are using the _______ Property of Addition.
5. When a number is added to each side of an equation and the two sides remain equal, it is the _______ Property of Equality.
6. When you change the way in which the factors are grouped, you are using the _____ Property of Multiplication.
7. When you multiply a number by 1 and the same number results, it is the ___________ Identity.

a. variable
b. inverse
c. Associative
d. Multiplicative
e. Commutative
f. Additive
g. Addition

Skills and Problem Solving

Identify the property shown by each equation. (Section 12.1)

8. $8 \cdot 7 = 7 \cdot 8$
9. $3 + 0 = 3$
10. $(5 + 9) + 3 = 5 + (9 + 3)$
11. $18 + 25 = 25 + 18$

Replace each ■ with a number. (Section 12.1)

12. $8 \cdot$ ■ $= 8$
13. $4 + 6 =$ ■ $+ 4$
14. $3 \cdot (5 \cdot 9) = ($■ $\cdot 5) \cdot 9$
15. ■ $+ 9 = 9$

Name the number you would add or subtract to each side to solve each equation. (Sections 12.2–12.3)

16. $x + 9 = 16$
17. $y - 12 = 25$
18. $z + 2.3 = 4.9$

Name the number you would multiply or divide each side by to solve each equation. (Section 12.4)

19. $12m = 84$
20. $\frac{a}{3} = 14$
21. $8 = 0.4x$

Solve each equation. Check your solution. (Sections 12.3–12.5)

22. $6 + p = 14$

23. $x - 2 = 11$

24. $4d = 20$

25. $\frac{x}{8} = 4$

26. $m + 12 = 25$

27. $0.5y = 10$

28. $14 = p - 5$

29. $h + 1\frac{1}{2} = 3\frac{3}{4}$

30. $\frac{d}{2.3} = 5.8$

31. $64 = 4p$

32. $12.7 + r = 10.9$

33. $4x - 10 = 6$

34. $15 = 3x + 6$

35. $\frac{r}{2} + 6 = 14$

36. $7 + 8d = 31$

Write an equation and solve. (Sections 12.4–12.6)

37. Three times a number increased by 8 is 14. What is the number?

38. The product of a number and 9 is 63. What is the number?

39. The difference between x and 7 is 12. What is the number?

40. Twenty-four is five times a number more than four. What is the number?

41. Porter can throw a football twice as far as his brother Parker. Porter can throw a football 25 yards. How far can Parker throw the ball?

42. Rachel bought three times as many apples as oranges. If she bought 21 apples, how many oranges did she buy?

43. The area of a parallelogram is 50 m^2. The base is 5 m. What is the height?

44. Gabriella went skiing. She paid $35 to rent skis and $15 an hour to ski. If she paid a total of $95, how many hours did she ski?

Chapter 12 Test

Identify the property shown by each equation.

1. $5 \bullet 14 = 14 \bullet 5$
2. $35 + 0 = 35$
3. $18 + 25 + 12 = 25 + 18 + 12$
4. $8 \bullet (9 \bullet 5) = (8 \bullet 9) \bullet 5$

Replace each ■ with a number.

5. $12 + 7 = 7 + ■$
6. $4 \bullet ■ = 4$
7. $8 \bullet (9 \bullet 3) = (8 \bullet ■) \bullet 3$
8. $14 \bullet 6 = ■ \bullet 14$

Find each sum or product mentally.

9. $14 + 8 + 6$
10. $4 \bullet 3 \bullet 25$
11. $24 + 6 + 27$

Solve each equation. Check your solution.

12. $8 + p = 17$
13. $m - 9 = 14$
14. $7s = 49$
15. $4t = 64$
16. $6.3 + a = 11.4$
17. $c - 11.9 = 15.3$
18. $t - 4\frac{2}{5} = 3\frac{2}{3}$
19. $\frac{h}{8} = 14$
20. $0.2h = 20$
21. $15 = 2x - 3$
22. $\frac{r}{5} - 4 = 11$
23. $4x + 12 = 22$

Write an equation and solve.

24. The product of a number and 7 is 28. What is the number?
25. A number times 5 added to 9 is 14. What is the number?
26. The lacrosse team held a bake sale. They earned $0.25 for every brownie they sold. If they earned $100, how many brownies did they sell?
27. Jane and three friends went to the movies. They bought tickets and a tub of popcorn. If the popcorn cost $2.50, and they spent a total of $34.50, how much did each ticket cost? Explain.
28. Jessa's age is twice her sister's age less two years. When Jessa is ten years old, how old will her sister be?
29. Howard collects stamps and coins. He has five more than three times as many stamps as coins. If he has 80 coins, how many stamps does he have?

Change of Pace

Sets

A set is a collection of elements. The elements, or members of a set, can be objects like coins and books or mathematical items like geometric figures or numbers.

Let set A equal the multiples of 8. Let set B equal the multiples of 10. These sets are listed in set notation.

$A = \{0, 8, 16, 24, 32, 40, 48, \ldots\}$ Read this as A equals the set containing 0, 8, 16, 24, 32, 40, 48, and so on.

$B = \{0, 10, 20, 30, 40, 50, \ldots\}$

You can use the intersection of the sets to find the common multiples and the least common multiple. The intersection of the two sets is the sets of elements common to both sets. The symbol for intersection is $\cap$.

$A \cap B = \{0, 40, 80, 120, \ldots\}$ Read this as A intersection B equals the set containing 0, 40, 80, 120, and so on.

The LCM of 8 and 10 is 40.

The union of two sets is the set of elements in either set or in both sets. The symbol for union is $\cup$.

Let set $C = \{0, 2, 4, 6, 8\}$ and set $D = \{0, 3, 6, 9\}$. Find $C \cup D$.

$C \cup D = (0, 2, 3, 4, 6, 8, 9\}$ Read this as C union D equals the set containing 0, 2, 3, 4, 6, 8, and 9.

Find the intersection and union of each pair of sets.

1. $M = \{4, 5, 6, 7, 8\}$
$N = \{0, 1, 2, 3, 4\}$

2. $R = \{0, 2, 4, 6, 8\}$
$S = \{1, 3, 5, 7, 9\}$

3. $K = \{0, 2, 4, 6, 8\}$
$L = \{0, 3, 6, 9, 12\}$

4. Let set X contain all the factors of 20. Let set Y contain all the factors of 45. Find $X \cap Y$. Give the GCF of 20 and 45.

5. Let $E = \{0, 2, 4, 6, 8, \ldots\}$ and $O = \{1, 3, 5, 7, 9, \ldots\}$. Find $E \cup O$. Describe in words sets E, O, and $E \cup O$.

Suppose F = {rectangles}, G = {rhombi}, H = {squares}, and J = {parallelograms}. Find each of the following.

6. $F \cup H$

7. $G \cap J$

8. $G \cup H$

9. $H \cup J$

10. $F \cap G$

11. $F \cup J$

Cumulative Test

1. Find $y + 23.65$, if $y = 9.08$.
   a. 9.08
   b. 14.5
   c. 23.65
   d. 32.73

2. Find p if $p - 45 = 12$.
   a. 33
   b. 45
   c. 57
   d. 69

3. What is the prime factorization of 36?
   a. $2^4 \times 3$
   b. $2^2 \times 3^2$
   c. $3^2 \times 5$
   d. none of the above

4. This is a drawing of a ____.
   a. quadrilateral
   b. circle
   c. triangle
   d. none of the above

5. What is the rule?

Input	6	12	18
Output	12	24	36

   a. add 6
   b. multiply by 0
   c. multiply by 2
   d. subtract 12

6. Multiply.
   $6\frac{3}{7} \times 1\frac{1}{9} =$ ____
   a. $6\frac{1}{21}$
   b. $7\frac{1}{4}$
   c. $8\frac{2}{3}$
   d. none of the above

7. In a classroom discussion, Emile says that subtracting 3 from 7 is the same as subtracting 7 from 3. This statement is ____.
   a. false
   b. false, only if you are subtracting fractions
   c. true
   d. true, only if you are subtracting whole numbers

8. Marta needs 450 points to get an A in math. On her first 4 exams, her grades were 87, 89, 91, and 93. What score must she earn on the last exam to get an A in math?
   a. 88
   b. 90
   c. 92
   d. 94

9. Mr. Salinas is planning a trip. What is one of the things he needs to know?
   a. the color of his sofa
   b. the price of gasoline
   c. the price of onions
   d. the weight of his children

10. Suppose you drove for $3\frac{1}{2}$ hours at 52 mph. Which formula would you use to determine how many miles you drove?
   a. $r = d \div t$
   b. $m = s \times g$
   c. $d = r \times t$
   d. none of the above

Integers

Sarah Mesrobian
Virginia

13.1 Integers

Objective: to find the opposite and absolute value of integers; to compare and order integers

The temperature 5 degrees *below* zero is written -5. Similarly, the temperature 5 degrees *above* zero can be written as +5. Numbers such as -5 and +5 are called **integers**. The set of integers is listed below.

… -5, -4, -3, -2, -1, 0, 1, 2, 3, 4, 5, …

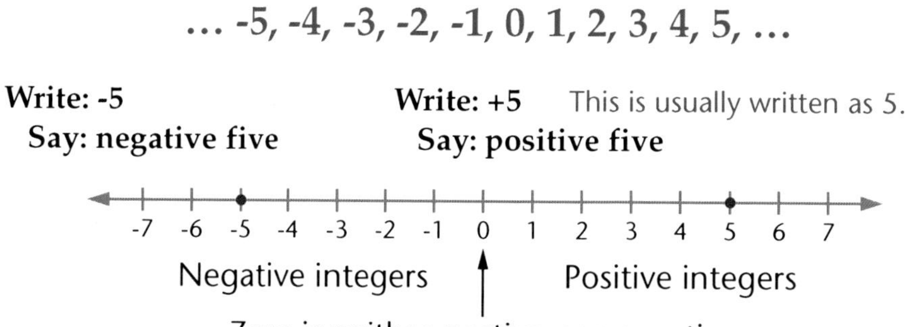

The numbers -5 and 5 are **opposites**. Two numbers are opposites if they are the same distance from zero in opposite directions. The **absolute value** of a number is its distance from 0 on a number line. You can write the "absolute value of negative five" as $|-5|$. Opposites have the same absolute value.

Examples

A. Find the opposite of 4.

B. Find $|-3|$.

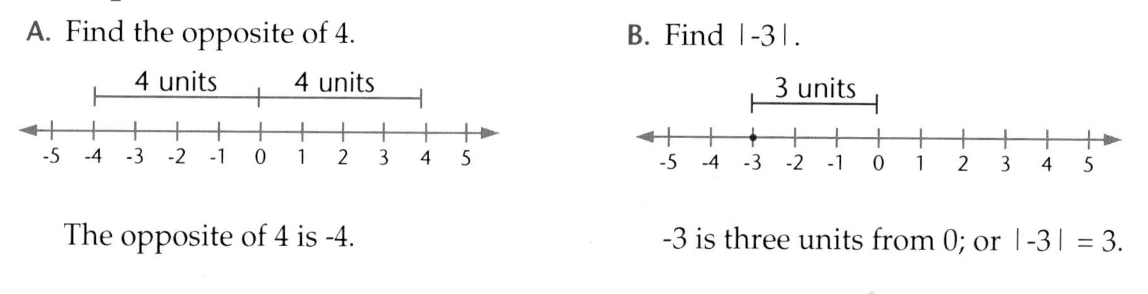

The opposite of 4 is -4.

-3 is three units from 0; or $|-3| = 3$.

You can also compare and order integers. On a number line, the greater integer is to the right of the lesser integer.

C. Compare 2 ■ -5.

2 is to the right of -5, so 2 > -5.

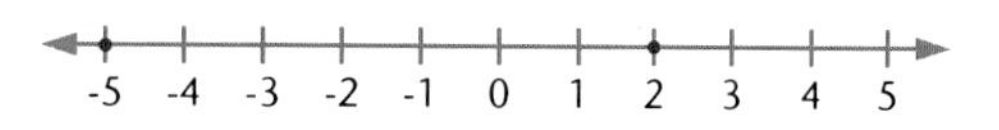

D. Compare -7 ■ -4.

-7 is to the left of -4, so -7 < -4.

-7 -6 -5 -4 -3 -2 -1 0 1 2 3 4 5 6 7

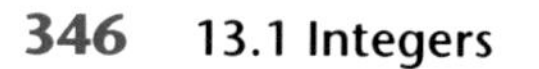

E. Order 3, -5, 0, -2, and 6 from least to greatest.

You can look on a number line to help you order the integers.

-10 -9 -8 -7 -6 -5 -4 -3 -2 -1 0 1 2 3 4 5 6 7 8 9 10

The order from least to greatest is -5, -2, 0, 3, and 6.

Try THESE

Write an integer to describe each situation.

1. deposited $25 into your bank account
2. 12 degrees below zero

Exercises

Name the opposite of each integer.

1. -11 2. -23 3. 41 4. 100 5. 6 6. -5

Name the absolute value of each integer.

7. 9 8. -10 9. 2 10. -15 11. 12 12. -9

Compare using $<$, $>$, or $=$.

13. -5 ● -7 14. 5 ● 7 15. -16 ● -33 16. 57 ● -96

Order from least to greatest.

17. 4, 10, 8
18. 7, -3, -6,
19. -5, -9, -1
20. -2, -11, -4, 3, 11, -8
21. 23, 15, -17, -19, -14, 13

Order from greatest to least.

22. 3, 0, 5
23. -8, -6, -12
24. -7, 2, -9, -4
25. -18, -14, 15, 14, -8
26. 3, -3, 13, 0, -13

Write *true* or *false*.

27. $-7 > -9$ 28. $12 > 14$ 29. $-6 > -4$ 30. $0 > -15$

Problem SOLVING

31. If -16 stands for 16 meters in one direction from a point, then what integer stands for 25 meters in the opposite direction from the same point?

32. If -7 represents 7 meters south, what does +9 represent?

33. What is the absolute value of zero?

34. Can a number have a negative absolute value? Explain your reasoning.

35. Golf scores can be positive or negative, but the lowest score wins. A negative score means that many shots under par and a positive score means that many shots above par. Put the following golf scores in order from least to greatest: -8, 4, -2, 3, 0, -6, 5.

★36. A dear grandfather named Dunn is two times as old as his son. Twenty-five years ago their age ratio was actually three to one. How old is grandfather?

37. The perimeter of a square-shaped yard is 68.4 feet. What is the area of the yard?

38. Order 4, -6, 0, and -3 from least to greatest.

a. 4, 0, -3, -6 b. -6, -3, 0, 4 c. 0, -3, 4, -6 d. 4, -3, 0, -6

39. Which statement is *not* true?

a. $-4 < -6$ b. $2 > -3$ c. $-2 > -3$ d. $-7 < 5$

MIXED REVIEW

Add or subtract. Write each sum or difference in simplest form.

40. $\frac{1}{9} + \frac{5}{7}$ 41. $\frac{1}{3} + \frac{5}{6}$ 42. $\frac{3}{5} - \frac{1}{3}$ 43. $\frac{11}{12} - \frac{2}{3}$

44. $4\frac{1}{8} + 3\frac{5}{8}$ 45. $6\frac{2}{7} + 8\frac{3}{14}$ 46. $8\frac{7}{9} - 1\frac{4}{9}$ 47. $5\frac{8}{9} - 4$

Cumulative Review

Compute.

1. $\begin{array}{r} 9{,}301 \\ +\ 789 \\ \hline \end{array}$

2. $\begin{array}{r} 50.07 \\ -\ 27.99 \\ \hline \end{array}$

3. $\begin{array}{r} 432 \\ \times\ 48 \\ \hline \end{array}$

4. $3.2\overline{)6.624}$

Rename each decimal as a fraction or mixed number in simplest form.

5. 0.7
6. 0.53
7. 0.4
8. 1.5
9. 1.25

Compute. Write each answer in simplest form.

10. $\frac{1}{5} + \frac{1}{5}$
11. $1\frac{1}{8} + \frac{3}{8}$
12. $\frac{5}{7} - \frac{3}{7}$
13. $\frac{8}{9} - \frac{5}{9}$
14. $\frac{2}{3} \times \frac{1}{5}$
15. $\frac{3}{14} \times \frac{7}{9}$
16. $\frac{1}{2} \div \frac{1}{8}$
17. $1\frac{1}{2} \div \frac{2}{3}$

Replace each ■ so the ratios are equivalent.

18. $\frac{1}{2} = \frac{■}{8}$
19. $\frac{1}{3} = \frac{3}{■}$
20. $\frac{10}{15} = \frac{■}{3}$
21. $\frac{8}{24} = \frac{■}{3}$

Find the area. Use 3.14 for π.

22. 23. 24.

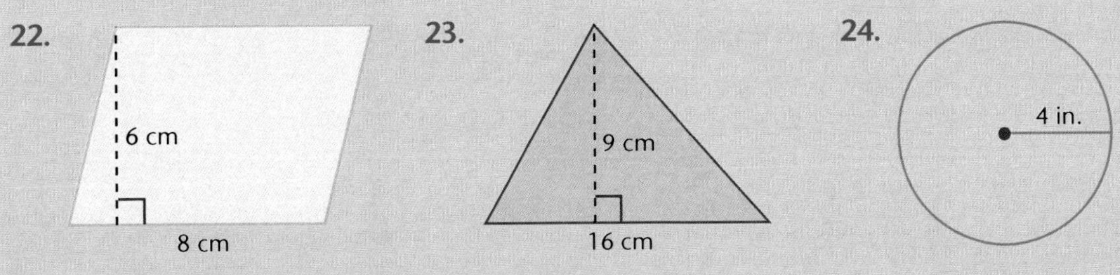

Solve.

25. Sharon bowled three games. Her scores were 117, 125, and 142. What was her average score?

26. William needs $1\frac{3}{8}$ yards of wool for pants. He needs $2\frac{1}{4}$ yards for a jacket. How many yards does he need in all?

27. Twelve cans of soup cost $5.40. What is the cost of four cans?

28. A bag contains 2 red marbles and 3 yellow marbles. What is the probability of choosing a yellow marble?

13.2 Adding Integers

Objective: to add integers

Exploration Exercise

1. Use the red counters to represent positive integers and the yellow counters to represent negative integers. Each red counter represents +1 and each yellow counter represents -1.

   (+) = +1

   (−) = -1

2. Place three red counters on your desk. What integer is represented by the counters?

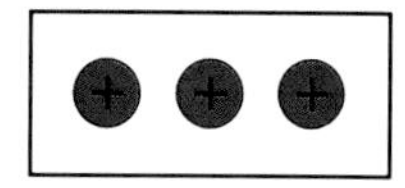

3. Place three yellow counters on your desk. What integer is represented by the counters?

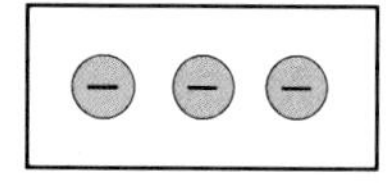

4. Pair up each red counter on your desk with a yellow counter. Each pair is equal to zero. It is called a zero pair.

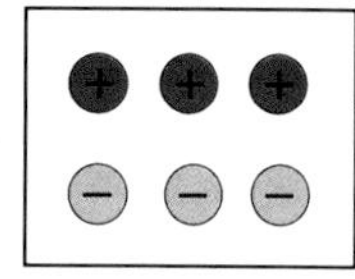

5. Find 5 + 3. Place five red counters on your desk. Add three red counters. How many red counters do you have? What integer does this represent?

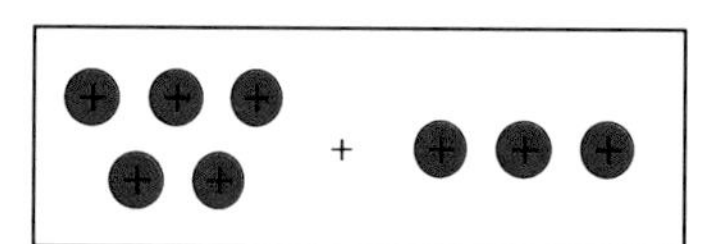

6. Find -5 + -3. Place five yellow counters on your desk. Add three yellow counters. How many yellow counters do you have? What integer does this represent?

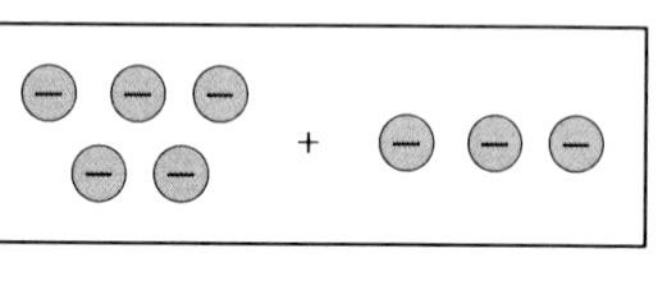

7. Find 5 + -3. Place five red counters on your desk. Add three yellow counters. Match up the yellow and red counters. Remove the zero pairs. How many red counters are left?

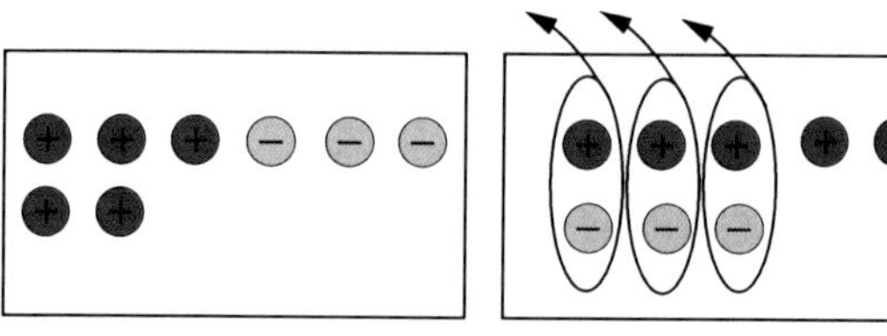

8. Find -5 + 3. Place five yellow counters on your desk. Add three red counters. Match up the yellow and red counters. Remove the zero pairs. How many yellow counters are left?

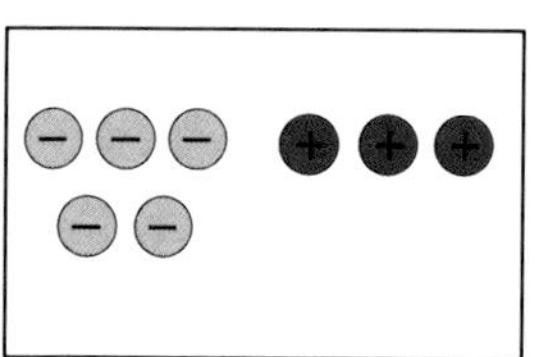

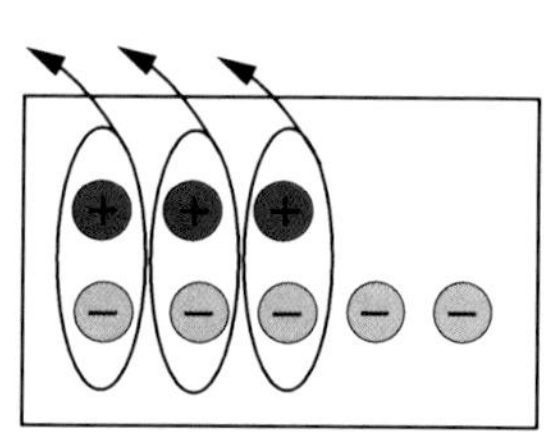

As you saw from the exploration you can use counters to add integers. You can also use a number line to add integers.

Add +3 + +4.

Step 2: You move **right** 4 units.

Step 1:
You start at 0 and move **right** 3.

Start

End

0 1 2 3 4 5 6 7 8

Step 3:
You end at 7. So 3 + 4 = 7.

Add -2 + -4.

Then you move **left** 4 units.

You start at 0 and move **left** 2.

End

Start

-6 -5 -4 -3 -2 -1 0 1 2

You end at -6. So -2 + -4 = -6.

Adding Integers with the Same Sign

When you add two positive integers, the sum is positive.

$$1 + 3 = 4$$

When you add two negative integers, the sum is negative.

$$-3 + -6 = -9$$

Add 1 + -3.

Then you move **left** 3 units.

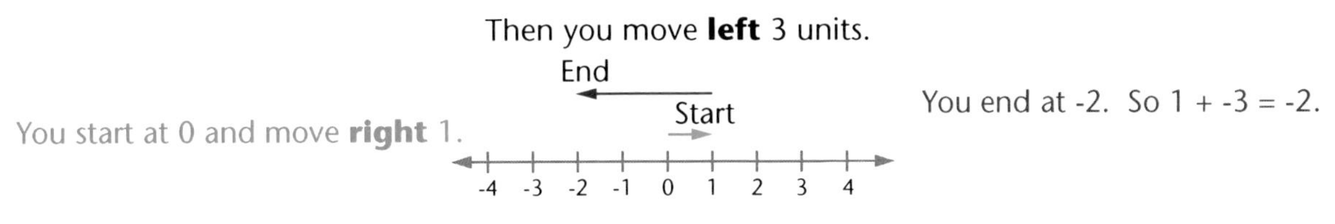

Add –4 + 5.

Then you move **right** 5 units.

End

Start

You start at 0 and move **left** 4.

-4 -3 -2 -1 0 1 2 3 4

You end at 1. So -4 + 5 = 1.

Adding Integers with Different Signs

To add integers with different signs, find the absolute value of each integer. Then subtract the smaller absolute value from the larger absolute value. The sum has the sign of the integer with the greatest absolute value.

Examples: -7 + 5 = -2 6 + -1 = 5

Examples

A. Add -7 + -3.

-7 + -3 = -10 The sum of two negative integers is negative.

B. Add -8 + 6.

| -8 | = 8
| 6 | = 6 Find the absolute value of each integer.

8 – 6 = 2 Subtract the absolute values.

-8 + 6 = -2 The sum has the sign of the integer with the greater absolute value.

Try THESE

Write an addition sentence for each set of counters.

1. 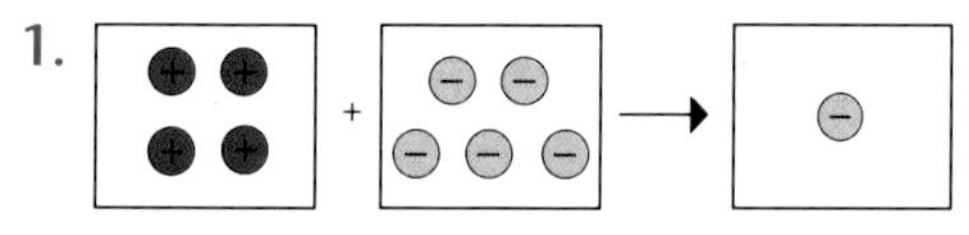

2. 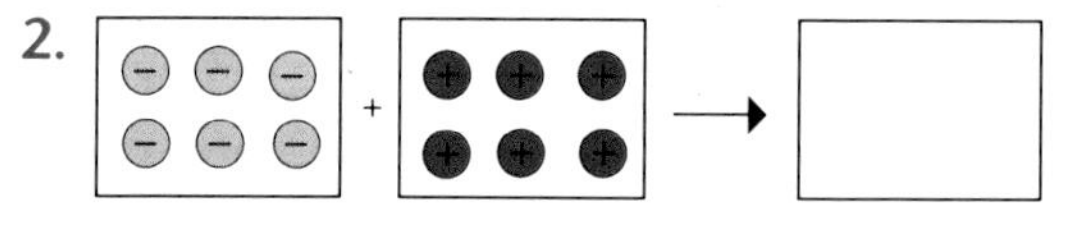

Exercises

Use counters to find each sum.

1. -3 + -3
2. -5 + -4
3. -8 + 5
4. 6 + -4
5. 3 + -9
6. 2 + 5
7. -3 + -2
8. -5 + -1

Use a number line to find each sum.

9. 4 + 8
10. -3 + 2
11. -2 + 3
12. -5 + 8
13. 6 + -3
14. -1 + -7
15. 8 + -3
16. 2 + 9

Add.

17. $-5 + 2$
18. $12 + -12$
19. $5 + -2$
20. $-10 + -10$
21. $-8 + 0$
22. $-10 + -4$
23. $-12 + 9$
24. $-5 + -7$
25. $14 + -6$
26. $-3 + -10$
27. $-23 + -18$
28. $-8 + -12$
29. $24 + -7$
30. $-18 + 6$
31. $15 + -18$
32. $-12 + 15$
33. $-4 + 7 + -1$
34. $2 + 8 + -5$
35. $-5 + -4 + -6$
36. $-5 + 3 + 2$

Problem SOLVING

37. A golfer finishes 9 holes of golf at 3 under par. On the second 9 holes he is 2 under par. How far under par is he for the whole 18 holes?

38. A summer drought causes a lake to drop 6 inches below its normal level. If a storm adds 2 inches to the lake's level, how far below level is it?

MixeD REVIEW

39. $\frac{1}{4} \times \frac{4}{5}$
40. $\frac{1}{2} \times 1\frac{1}{3}$
41. $10 \times 2\frac{4}{5}$
42. $10\frac{1}{7} \times 10\frac{5}{9}$
43. $\frac{9}{10} \div \frac{1}{3}$
44. $\frac{5}{8} \div \frac{15}{16}$
45. $5 \div \frac{3}{4}$
46. $3\frac{3}{7} \div \frac{3}{4}$

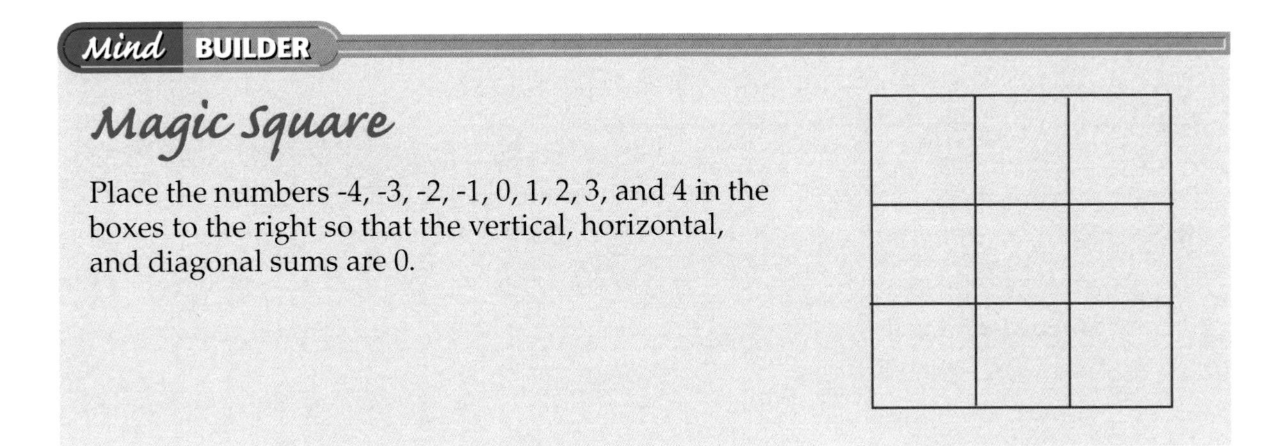

Mind BUILDER

Magic Square

Place the numbers -4, -3, -2, -1, 0, 1, 2, 3, and 4 in the boxes to the right so that the vertical, horizontal, and diagonal sums are 0.

13.3 Subtracting Integers

Objective: to subtract integers

Exploration Exercise

We will again use red counters for positive integers and yellow counters for negative integers in our exploration exercise.

1. Find 5 + -2. Place five red counters on your desk. Add two yellow counters. Match up the zero pairs and remove them. How many counters are left? What integer does this represent?

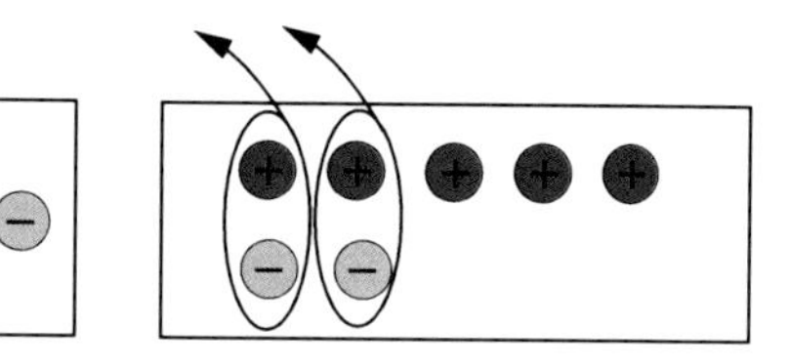

2. Find 5 – 2. Place five red counters on your desk. Take two counters away. How many counters are left? What integer does this represent?

3. Find -5 + -2. Place five yellow counters on your desk. Add two yellow counters. How many counters are left? What integer does this represent?

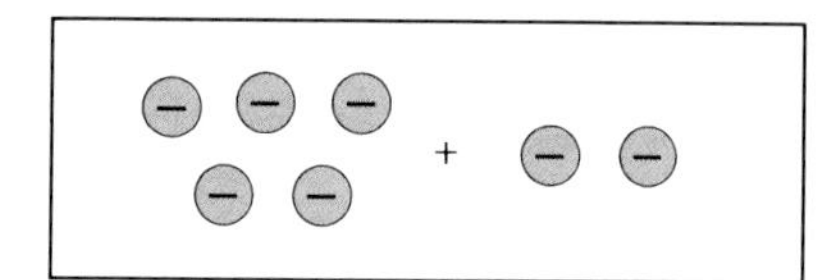

4. Find -5 – 2. Place five yellow counters on your desk. Add two zero pairs since there are no positive counters. Take away two red counters. How many counters are left? What integer does this represent?

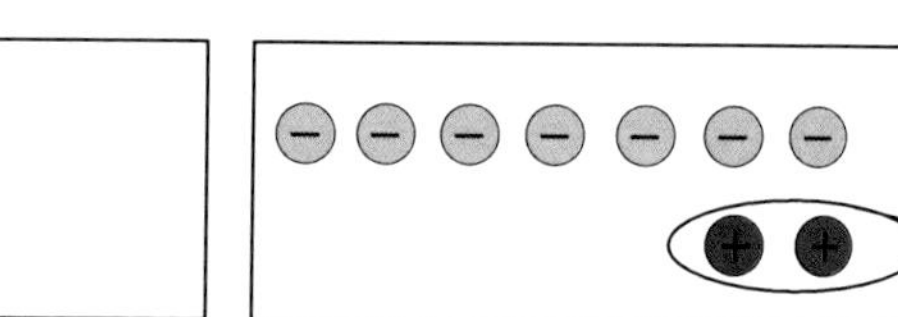

In the exploration exercise you find that subtracting two numbers is the same as adding the opposite.

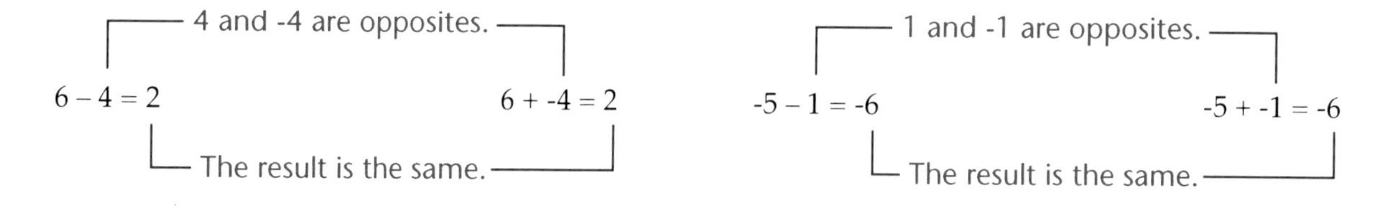

Subtracting Integers

To subtract an integer, add its opposite.

Example: $8 - 6 = 2$ $8 + -6 = 2$

Examples

A. Find -8 – 3.

-8 – 3 = -8 + -3 To subtract 3, add its opposite.

= -11 Simplify.

B. Find -11 – -6.

-11 – -6 = -11 + 6 To subtract -6, add its opposite.

= -5 Simplify.

Try THESE

Use counters to model each subtraction problem. Then write each difference.

1. 7 – 3
2. -4 – -2
3. -6 – -3
4. -2 – -1
5. -5 – 2
6. -3 – -4
7. 2 – -3
8. -3 – -1

Exercises

Subtract.

1. 9 – 4
2. 5 – 7
3. -6 – 8
4. -3 – -7
5. 7 – -4
6. -11 – 9
7. -7 – 0
8. 4 – 10
9. -5 – -2
10. 9 – 12
11. -5 – 7
12. 3 – 10
13. 2 – -3
14. -8 – 9
15. 14 – 6
16. -10 – -10
17. 5 – -1
18. -4 – -6
19. 8 – -3
20. 8 – 20
21. 14 – 28
22. 9 – -23
23. -25 – 16
24. -18 – 21

Evaluate each expression for $a = -2$, $b = 5$, and $c = -4$.

25. $a - c$
26. $a - b$
27. $b - a$
28. $c - a$

Tell whether each statement is *always*, *sometimes*, or *never* true. Give an example or counter example for each.

29. positive – positive = positive
30. positive – negative = positive
31. negative – positive = negative
32. negative – negative = negative

13.4 Multiplying Integers

Objective: to multiply integers

Exploration Exercise

What happens when you multiply a positive number by a negative number? What happens when you multiply two negative numbers?

1. Imagine walking on a giant number line. Multiplying 2 • 3 is like stepping three spaces forward two times. First you step from 0 to 3, and then to 6. So 2 • 3 = 6.

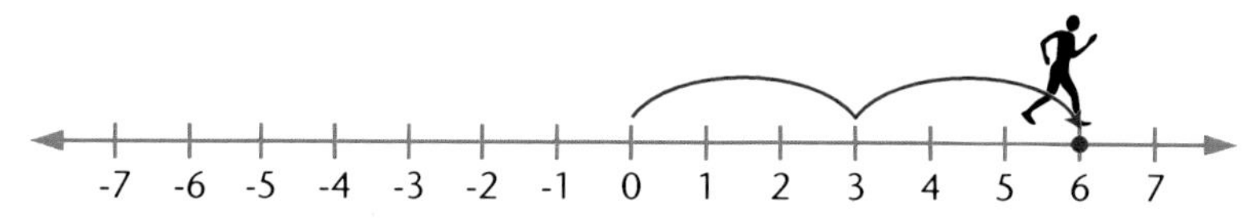

2. Now consider multiplying 2 • -3. You take two steps that are 3 spaces long, but you step in the negative direction, or backward. Step from 0 to -3 and then to -6. What is 2 • -3?

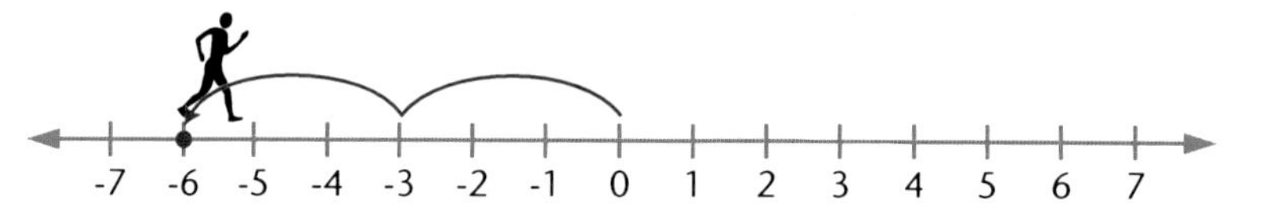

3. How can you multiply -2 • 3? Your steps are 3 spaces long and forward, but since you take -2 steps, you turn around and face the other way. Heading forward, you step from 0 to -3 to -6. So -2 * 3 = -6.

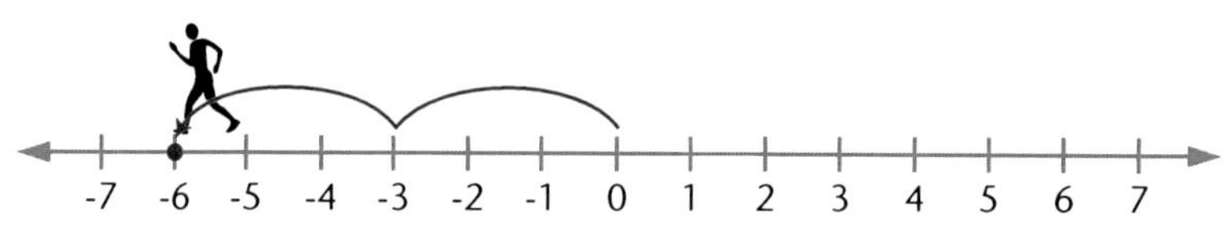

4. What is -2 • -3? Your steps are 3 spaces long and backward, but since you are taking -2 steps, you face the other way. So you step backward from 0 to 3 to 6. What is -2 • -3?

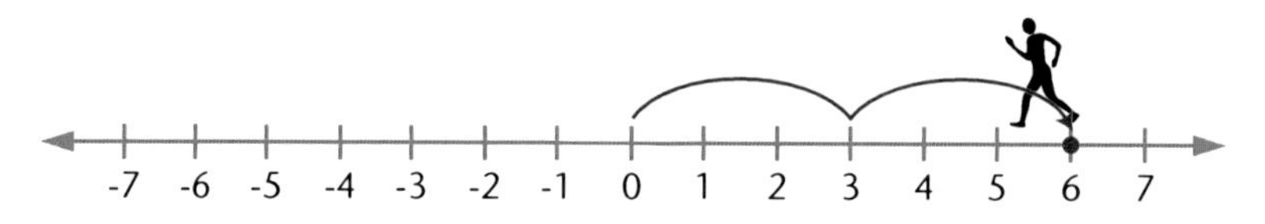

5. What pattern do you see?

To multiply integers the following rules apply.

Multiplying Integers

The product of two integers with the **same** sign is **positive.**

Examples: 4 • 6 = 24 -3 • -6 = 18

The product of two integers with **different** signs is **negative.**

Examples: -2 • 6 = -12 5 • -3 = -15

Examples

A. Find 5 • -8.

5 • -8 = -40

different signs = negative product

B. Find -6 • -4.

-6 • -4 = 24

same signs = positive product

Try THESE

Use counters to model each multiplication problem. Then write the product.

1. 3 • -5
2. 4 • -3
3. -2 • -2
4. -3 • 1

Exercises

Multiply.

1. 8 • -9
2. -5 • -3
3. -6 • 0
4. -5 • 9
5. 8 • -5
6. 4 • -2
7. -9 • 4
8. 7 • 6
9. -6 • -9
10. -8 • -6
11. -12 • 3
12. 7 • -18
13. -5 • 6
14. -1 • -8
15. -10 • 5
16. 11 • -6

Evaluate if $x = -3$, $y = -8$, and $z = 7$.

17. $4y$
18. $5x$
19. xy
20. yz

Problem SOLVING

21. Explore what happens when you multiply more than two integers.
    a. Is the product of three positive integers positive or negative?
    b. Is the product of three negative integers positive or negative?
    c. Is the product of four negative integers positive or negative?

13.5 Dividing Integers

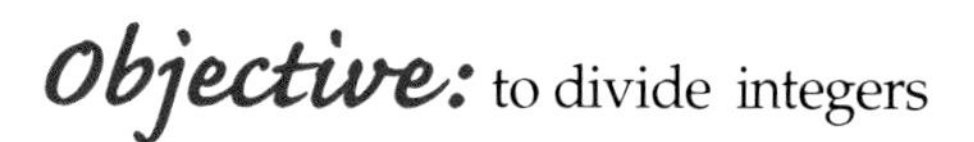 to divide integers

Exploration Exercise

1. Complete each division problem.

a. $2 \bullet 3 = 6$
   $6 \div 3 = \blacksquare$
   $6 \div 2 = \blacksquare$

b. $2 \bullet -3 = -6$
   $-6 \div -3 = \blacksquare$
   $-6 \div 2 = \blacksquare$

c. $-2 \bullet -3 = 6$
   $6 \div -3 = \blacksquare$
   $6 \div -2 = \blacksquare$

d. $-2 \bullet 3 = -6$
   $-6 \div 3 = \blacksquare$
   $-6 \div -2 = \blacksquare$

2. Write a statement that explains how the sign of the quotient is related to the signs of the dividend and divisor.

The above exploration suggests the following rule.

Dividing Integers

The quotient of two integers with the **same** sign is **positive.**

Examples: $10 \div 2 = 5$ $-12 \div -3 = 4$

The quotient of two integers with **different** signs is **negative.**

Examples: $-14 \div 7 = -2$ $24 \div -6 = -4$

Examples

A. Find $-18 \div -3$.

$\mathbf{-18 \div -3 = 6}$ same signs = postive quotient

B. Find $-36 \div 9$.

$\mathbf{-36 \div 9 = -4}$ different signs = negative quotient

Try THESE

Choose the correct value for each expression.

1. $-20 \div 4$ a. -5 b. 5
2. $-56 \div -8$ a. -7 b. 7
3. $18 \div -6$ a. -3 b. 3
4. $25 \div -5$ a. -5 b. 5
5. $-48 \div -8$ a. -6 b. 6
6. $-21 \div 3$ a. -7 b. 7

Exercises

Divide.

1. $18 \div 2$
2. $-36 \div -6$
3. $81 \div -9$
4. $54 \div -9$
5. $-8 \div 2$
6. $-12 \div 3$
7. $-45 \div -9$
8. $35 \div 7$
9. $-63 \div -7$
10. $-32 \div 8$
11. $-25 \div -5$
12. $6 \div 2$
13. $-30 \div -6$
14. $64 \div -8$
15. $-112 \div 4$
16. $-85 \div -17$

Use the order of operations to simplify each expression.

17. $-12 \div 4 + 2$
18. $3 \bullet -5 - 6$
19. $5 + 9 \div -3$
20. $-24 \div -6 + 3 \bullet -2$
21. $-8 + 3 + -2 \bullet 6$
22. $32 \div -8 + -4 \bullet -3$

Problem SOLVING

23. From 9:00 P.M. until midnight, the temperature change was -9°. Find the average change per hour.

★24. List all of the factors of -18.

Test PREP

25. What is the quotient of -48 divided by -6?

a. -8 b. 54 c. -54 d. 8

26. Which expression has the greatest quotient?

a. $-20 \div -5$ b. $8 \div -4$ c. $-6 \div -3$ d. $12 \div -3$

Mid-Chapter REVIEW

Name the absolute value of each integer.

1. 6
2. -9
3. -21
4. 35

Add or subtract.

5. $-3 + -9$
6. $12 - 15$
7. $-7 + 8$
8. $-5 - -10$

Multiply or divide.

9. $-4 \bullet -7$
10. $-63 \div 9$
11. $8 \bullet -9$
12. $-42 \div -6$

13.6 Solving Equations with Integers

Objective: to solve equations with integers

The same rules used to solve equations with whole numbers are used to solve equations with integers.

Examples

Solve each equation. Check your solution.

A. $x + -3 = 7$

$x + -3 - -3 = 7 - -3$ Subtract -3 from each side.

$x + -3 + 3 = 7 + 3$

$x + 0 = 10$

$x = 10$

Check:

$x + -3 = 7$

$10 + -3 = 7$ Replace x with 10.

$7 = 7$ ✔ The solution is 10.

B. $y - 5 = -8$

$y - 5 + 5 = -8 + 5$ Add 5 to each side.

$y = -3$

Check:

$y - 5 = -8$

$-3 - 5 = -8$ Replace y with -3.

$-3 + 5 = -8$

$-8 = -8$ ✔ The solution is -3.

C. $-2n = -8$

$\frac{-2n}{-2} = \frac{-8}{-2}$ Divide each side by -2.

$n = 4$

Check:

$-2n = -8$

$-2 \times 4 = -8$ Replace n with 4.

$-8 = -8$ ✔ The solution is 4.

D. $\frac{k}{7} = -3$

$\frac{k}{7} \times 7 = -3 \times 7$ Multiply each side by 7.

$k = -21$

Check:

$\frac{k}{7} = -3$

$\frac{-21}{7} = -3$ Replace k with -21.

$-3 = -3$ ✔ The solution is -21.

Try THESE

State how to solve each equation.

1. $y + 3 = -1$
2. $a - -4 = 8$
3. $-4x = 16$
4. $\frac{m}{2} = -6$

Exercises

Solve each equation. Check your solution.

1. $y + 2 = -1$
2. $x + -2 = 5$
3. $r - 3 = -2$
4. $y - 5 = -9$
5. $b - 3 = -4$
6. $a + -4 = -9$
7. $-3k = 18$
8. $4m = -20$
9. $\frac{b}{5} = -8$
10. $\frac{m}{9} = 3$
11. $5a = -35$
12. $\frac{n}{-4} = 20$
13. $k + -5 = -2$
14. $n + -1 = 5$
15. $x - -1 = 5$
16. $m - 6 = -4$
17. $-7x = -56$
18. $-9k = 36$
19. $\frac{r}{-8} = -15$
20. $\frac{a}{-6} = -6$

Problem SOLVING

21. The product of a number and -20 is 60. Find the number.
22. The sum of a number and -30 is -70. Find the number.
23. When 10 is subtracted from a number, the result is -90. Find the number.
24. When a number is divided by -10, the result is 80. Find the number.

★25. Find the next term of the sequence -18, 6, -2, $\frac{2}{3}$, ■.

Mind BUILDER

Puzzles

Copy each diagram. Replace each ■ with the product of the numbers in the row or column. Then replace each ? with the product of the numbers in the bottom row or the column at the right.

1.

5	-4	■
-3	2	■
■	■	?

2.

-1	-6	■
5	-2	■
■	■	?

3.

-3	-5	■
-6	-8	■
■	■	?

4. Make a puzzle of your own.

Objective: to solve inequalities

To ride the roller coaster at the Kiddie Land Amusement Park you must be at least 45 inches tall.

In the last chapter you learned that an equation is a sentence with an equal sign. Another type of mathematical sentence is an **inequality**. An inequality contains $<$, $>$, $\leq$, $\geq$, or $\neq$. Each symbol has a special meaning.

Symbol	Meaning
$>$	Greater than
$<$	Less than
$\geq$	Greater than or equal to
$\leq$	Less than or equal to
$\neq$	Not equal to

Examples

A. Write an inequality expressing the minimum height to ride a roller coaster at the Kiddie Land Amusement Park.

Let h = a person's height.

person's height	is greater than, or equal to	45 inches
h	$\geq$	45 in.

If Jessica is 46.5 inches tall; can she ride the roller coaster?

Yes, $46.5 \geq 45$

You can also graph an inequality on a number line. The graph of an inequality shows all of the solutions of that inequality. The **solution of an inequality** is any number that makes the inequality true.

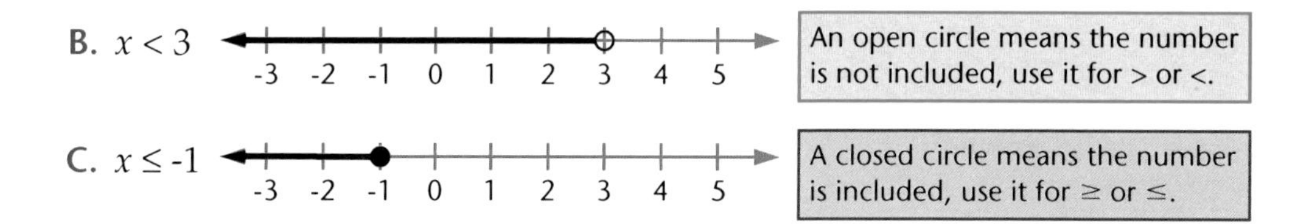

You can solve an inequality and graph its solution set. To solve an inequality, you use inverse operations to get the variable alone on one side.

To undo addition, use subtraction.
To undo subtraction, use addition.
To undo multiplication, use division.
To undo division, use multiplication.

D. Solve $n - 4 < 2$.

$n - 4 + 4 < 2 + 4$ Add 4 to both sides to undo the subtraction.

$n < 6$ Simplify.

Graph the solution.

-1 0 2 3 4 5 6 7 8

Use an open circle and shade to the left for less than.

Try THESE

Write an inequality for each situation.

1. x is less than seven.
2. A number is more than five.
3. Carol is at least twelve years old.
4. There are more than five students in a class.

Exercises

Solve each inequality. Graph each solution on a number line.

1. $x + 7 \leq 10$
2. $g - 4 > 1$
3. $m - 3 \geq -5$
4. $d - 6 > -4$
5. $3m \leq 12$
6. $s - 4 < 8$
7. $g + 5 \geq -7$
8. $w + 3 < -6$
9. $4d > 16$
10. $\frac{x}{5} > 3$
11. $x - 7 > 3$
12. $6 > x + 1$

Problem SOLVING

13. Compare and contrast solving an equation and solving an inequality.
14. You have \$5.00 for lunch. You bought a turkey sandwich for \$2.25. Write an inequality and solve it to show how much more money you can spend.

13.8 Problem-Solving Strategy: Use a Variety of Strategies

Objective: to solve word problems using a variety of strategies

A weather balloon is located 30,000 feet above sea level and a submarine is located 400 feet below sea level. What is the distance between the balloon and submarine?

Sometimes you can use more than one strategy to solve the same problem.

Use Logic	Make a List	Act It Out
Use Matrix Logic	Solve a Simpler Problem	Make a Diagram
Guess and Check	Work Backward	Use Venn Diagrams

You need to find the distance between the balloon and the submarine. You know the height of the balloon and the depth of the submarine.

Make a Diagram is a good choice because you can see the problem visually. **Solve a Simpler Problem** is a second choice because you can work with smaller numbers.

3. SOLVE

Make a Diagram

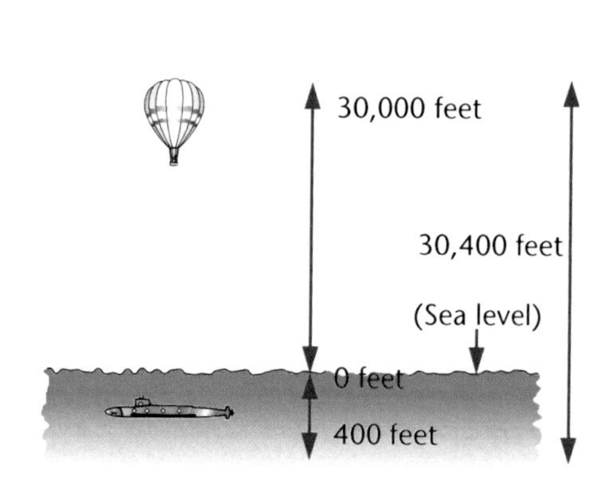

Since the balloon is 30,000 feet above sea level and the submarine is 400 feet below sea level, the distance between them is 30,400 feet.

Solve a Simpler Problem

If a balloon is 3 feet above the water and a rock is 2 feet below the water, what is the distance between them?

$$\begin{array}{rl} 3 & \text{feet above the water} \\ +\ 2 & \text{feet below the water} \\ \hline 5 & \text{feet between the balloon and the rock} \end{array}$$

By solving a simpler problem, it is easier to understand that there are 5 feet between the balloon and the rock. The answer makes sense.

The distance from 30,000 to 0 and from 0 to -400 is 30,400 so the answer is reasonable.

Try THESE

Solve.

1. The temperature on the ground is 31°F. The temperature on the mountaintop is -7°F. What is the difference in temperature?

2. An elevator on the first floor goes up 10 floors, down 8 floors, and then up 3 floors. On what floor is the elevator?

Solve

1. Mr. and Mrs. Msumba have 5 daughters, and each daughter has a brother. How many people are in the Msumba family?

2. Jeff and Kim ate $\frac{1}{3}$ of the cookies. Then Doug ate 6 cookies, which was $\frac{1}{3}$ of the remaining cookies. How many cookies were there at first?

3. Some number added to -8 is equal to 13. What is the number?

4. Complete.
   8, 7, 5, 2, -2, -7, __?__, __?__, __?__

5. The temperature was 4°C. Then the temperature fell 7°C. What was the final temperature?

6. If the temperature decreases 3.6° for every 1,000 feet in elevation, how much cooler will it be at the top of a 2,500-foot mountain?

7. A number is multiplied by -3, and then 5 is added to the product. The result is -1. What is the number?

8. Claire needs to get to school by 8:00 A.M. It takes her 25 minutes to drive to school, 14 minutes to eat breakfast, and 23 minutes to get ready. What is the latest time she can get up?

MIXED REVIEW

Simplify.

9. 4 + 3 • -2

10. -5 • -4 + 7

11. -9 ÷ 3 + 5

12. -2 • -2 • -2

13. -6 + 2 + 1 − 5

14. 2 + 14 ÷ -2

Chapter 13 Review

Language and Concepts

Write *true* or *false*. If false, replace the underlined word to make a true statement.

1. The absolute value of a number is always negative.
2. A combination of a positive integer and a negative integer of the same value is the same as zero.
3. You would read -8 as eight.
4. Negative nine is greater than negative four.
5. You do not have to write the positive sign if the number is +3.
6. The opposite of a number is always positive.

Skills and Problem Solving

Name the opposite of each integer. (Section 13.1)

7. -9 8. 15 9. -31 10. -28

Name the absolute value of each integer. (Section 13.1)

11. 14 12. -5 13. -16 14. 81

Compare using <, >, or = . (Section 13.1)

15. -2 ■ -4 16. 1 ■ 3 17. -5 ■ -7 18. 8 ■ 11

19. -2 ■ 2 20. 0 ■ -6 21. -15 ■ -15 22. 38 ■ -83

Order from least to greatest. (Section 13.1)

23. -2, 6, -9 24. -3, 0, -10 25. 8, -4, -8, 4

Write an equation for each diagram. (Section 13.2)

26. 27.

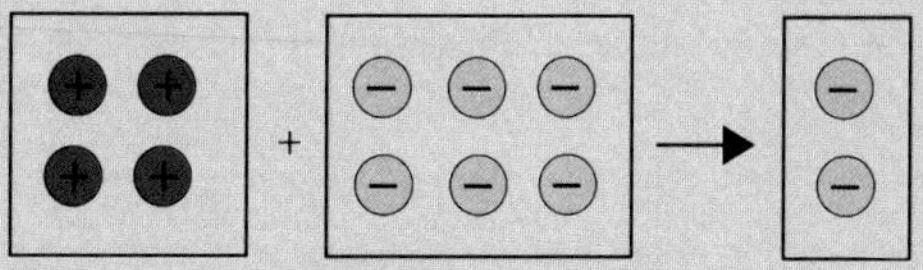

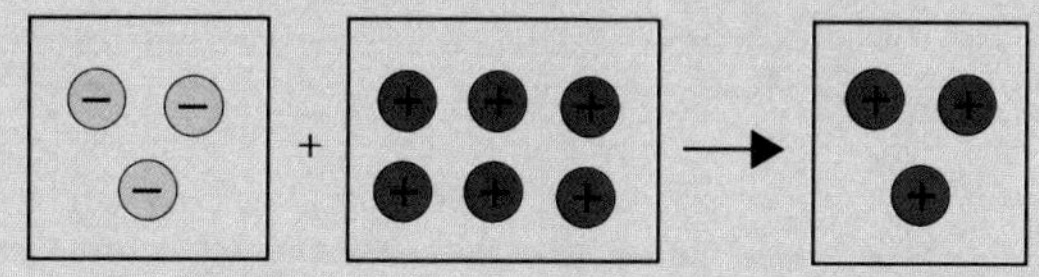

Add or subtract. (Sections 13.2–13.3)

28. -3 + -8 29. 3 – 10 30. 6 + -3 31. -4 + -3

32. -4 + 2 33. -2 – 8 34. -4 – -7 35. 12 + -5

Multiply or divide. (Sections 13.4–13.5)

36. $-8 \bullet 3$
37. $28 \div -7$
38. $-12 \div -4$
39. $-6 \bullet -5$
40. $-3 \bullet 15$
41. $-64 \div 8$
42. $-10 \bullet -5$
43. $36 \div -3$

Solve each equation. (Section 13.6)

44. $x + 7 = -2$
45. $y - 6 = -3$
46. $-4k = 32$
47. $4 + n = 9$
48. $\frac{n}{2} = -9$
49. $-6m = -72$
50. $14 = x - -5$
51. $16 = \frac{w}{-2}$

Solve each inequality. Graph each solution. (Section 13.7)

52. $x + 5 > 7$
53. $g - 3 \leq -8$
54. $r - 8 \geq 7$
55. $h + 4 < 3$

Solve. (Sections 13.2, 13.3, 13.8)

56. In the football game, a penalty of -15 yards was marked off a gain of 43 yards. What was the net gain?
57. The temperature was -8°C. Then the temperature rose 12°. What was the final temperature?
58. Maria's game scores were 40, -20, and 10. Find her total score.
59. The temperature at 8:00 A.M. was -2°. At 2:00 P.M. the temperature was 22°. What was the temperature change?
60. During a three-day period, the stock of Money, Inc. gained 6 points, lost 10 points, and gained 2 points. What was the total gain or loss for the three days?

Chapter 13 Test

Compare using <, >, or = .

1. -9 ■ -8
2. 16 ■ -18
3. -21 ■ 0
4. 67 ■ -76
5. 4 ■ 18
6. 21 ■ -21
7. -13 ■ 10
8. -8 ■ -12

Add or subtract.

9. $-6 + -8$
10. $0 + -12$
11. $-2 - 10$
12. $-4 - -2$
13. $8 + -3$
14. $9 + -9$
15. $-12 - 7$
16. $5 - 6$

Multiply or divide.

17. $5 \bullet -7$
18. $-2 \bullet 7$
19. $-9 \div 3$
20. $9 \bullet -1$
21. $-5 \div -5$
22. $-8 \bullet 9$
23. $-21 \div -3$
24. $-56 \div 8$

Solve each equation.

25. $n + 4 = -5$
26. $k - 7 = -1$
27. $-6m = -48$
28. $\frac{x}{5} = -30$

Solve each inequality.

29. $k - 6 \leq 3$
30. $x - 5 < -2$
31. $y + 7 \geq -3$
32. $4x \leq 4$

Solve.

33. The Oakville Bears had a first down on the 40-yard line. In the next three plays, they lost 2 yards, gained 13 yards, and lost 15 yards. How many yards did they gain or lose?
34. Ted's scores in four games were 20, -16, 12, and 16. What was his average score?
35. Mrs. Galvoz lost 3 pounds a month for 5 months. What integer represents her total weight lost?
36. On a math quiz worth 25 points, Lily missed 3 points in the first section, 2 points in the second section, and 2 points in the third section. She then earned 4 points on the extra credit. What was Lily's score on the quiz? Explain your reasoning using your knowledge of positive and negative integer operations.

Solving Equations Using the Distributive Property

The Distributive Property states that the product of a number and a sum is equal to the sum of the products.

$6 \times (7 + 20) =$	The problem is asking for the product of a number and a sum.
$(6 \times 7) + (6 \times 20) =$	The answer can be found by finding the sum of the products.
$42 + 120 = 162$	Finally, find the sum of these products.

This property is also used in algebra to solve equations.

Examples

A.

$$3 \times (a + 5) = 21$$
$$(3 \times a) + (3 \times 5) = 21$$
$$3a + 15 = 21$$
$$3a + 15 - 15 = 21 - 15$$
$$3a = 6$$
$$3a \div 3 = 6 \div 3$$
$$a = 2$$

B.

$$6(2b - 5) = 18$$
$$(6 \times 2b) - (6 \times 5) = 18$$
$$12b - 30 = 18$$
$$12b - 30 + 30 = 18 + 30$$
$$12b = 48$$
$$12b \div 12 = 48 \div 12$$
$$b = 4$$

Replace each ■ with a number or variable to make the equation true.

1. $7 \times (c + 5) = (7 \times ■) + (7 \times ■)$
2. $12 \times (d - 2) = (12 \times ■) - (■ \times ■)$
3. $5\ (r + 6) = (5r + ■)$
4. $-6\ (w + 5) = (■ + -30)$

Use the Distributive Property to solve the following equations.

5. $5\ (m + 3) = 25$
6. $2\ (n - 5) = 6$
7. $3\ (2h + 4) = 42$
8. $8\ (3t - 4) = 112$
9. $-6\ (s + 7) = -60$
10. $-3\ (2x - 5) = 27$

Cumulative Test

1. Estimate.

$$\begin{array}{r} 223 \\ \times\ 498 \\ \hline \end{array}$$

a. 10,000
b. 100,000
c. 1,000,000
d. none of the above

2. Which of the following is the standard form for twenty-seven?
a. 27
b. 72
c. 207
d. none of the above

3. Find the sum of 49 and 38.
a. 87
b. 89
c. 97
d. none of the above

4. If $4^4 = 4 \times 4 \times 4 \times 4$, then what is 4^3?
a. 4×4
b. $4 \times 4 \times 4$
c. 4×3
d. $3 \times 3 \times 3 \times 3$

5. Which is true?
a. $7{,}452 = 7{,}652$
b. $7{,}452 > 7{,}652$
c. $7{,}452 < 7{,}652$
d. none of the above

6.

$$\begin{array}{r} 5{,}802 \\ \times\ \ \ 15 \\ \hline \end{array}$$

a. 87,000
b. 87,030
c. 87,130
d. none of the above

7. The odometer on Mrs. Pratt's car reads 58,000 miles. During May, June, and July, she drove 3,411 miles. What was the reading before May?
a. 54,589
b. 54,599
c. 55,119
d. 61,411

8. Between which two amounts is the product $84 × 40?
a. $3,000 and $3,100
b. $3,200 and $3,600
c. $3,600 and $3,800
d. $3,600 and $4,000

9. Which fact do you need in order to solve this problem?

Murry is 6 inches shorter than Barry. How tall is Murray?

a. Barry's weight
b. Murray's age
c. Murray's weight
d. none of the above

10. Kelly marked a number chart like this.

(1) 2 (3) 4 5 (6) 7
8 9 (10) 11 12 13 14
(15) 16 17 18 19 20 (21)

Using this pattern, which number would she circle next?
a. 26 b. 27
c. 28 d. 29

CHAPTER 14

Coordinate Graphing

Joy Williams
Panama

14.1 The Coordinate Plane

Objective: to locate points in the coordinate plane

The weather bureau keeps track of weather balloons by means of a **coordinate plane** and **ordered pairs**.

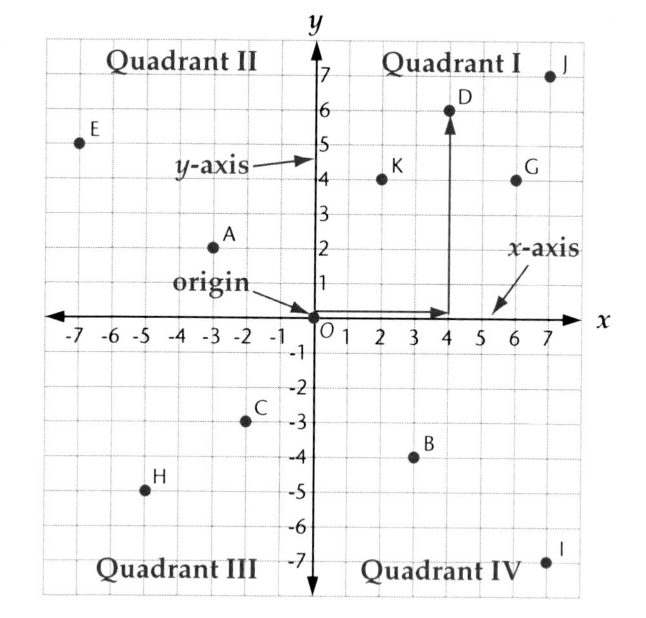

The location of point D is given by the number pair (4, 6). The point for (6, 4) is G. Such number pairs are called ordered pairs. Note that the **different** orders, (4, 6) and (6, 4), locate **different** points.

The coordinate plane has two perpendicular number lines. They intersect at a point called the **origin,** labeled *O*.

Locate points on the grid as follows.

- Start at the origin. The ordered pair for point *O* is (0, 0).
- The first number tells how many units to move **right** (+) or to move *left* (-) on the *x*-axis, **horizontal** number line.
- The second number tells how many units to move **up** (+) or move **down** (-) on the *y*-axis, **vertical** number line.

Examples

A. Locate the point for (2, 4).

Start at *O*. Move right 2 units. → **(2, 4)** ↑ Move up 4 units.

The point for (2, 4) is K.

B. Locate the point for (-3, 2).

Start at *O*. Move left 3 units. ← **(-3, 2)** ↑ Move up 2 units.

The point for (-3, 2) is A.

C. Locate the point for (-5, -5).

Start at *O*. Move left 5 units. ← **(-5, -5)** ↓ Move down 5 units.

The point for (-5, -5) is H.

D. Locate the point for (3, -4).

Start at *O*. Move right 3 units. → **(3, -4)** ↓ Move down 4 units.

The point for (3, -4) is B.

Exercises

Name the ordered pair for each point.

1. A
2. B
3. C
4. D
5. E
6. F
7. G
8. H
9. I
10. J
11. K
12. L
13. M
14. N
15. O
16. P

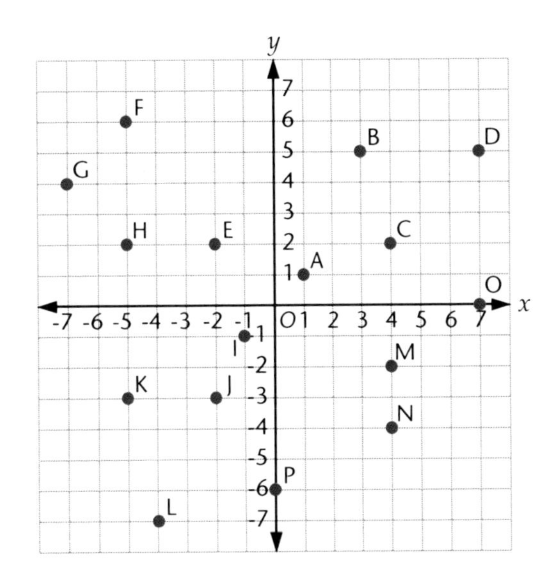

Use grid paper to draw a coordinate system like the one shown above. Locate the point for each ordered pair. Label each point with the letter given.

17. A (1, 2)
18. B (2, 1)
19. C (8, -2)
20. D (-2, 8)
21. E (-8, 2)
22. F (-8, -2)
23. G (9, -6)
24. H (9, 6)
25. I (0, 0)
26. J (0, -5)
27. K (5, -4)
28. L (-5, 4)
29. M (7, 7)
30. N (-7, -7)
31. O (-6, 0)
32. P (0, -8)

Problem SOLVING

Complete. Use grid paper.

33. Draw segments from (0, 0) to (2, 2) to (5, 0) to (0, 0). Name the figure.

34. Draw segments from (4, 5) to (4, 6) to (1, 6) to (1, 5) to (4, 5). Name the figure.

★35. Is the order of the coordinates in an ordered pair important? Explain.

★36. A square has the following coordinates (0, 0), (2, 0), and (2, 2). What is the fourth coordinate?

MIXED REVIEW

Solve each proportion.

37. $\frac{n}{5} = \frac{3}{60}$
38. $\frac{6}{p} = \frac{9}{18}$
39. $\frac{4}{7} = \frac{t}{28}$
40. $\frac{21}{33} = \frac{7}{b}$
41. $\frac{2}{19} = \frac{a}{57}$
42. $\frac{10}{45} = \frac{8}{z}$
43. $\frac{5.1}{s} = \frac{25.5}{11}$
44. $\frac{x}{6.09} = \frac{1}{0.87}$

Objective: to translate figures in the coordinate plane

Exploration Exercise

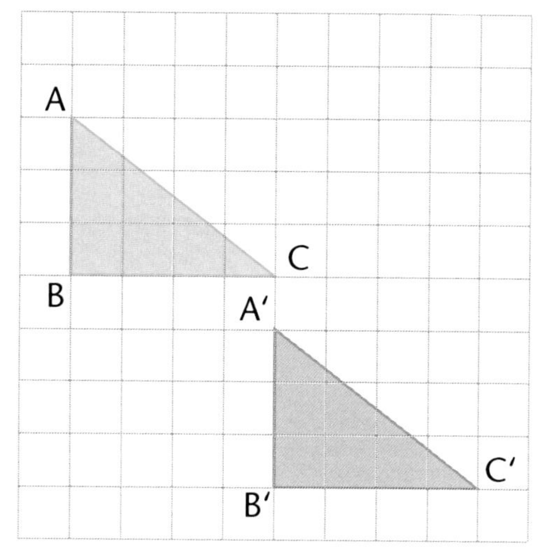

1. Look closely at the triangles on the grid. △ABC shows the original figure. △A′B′C′ shows the figure after it was moved. △A′B′C′ is called the **image** of △ABC. Did the size or shape of the figure change after it was moved? What can you say about the figure and its image?
2. Trace △ABC and cut it out. Place your tracing over the original figure. Then move your tracing around on the grid to see if you can tell how the figure was moved to make the image.
3. If you had not traced the figure, how would you use the letters A′, B′, and C′ to find out how the figure was moved?

The triangle above was translated. A **translation**, or slide, is a transformation that moves every point on a figure the same distance in the same direction. In the above exploration △A′B′C′ is an image of △ABC after it was translated.

Examples

A. Triangle DEF is a translation of △ABC.

Think of an object being pushed from one position to another.

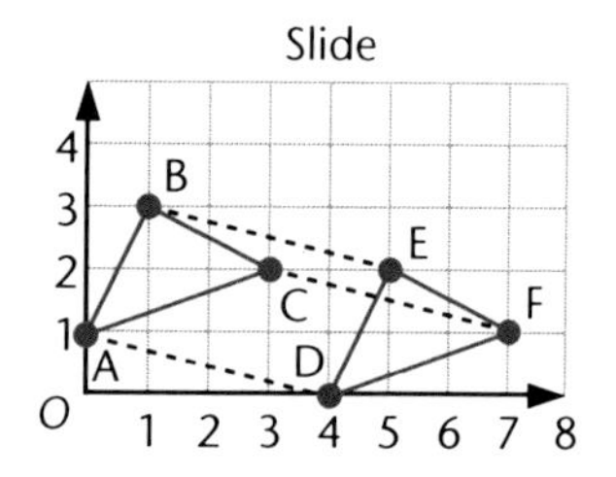

Move each point 4 units right and 1 unit down.

Point A (0, 1) is moved to point D (4, 0).
Point B (1, 3) is moved to point E (5, 2).
Point C (3, 2) is moved to point F (7, 1).

B. Move parallelogram RSTU 2 units right. Label the image R′S′T′U′.

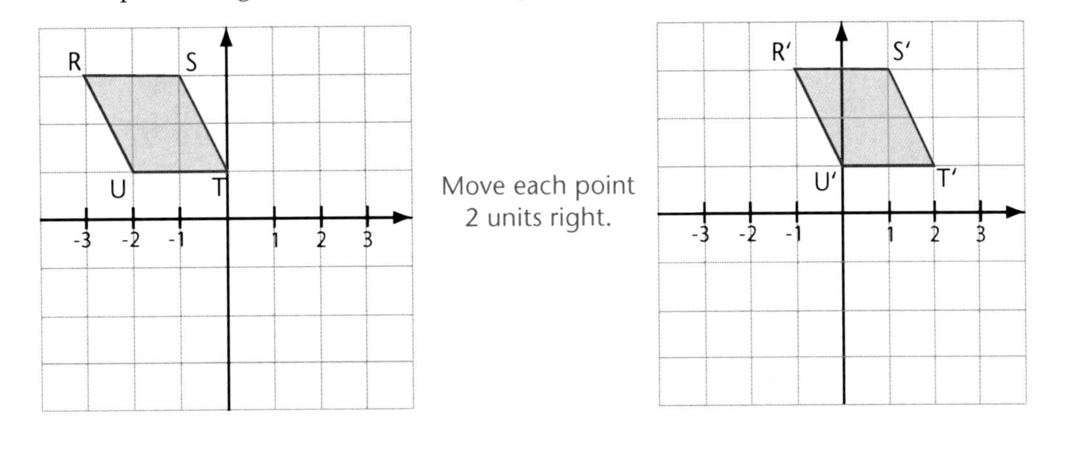

State how each figure below was translated.

1.

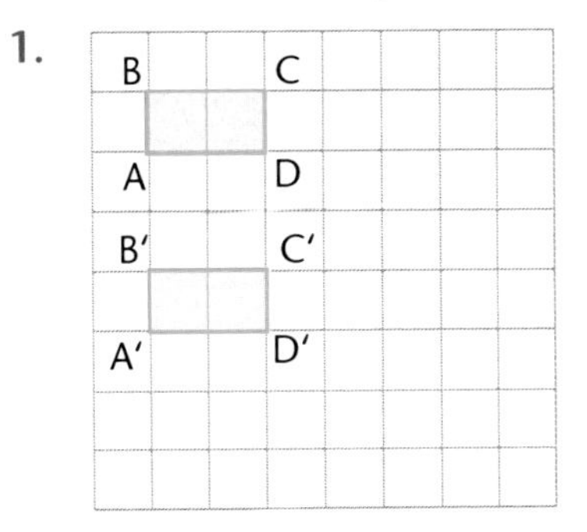

2.

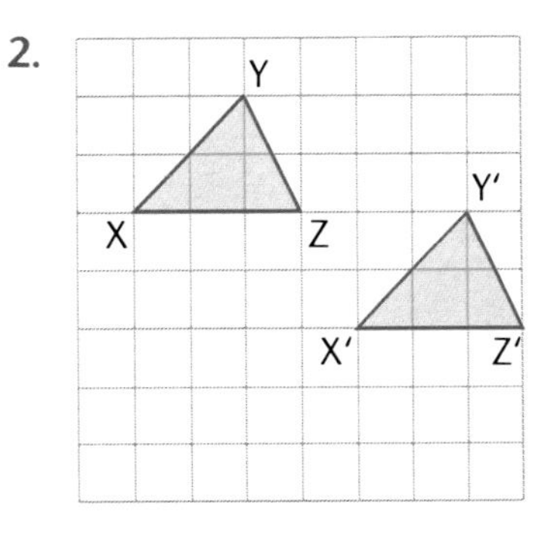

Is the image a translation of the figure? Write *yes* or *no*.

1.

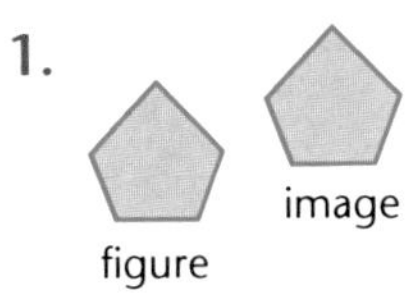

2.

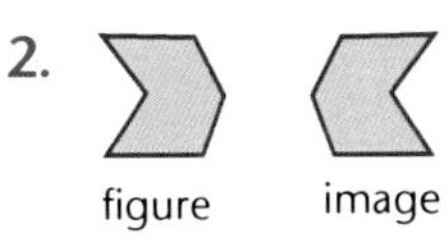

3.

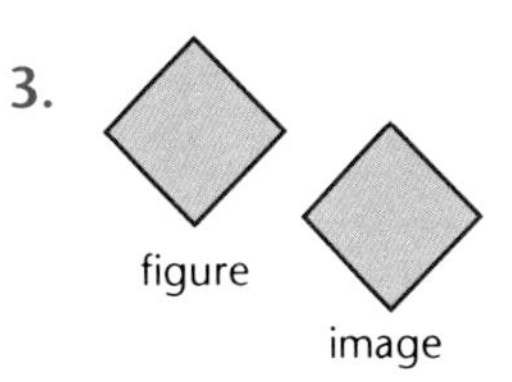

Trace △ABC on grid paper and use it for problems 4–6.

4. Translate △ABC 2 units up.
5. Translate △ABC 3 units left.
6. Translate △ABC 5 units right.

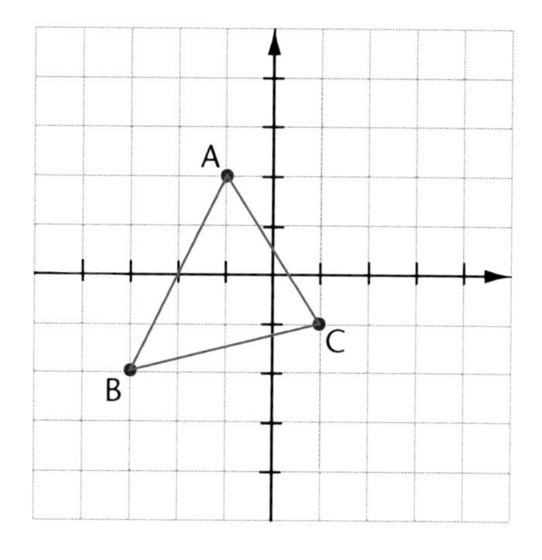

Trace parallelogram DEFG on grid paper and use it for problems 7–9.

7. Translate parallelogram DEFG 3 units down.
8. Translate parallelogram DEFG 1 unit right.
9. Translate parallelogram DEFG 2 units right and 2 units up.

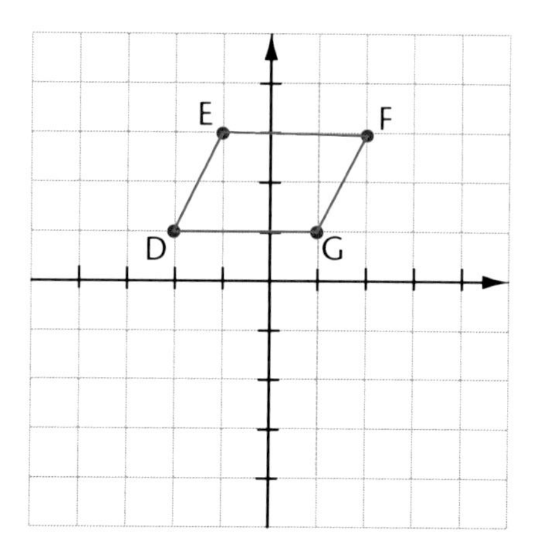

Problem SOLVING

10. Draw a triangle on grid paper and label it EFG. Move △EFG on the grid paper so that it shows a translation.

★11. Draw △RST with vertices R(2, 1), S(4, 4), and T(6, 1). Translate △RST 4 units left and 2 units down.

How Will Kim Get Home?

On her way back from the bus station, Kim got lost and could not find her way home (U). What Kim does not know is that there are many different ways for her to get home. Can you name all of the routes she could take? (Name routes like this: A – B – C – G – H, and so on.)

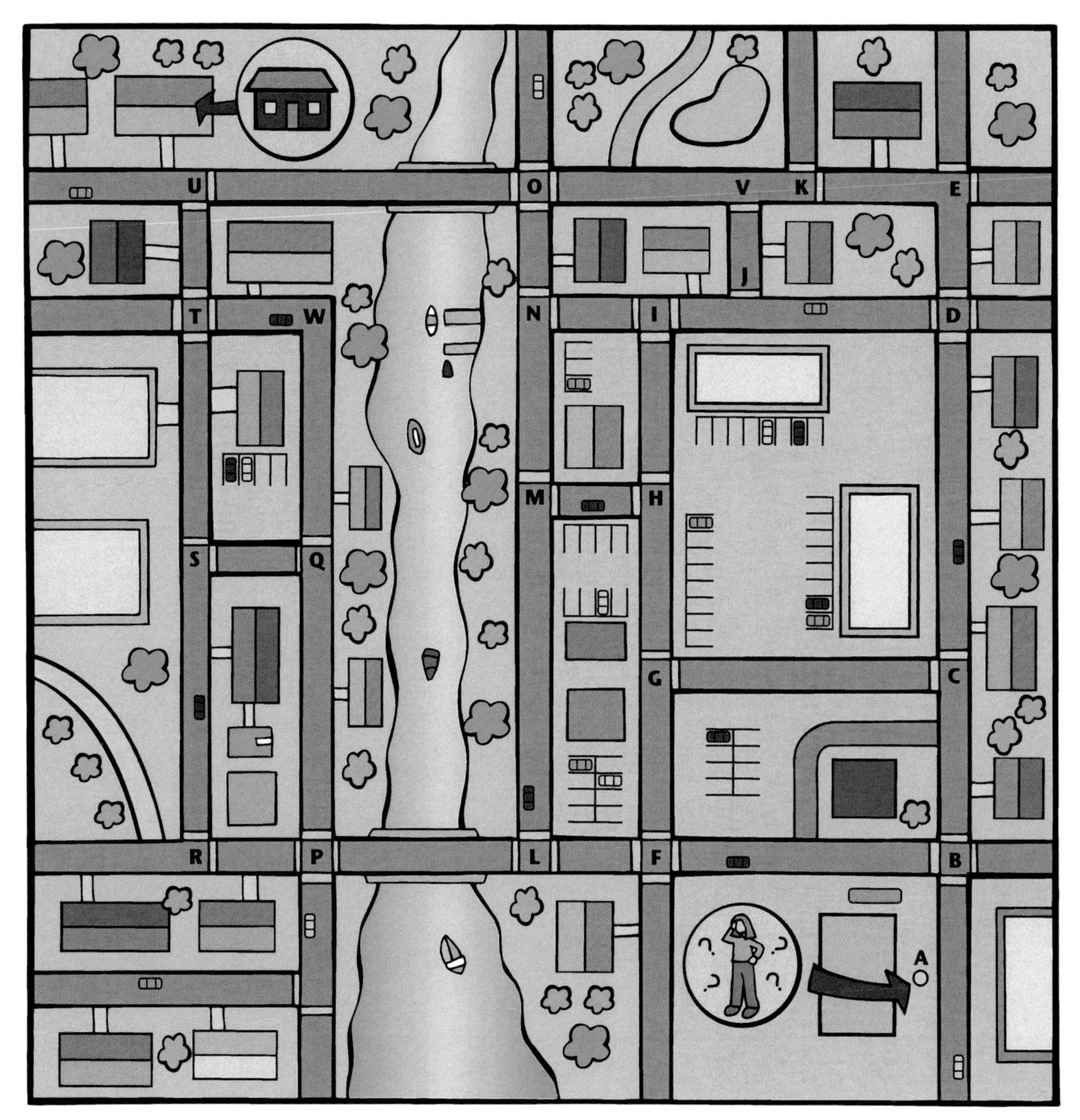

Extension

Place a 0.5 cm grid over the paths and count the units in each path. What did you discover?

14.3 Reflections and Rotations

Objective: to reflect and rotate figures

On a windy day, Zack watched a windmill as it turned. He noticed the various positions it took. Although the position changed, the sail was always the same size and shape.

When any figure is rotated or reflected, its position changes, but it keeps its shape.

A **reflection**, or flip, is a transformation that flips a figure over a line. The line is the **line of reflection**. A **rotation**, or turn, is a transformation that turns a figure about a fixed point. The point is called the **center of rotation**.

Examples

A. Triangle GHI is a reflection of $\triangle ABC$.

Think of an object and its mirror image.

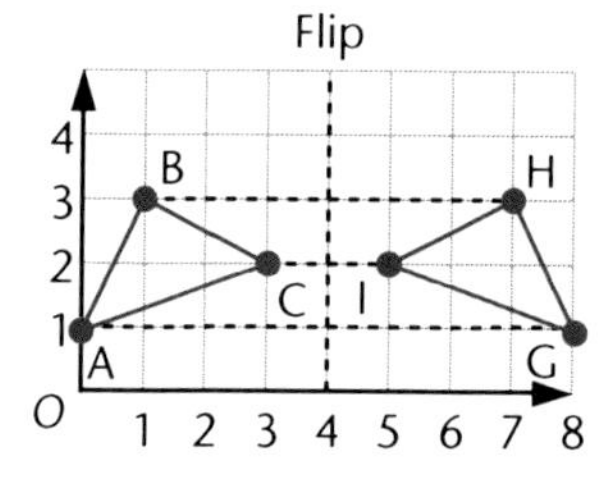

B. Triangle JKL is a rotation of $\triangle ABC$.

Think of an object as being on a wheel that turns around a center point.

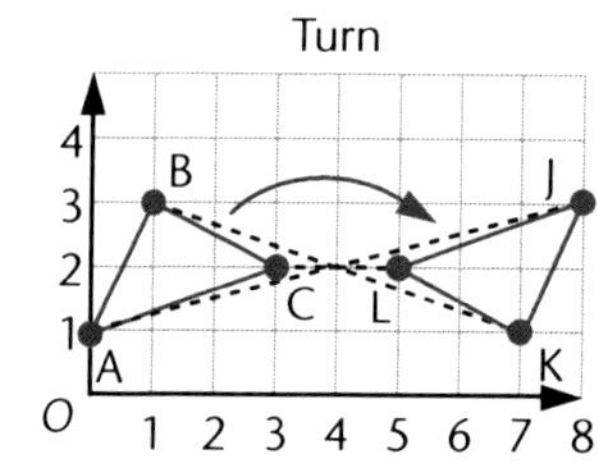

The center of rotation is (4, 2).

Copy each figure and draw its reflection over the given line of reflection.

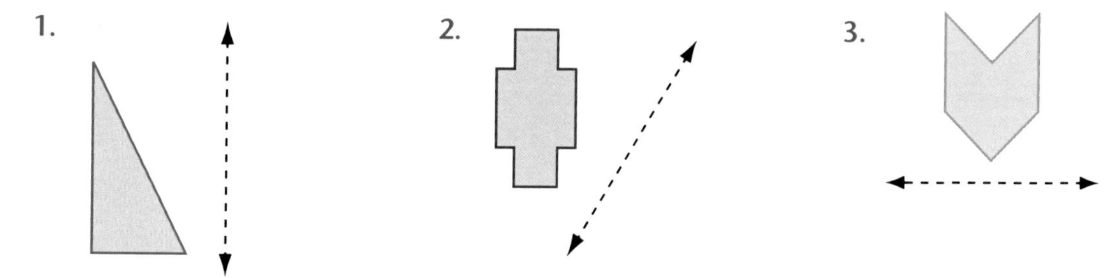

Exercises

Copy each grid. Then reflect each figure over the line of reflection.

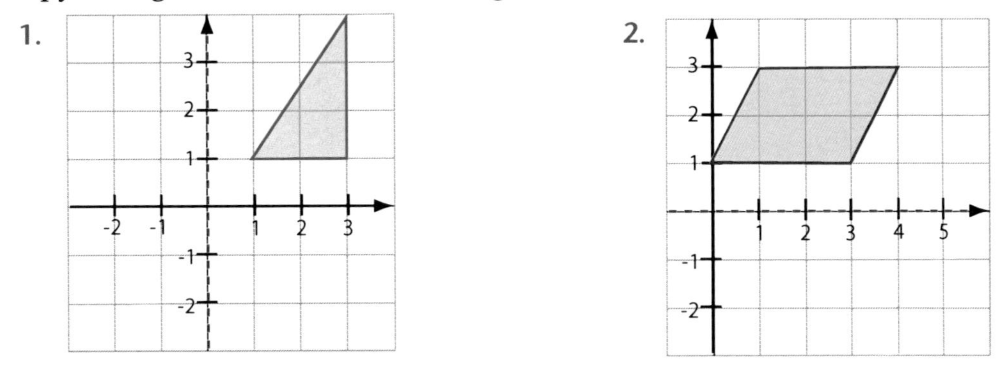

Tell whether the image is a rotation of the figure. Write *yes* or *no*.

Copy the figures onto grid paper. Then draw the line of reflection.

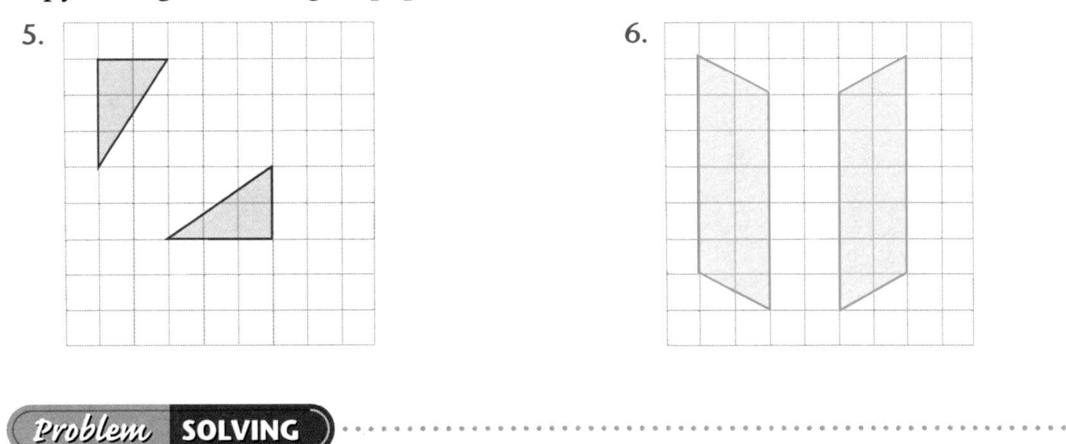

Problem SOLVING

7. Draw a triangle on grid paper and label it EFG. Move $\triangle$EFG on the grid so that it shows a reflection. Label the line of reflection.
8. Draw a triangle on grid paper and label it MNO. Move $\triangle$MNO on the grid so that it shows a rotation. Label the center of rotation.
9. Draw a triangle on grid paper and label it RST. Move $\triangle$RST on the grid so that it shows translation.

14.4 Problem-Solving Strategy: Look for a Pattern

Objective: to solve problems by finding a pattern

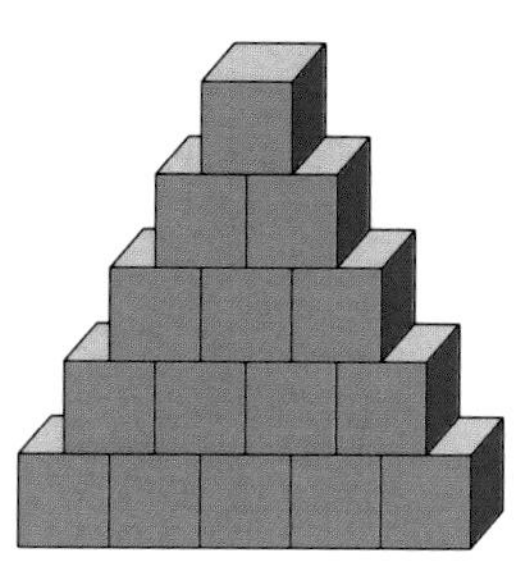

Karen wants to build a pyramid like the one shown at the right. She has 45 blocks. What is the largest pyramid she can build?

You need to find how many rows are in the pyramid. You know that the number of blocks in each row increases by one.

Draw a diagram to show what you know. Look for a pattern and continue the diagram.

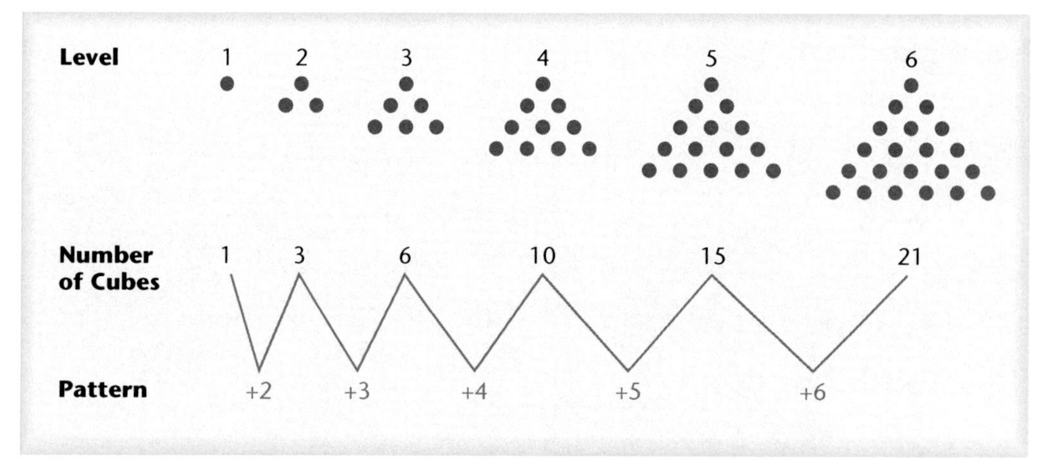

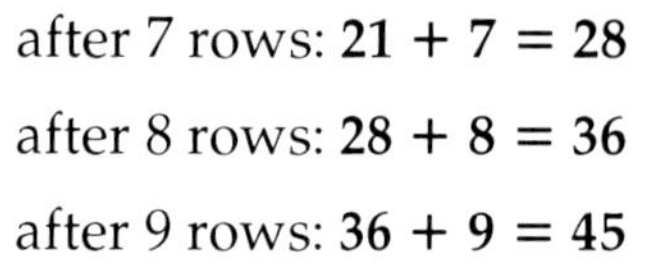

after 7 rows: **21 + 7 = 28**

after 8 rows: **28 + 8 = 36**

after 9 rows: **36 + 9 = 45**

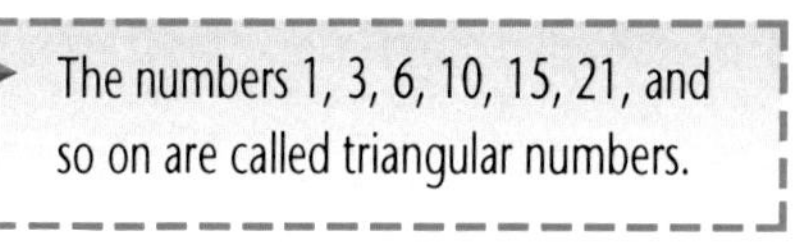

The numbers 1, 3, 6, 10, 15, 21, and so on are called triangular numbers.

The largest pyramid she could build would have 9 rows.

After 9 rows there should be 1 + 2 + 3 + 4 + 5 + 6 + 7 + 8 + 9 blocks. This sum is 45, so the solution is correct.

Solve. Look for a pattern.

1. One person enters an auditorium the first minute, 2 people enter the second minute, 3 the third minute, and so on. How many minutes will it take for a total of 105 people to be in the auditorium?

2. How many people will be in the auditorium after 17 minutes?

Solve

1. Describe how the pyramid at the right differs from the pyramid on the previous page.

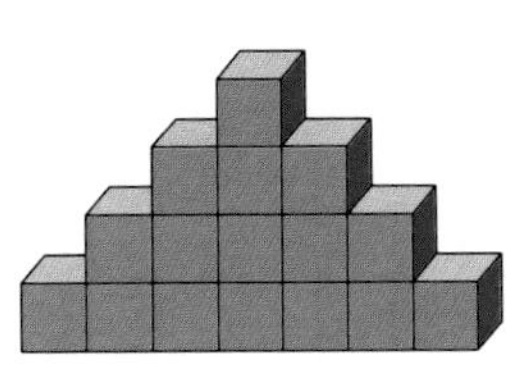

2. Find the number of blocks needed to build this kind of pyramid. Copy and complete this table.

Number of rows	1	2	3	4	5	6
Number of blocks	1	4	9			

3. Describe the pattern in the number of blocks needed.

4. How many blocks are needed to build a pyramid with 9 rows?

Find the fifth and sixth terms for each pattern.

5. , , , , . . .

6. 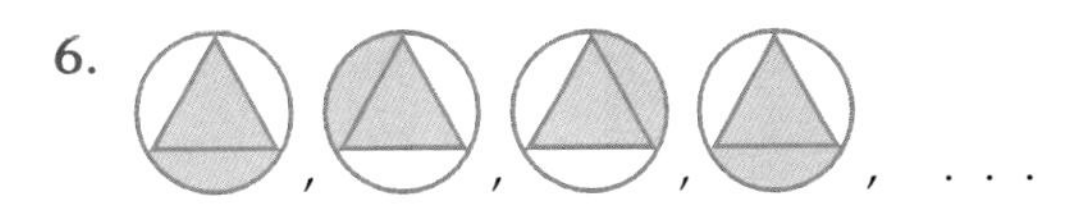

7. AC, BD, CE, DF, . . .

8. 10, 11, 13, 16, . . .

9. 8, 16, 32, 64, . . .

10. 6, 11, 21, 36, 56, . . .

Mid-Chapter REVIEW

1. Name the ordered pair for each point.

   a. A b. B

   c. C d. D

2. Translate ΔABC up 3 units.

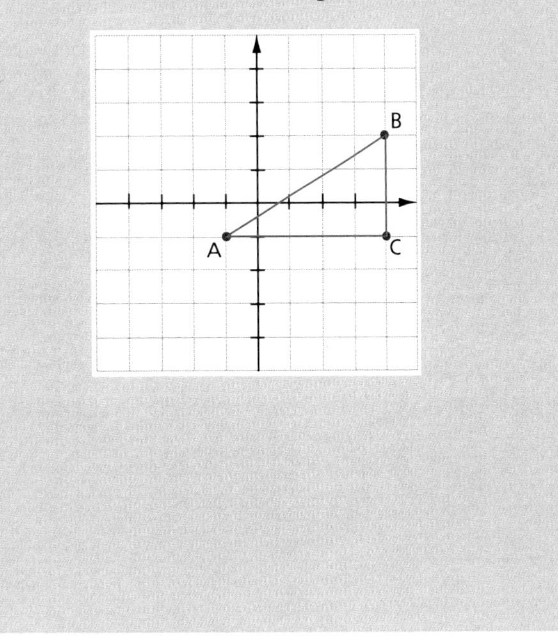

Objective: to find the missing terms in a sequence

Sharon is a doll collector. She has collected 5 dolls. If she is able to collect 3 dolls a year, how many dolls will she have in 6 years?

The pairs of numbers below form a function called a **sequence**. A sequence is a set of numbers called terms that are in a specific order. This sequence is called an **arithmetic sequence**. Each term of an arithmetic sequence can be found by adding or subtracting the same number to the term before it.

Sharon needs to add to complete the sequence.

now	1 year	2 years	3 years	4 years	5 years	6 years
5	**8**	**11**	**14**	**17**	**20**	**23**
	+3	+3	+3	+3	+3	+3

Sharon will have 23 dolls in 6 years.

Another Example

Another type of sequence is called a **geometric sequence**. Each term of a geometric sequence can be found by multiplying the term before it by the same number or by dividing the term before it by the same number.

Find the next two terms in the sequence.

48, 24, 12, 6, __?__, __?__, . . .

Divide by 2.
$6 \div 2 = 3$
$3 \div 2 = 1.5$

The next two terms in the sequence are 3 and 1.5.

Try THESE

Find the number that was added to or subtracted from each term.

1. 2, 4, 6, 8, . . .
2. 24, 20, 16, 12, . . .
3. 25, 50, 75, 100, . . .

Find the number each term is multiplied by or divided by.

4. 3, 6, 12, 24, . . .
5. 54, 18, 6, 2, . . .
6. 6, 30, 150, 750, . . .

Exercises

State whether each sequence is *arithmetic* or *geometric*.

1. 36, 30, 24, . . .
2. 1.1, 3.3, 9.9, . . .
3. 100, 10, 1, . . .
4. 9, 12, 15, . . .
5. 85, 76, 67, . . .
6. 27, 4.5, 0.75, . . .

Find the next three terms in each arithmetic sequence.

7. 10, 20, 30, __?__, __?__, __?__, . . .
8. 95, 80, 65, __?__, __?__, __?__, . . .
9. 2.5, 5, 7.5, __?__, __?__, __?__, . . .
10. 9, $8\frac{1}{2}$, 8, $7\frac{1}{2}$, __?__, __?__, __?__, . . .
11. 19, 17, 15, __?__, __?__, __?__, . . .
12. 71, 68, 65, __?__, __?__, __?__, . . .

Find the next three terms in each geometric sequence.

13. 3, 9, 27, __?__, __?__, __?__, . . .
14. 128, 64, 32, __?__, __?__, __?__, . . .
15. 1, 5, 25, __?__, __?__, __?__, . . .
16. 729, 243, 81, __?__, __?__, __?__, . . .
17. 400, 200, 100, __?__, __?__, __?__, . . .
18. 0.5, 3.5, 24.5, __?__, __?__, __?__, . . .

★19. Find the fifth term for the sequence 9, 18, 27, . . .

★20. Find the next two terms for the sequence $\frac{1}{4}, \frac{1}{8}, \frac{1}{16}, \ldots$

Problem SOLVING

21. An apartment rents for $360 a month. Each year the monthly rent increases $15. What will be the monthly rent in 5 years?

22. Suppose you have 40 tickets to sell. You sell $\frac{1}{2}$ of them each day. How many tickets will you have sold by the third day?

★23. Henry and Joey are finding the next term in the sequence 5, $7\frac{1}{2}$, 10, ___. Who is correct? Explain.

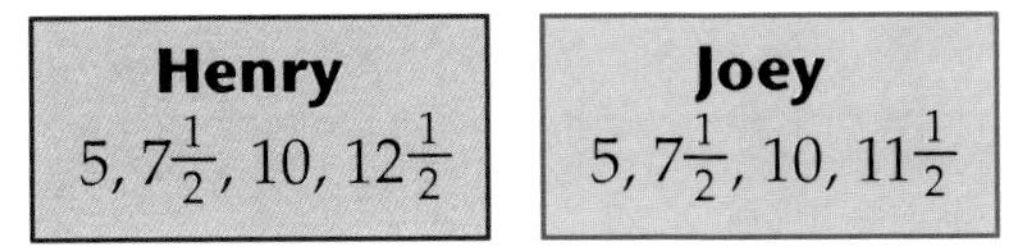

MiXeD REVIEW

Solve.

24. $6 \times \frac{2}{3}$
25. $\frac{5}{8} \times \frac{2}{3}$
26. $7 \times 3\frac{4}{9}$
27. $4\frac{1}{2} \times 5\frac{1}{3}$
28. $4 \div \frac{3}{4}$
29. $\frac{7}{9} \div \frac{2}{5}$
30. $3\frac{1}{2} \div 1\frac{3}{4}$
31. $1\frac{3}{8} \div 5$

14.6 Functions

Objective: to complete function tables; to find function rules

Kevin collects old and new comic books. If Kevin buys one old comic book, he will pay the seller $2.00. How much would three old comic books cost?

The relationship between the cost of the comic books and the number of comic books purchased is called a **function**.
A function pairs each number of one set with a number from another set. You can say the relationship between a comic book Kevin buys and what he pays is 1 to 2 (in other words, 1 to $2.00). The relationship between what he pays and a comic book would be 2 to 1. Since there is an exact relationship between comics and money (1 to 2) you can say 1 to 2 is an ordered pair.

A function is usually described with a rule.

> A function is a set of ordered pairs (x, y) in which each value of x is paired with exactly one value of y.

The **input**, or number of books purchased, is 3. The function is $\boxed{\times 2}$.

You can show this function with a function machine.

Input → 3 | × 2 | 6 → Output

The **output** is 6. So, three books cost $6.00.

Examples

There can be more than one number for the input.
Then a chart can be used to represent a function.

A.

Rule: + 3	
Input	**Output**
4	7
6	9
8	11
9.7	12.7

If 4 is the input, then 7 is the output. Explain.

B.

Rule: ÷ 6	
Input	**Output**
12	2
24	4
36	6
48	8

If 24 is the input, then 4 is the output. Explain.

Find the function rule for each chart. Use m to represent the variable.

C.

Input	Output
32	4
56	7
96	12

The output is less than the input.
So either subtract or divide.

32 – 28 = 4 yes **32 ÷ 8 = 4 yes**
56 – 28 = 28 no **56 ÷ 8 = 7 yes**
96 – 28 = 68 no **96 ÷ 8 = 12 yes**

The function rule is $m \div 8$.

D.

Input	Output
9	151
28	170
147	289

Is the output greater or less than the input? What operations would you use? How would you check your results?

The function rule is $m + 142$ or $142 + m$.
Explain how you know that this function rule is correct.

Complete.

1.

Rule: − 9	
Input	**Output**
9	?
13	?
28	?

2.

Rule: × 10	
Input	**Output**
2	?
4	?
7	?

3.

Rule: ÷ 5	
Input	**Output**
5	?
15	?
40	?

Exercises

Complete.

1.

Rule: × 3	
Input	**Output**
0	?
6	?
18	?
50.9	?

2.

Rule: ÷ 7	
Input	**Output**
7	?
?	4
350	?
4,368	?

3.

Rule: + 741	
Input	**Output**
0	?
59	?
62.9	?
?	1,482

4.

Rule: + 18	
Input	**Output**
5	?
?	64
60	?
72	?
125	?

5.

Rule: × $\frac{1}{2}$	
Input	**Output**
1	?
6	?
85	?
?	51
254	?

6.

Rule: − 49	
Input	**Output**
49	?
98	?
107	?
?	274
?	25

★7.

Rule: + 85				
Input	4	?	8	9.1
Output	?	91	?	?

★8.

Rule: × 0.8				
Input	0.4	0.8	?	8.9
Output	?	?	1.28	?

Find the function rule for each chart. Use any letter to represent the variable.

9.

Input	Output
45	32
36	23
24	11
13	0

10.

Input	Output
3	27
6	54
9	81
12	108

11.

Input	Output
50	50
40	60
25	75
10	90

12.

Input	Output
10	30
20	40
45	65
80	100

13.

Input	Output
15	3
20	4
30	6
45	9

14.

Input	Output
6	10.5
11	15.5
24	28.5
40	44.5

Problem SOLVING

15. For every two bags of building sticks you buy, you get one bag free. You have three free bags, how many bags did you buy? Write a function rule.

★16.

Input	Output
2	4
3	9
4	16

What is the function rule for this chart?

17. Can you find Millie's house number?

The number is less than 70,000.
The 5 is next to the 4.
The 5 is not next to the 6 or the 7.
The 8 is not next to the 6 or the 7.
The 4 is not next to the 6 or the 8.

Mind BUILDER

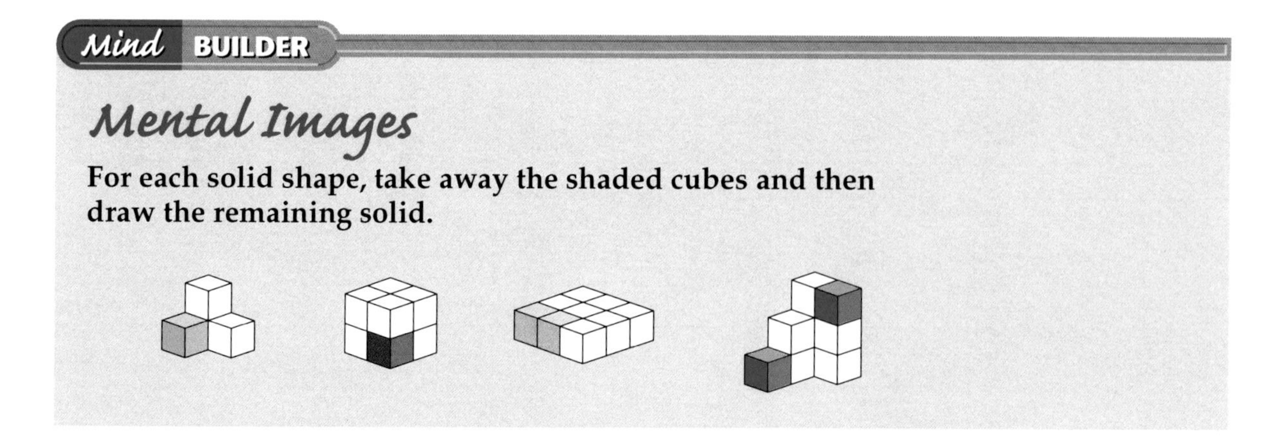

Mental Images

For each solid shape, take away the shaded cubes and then draw the remaining solid.

Cumulative Review

Write the place-value position of 4 in each number.

1. 3,405
2. 730,564
3. 29,341
4. 754,998
5. 4,337,205

Compare using <, >, or = .

6. 12,605 ● 12,506
7. 43,000 ● 430,000
8. 678,294 ● 678,249

Compute.

9. $\begin{array}{r} 27 \\ +\ 53 \\ \hline \end{array}$
10. $\begin{array}{r} 1{,}504 \\ +\ 7{,}389 \\ \hline \end{array}$
11. $\begin{array}{r} 71{,}302 \\ +\ 29{,}859 \\ \hline \end{array}$
12. $\begin{array}{r} \$94 \\ 52 \\ +\ 76 \\ \hline \end{array}$
13. $\begin{array}{r} 1{,}773 \\ 2{,}045 \\ +\ 7{,}362 \\ \hline \end{array}$
14. $\begin{array}{r} 95 \\ -\ 23 \\ \hline \end{array}$
15. $\begin{array}{r} \$845 \\ -\ 627 \\ \hline \end{array}$
16. $\begin{array}{r} 7{,}297 \\ -\ 5{,}468 \\ \hline \end{array}$
17. $\begin{array}{r} 6{,}007 \\ -\ 5{,}329 \\ \hline \end{array}$
18. $\begin{array}{r} 57{,}000 \\ -\ 48{,}254 \\ \hline \end{array}$
19. $\begin{array}{r} 43 \\ \times\ 100 \\ \hline \end{array}$
20. $\begin{array}{r} 578 \\ \times\ 6 \\ \hline \end{array}$
21. $\begin{array}{r} 5{,}000 \\ \times\ 90 \\ \hline \end{array}$
22. $\begin{array}{r} \$307 \\ \times\ 58 \\ \hline \end{array}$
23. $\begin{array}{r} 648 \\ \times\ 425 \\ \hline \end{array}$
24. $2\overline{)48}$
25. $8\overline{)168}$
26. $9\overline{)462}$
27. $4\overline{)654}$
28. $7\overline{)3{,}607}$

Find the average.

29. 53, 67, 43, 77
30. 1,306; 954; 1,211

Write using exponents.

31. $5 \times 5 \times 5 \times 5$
32. $9 \times 9 \times 9$
33. 1,000,000

Write an equation and solve. Write any facts not needed.

34. Suppose you rent 3 canoes at $12 each. Your canoe trip is 14 miles long. How much do you pay for the canoes?
35. It is 594 miles from Denver to Kansas City. Paula drives 159 miles in 3 hours. How much farther does she have to drive?
36. Kim weighs 83 pounds. Last year she weighed 78 pounds. Her older brother Carl weighs 110 pounds. How much more does Carl weigh?
37. Leslie Porter buys a used car for $4,650. She pays for the car in 6 equal installments. How much is each installment?

14.7 Graphing Functions

Objective: to graph functions

Anna earns $4 a week for doing her chores. She saves all of her money. Each week the amount of money in her savings increases. In the first week she has $4, in the second week she has $8, and so on. The amount of money she saves depends on the number of weeks. We can use the function below to represent Anna's savings.

amount saved → $y = 4x$ ← number of weeks

It is helpful to make a function table to find out how much money Anna saves. Use the function rule above.

Input (x)	Rule ($4x$)	Output (y)	Ordered Pair (x, y)
1	4 • 1	$4	(1, 4)
2	4 • 2	$8	(2, 8)
3	4 • 3	$12	(3, 12)
4	4 • 4	$16	(4, 16)

You can then use the function table to graph the function.

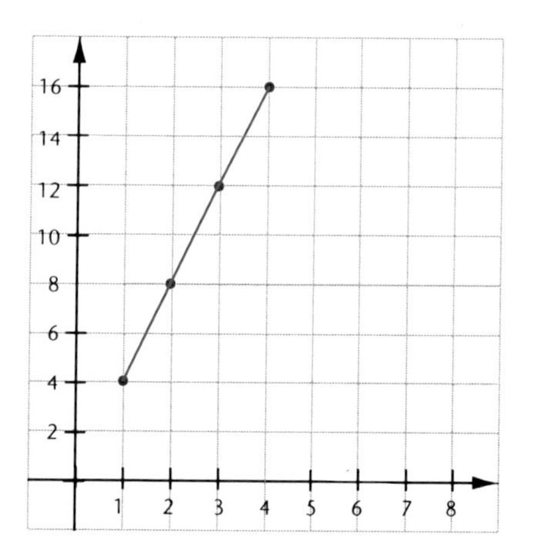

1. Graph each point.

   The x-coordinates represent the input values and the y-coordinates represent the output values.

2. Connect the points to form a line.

   The line is the graph of $y = 4x$.

Example

A. Make a table and graph the function $y = x + 1$. Use input values -2, -1, 0, 1, 2.

1. Make a table.

Input (x)	Rule (x + 1)	Output (y)	Ordered Pair (x, y)
-2	-2 + 1	-1	(-2, -1)
-1	-1 + 1	0	(-1, 0)
0	0 + 1	1	(0, 1)
1	1 + 1	2	(1, 2)
2	2 + 1	3	(2, 3)

2. Plot the ordered pairs on a graph.

3. Connect the points. They form a line.

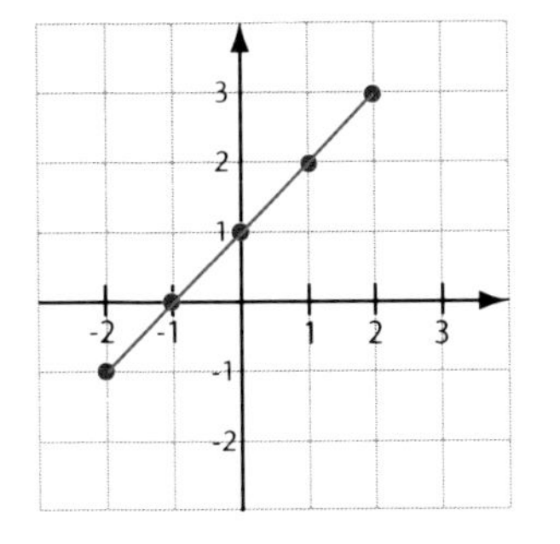

Try THESE

Make a table for each function rule. Use the input values -1, 0, 1, 2.

1. $y = x + 7$
2. $y = \frac{x}{2}$
3. $y = 2x + 4$

Exercises

Make a table for each function rule. Use the input values given. Then graph the function.

1. $y = x + 2$; -4, 0, 4
2. $y = 2x$; -2, 0, 2
3. $y = x - 3$; 6, 5, 4
4. $y = 2x + 1$; -2, 0, 2
5. $y = 3x - 4$; -1, 0, 1
6. $y = x + 5$; -3, 0, 3
7. $y = -3x$; -1, 0, 1
8. $y = \frac{x}{2} + 1$; -4, 0, 4

Problem SOLVING

9. Tamara works at an ice cream store where she makes $5.50 an hour. Use the function rule $y = 5.50x$, where x represents the number of hours worked, to find how much money she will make if she works 0, 2, 4, 6, and 8 hours. Make a table and graph.

10. The science class planted a tree in the school courtyard that was 6 inches tall. They are measuring its growth. It is growing 4 inches each week. Write a function rule for the growth of the tree.

Chapter 14 Review

Language and Concepts

Write the letter of the word or phrase that best matches each description.

1. pairs each number of one set with a number from another set
2. describes a function
3. found by adding or subtracting the same number to the term before it
4. found by multiplying or dividing the same number to the term before it
5. a pair of numbers in a specific order that indicates the location of a point on a coordinate plane
6. a transformation that moves every point on a figure the same distance and the same direction

a. reflection
b. arithmetic sequence
c. function
d. function rule
e. rotation
f. geometric sequence
g. ordered pairs
h. translation

Skills and Problem Solving

Name the point for each ordered pair. (Section 14.1)

7. (2, 0)
8. (-3, -3)
9. (1, -3)
10. (-1, -1)
11. (-2, 1)
12. (1, 3)

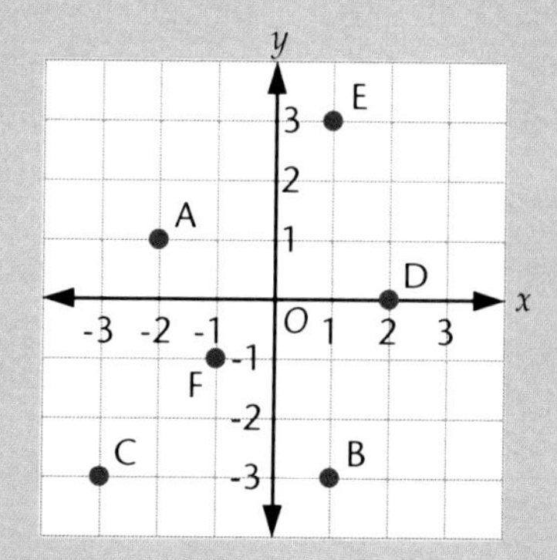

Copy each grid with triangle ABC. Then perform the indicated transformation. (Sections 14.2–14.3)

13.

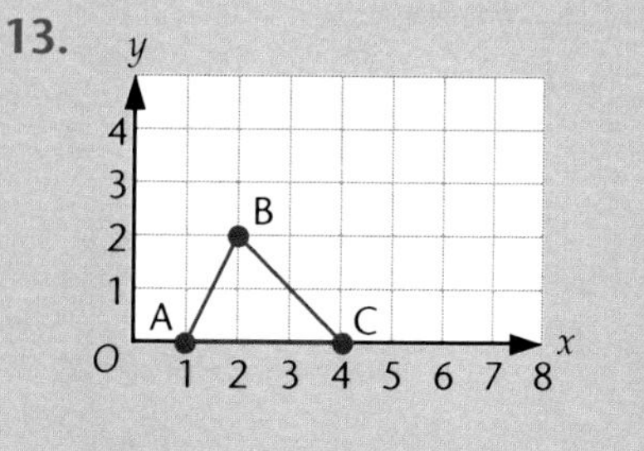

Move 3 units up.

14.

Move 4 units right.

15.

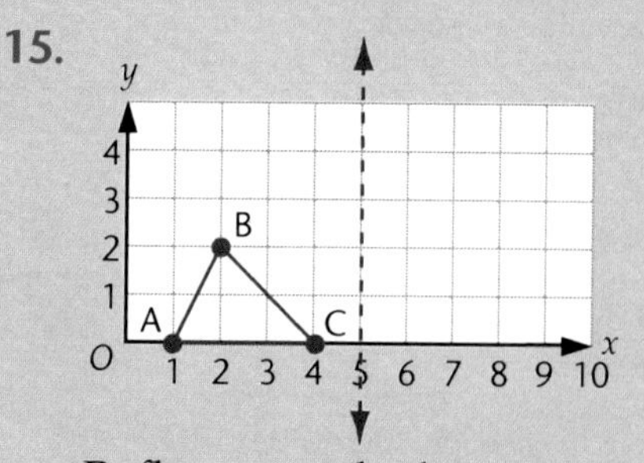

Reflect over the line of reflection.

Find the next three terms in each arithmetic sequence. (Section 14.5)

16. 22, 24, 26, __?__, __?__, __?__, . . .

17. 32, 28, 24, __?__, __?__, __?__, . . .

Find the next three terms in each geometric sequence. (Section 14.5)

18. 4, 16, 64, __?__, __?__, __?__, . . .

19. 128, 64, 32, __?__, __?__, __?__, . . .

Complete. (Section 14.6)

20.

21.

Rule: × 4	
Input	**Output**
7	?
9.3	?
?	24

22.

Rule: − 6	
Input	**Output**
24	?
36	?
?	54

Find the function rule for each chart. (Section 14.6)

23.

24.

Input (x)	Output (y)
3	12
5	20
7	28

25.

Input (x)	Output (y)
10	20
20	30
40	50

Solve. (Section 14.6–14.7)

26. It cost $7 for one student to attend the school play, $14 for two students to attend, and $21 for three students.

a. Write a function rule that relates the number of students to the cost.

b. How much will it cost five students to attend? Explain.

27. Using the function table to the right . . .

a. Graph the function on a coordinate plane.

b. Find the output values for input values 3 and 4.

c. Find the function rule.

Input (x)	Output (y)
-2	-3
0	1
1	3
2	5

Chapter 14 Test

Name the ordered pair for each point.

1. J
2. K
3. L
4. M
5. N
6. O

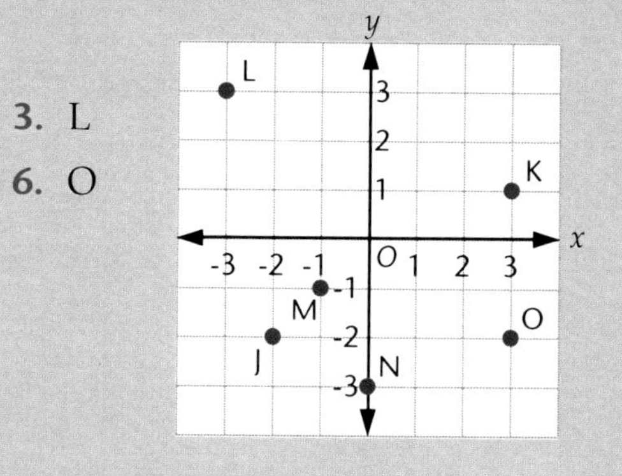

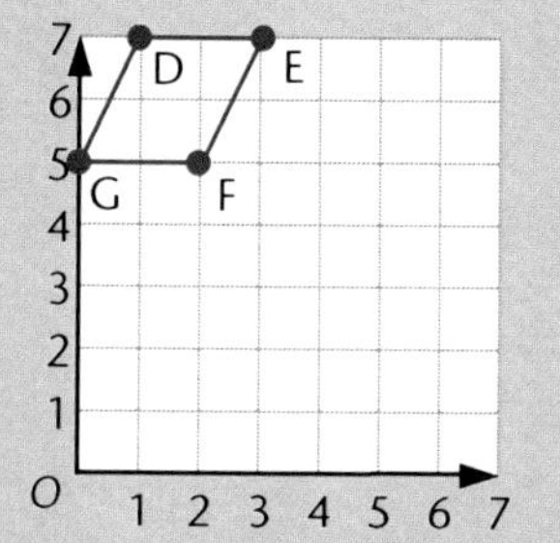

Copy the grid with rhombus DEFG. Then perform the indicated transformation.

7. Move 3 units down.
8. Move 3 units right.
9. Reflect over the *y*-axis.

Find the next three terms in each arithmetic sequence.

10. 5, 10, 15, __?__, __?__, __?__, . . .

11. 100, 99, 98, __?__, __?__, __?__, . . .

Find the next three terms in each geometric sequence.

12. 1.1, 2.2, 4.4, __?__, __?__, __?__, . . .

13. 1,024, 256, 64, __?__, __?__, __?__, . . .

Find the function rule for each chart.

14.

Input (x)	Output (y)
28	14
36	22
44	30

15.

Input (x)	6	8	10
Output (y)	18	24	30

Solve.

16. Tina, Kim, and Ed need to cross from one side of a creek to another in a canoe. The canoe can carry 190 pounds. Tina weighs 90 pounds, Kim weighs 95, and Ed weighs 105. How can all three cross safely?

17. The library charges $0.25 a day for overdue books.
    a. Write a function rule for the total amount due after *d* days.
    b. Make a function table and graph for 1, 2, 3, and 4 days late.
    c. If a person was charged $2.25, how many days late were the books? Explain your reasoning.

Sums and Terms of Arithmetic Sequences

You can find the sum of an arithmetic sequence without adding all the terms in the sequence.

Sequence: 56, 60, 64, 68, 72, 76

- Add the first and last terms.
- Divide by 2.
- Multiply by the number of terms.

$56 + 76 = 132$
$132 \div 2 = 66$
$66 \times 6 = 396$

$56 + 76 = 132$
$60 + 72 = 132$
$64 + 68 = 132$

The sum is 396.

When the first term and the common difference are known, any other term can be found.

Find the twentieth term.

Sequence: 45, 51, 57, . . .
Difference: 6

- Subtract 1 from the number of the term you are finding.
- Multiply by the difference.
- Add the first term.

$20 - 1 = 19$
$19 \times 6 = 114$
$114 + 45 = 159$

The twentieth term is 159.

Find the sum of each arithmetic sequence.

1. 4, 6, 8, 10, 12
2. 3, 6, 9, 12, 15, 18
3. 75, 78, 81, 84, 87
4. 46, 50, 54, 58
5. 50, 49, 48, . . . , 1
6. 1, 2, 3, . . . , 100

Find each term for the sequence 7, 14, 21 . . .

7. ninth
8. fifteenth
9. thirty-sixth
10. hundredth

Cumulative Test

1. Which is the lowest temperature?
   a. -5°C
   b. 5°C
   c. 15°C
   d. 25°C

2. 6 + -13 = ____
   a. 7
   b. -7
   c. -19
   d. none of the above

3. What is the least prime number greater than 100?
   a. 101
   b. 102
   c. 103
   d. 104

4. Murelle has 4 blouses, 3 skirts, and 2 jackets. How many outfits are possible?
   a. 4
   b. 12
   c. 24
   d. 48

5. Benito types 50 words per minute. How many words does he type in m minutes?
   a. $m + 50$
   b. $m - 50$
   c. $m \times 50$
   d. $m \div 50$

6. Which point is named by (-4, 3)?
   a. A
   b. B
   c. C
   d. D

7. The ten highest scores on the latest math test were 100, 98, 95, 92, 88, 88, 85, 82, 81, and 79. An average is 88.8. That average is the ____.
   a. mean
   b. mode
   c. median
   d. none of the above

8. A stop sign has a geometric shape. Name the shape.
   a. decagon
   b. hexagon
   c. octagon
   d. pentagon

9. Don drove 350 miles the first day of his vacation and 247 miles the second day. He told a gas station attendant that he was 597 miles away from home.

   Which of the following facts is relevant to the situation?
   a. He traveled slower on the second day of his trip.
   b. Each day he traveled in a straight line away from home.
   c. His car used more gas on the second day.
   d. His car averaged 37 mpg.

10. Judy got 90 percent correct on a science test and 85 percent correct on a math test. Inez said that Judy got more right answers on the science test than on the math test.

    Which of the following conclusions is correct?
    a. Inez's statement is always true.
    b. Inez's statement is never true.
    c. Inez's statement would not be true if Judy had gotten 85 percent correct in science.
    d. Inez's statement is true if the tests have the same number of questions.

Appendix

Mathematical Symbols

Symbol	Meaning	Symbol	Meaning	Symbol	Meaning
•	multiplication dot	$\cong$	is congruent to	$\overleftrightarrow{AB}$	line AB
$=$	is equal to	$\sim$	is similar to	$\overline{AB}$	line segment AB
$\neq$	is not equal to	$\perp$	is perpendicular to	$\overrightarrow{AB}$	ray AB
$>$	is greater than	$\parallel$	is parallel to	π	pi
$\geq$	is greater than or equal to	4:3	ratio of 4 to 3	$\sqrt{}$	square root
$<$	is less than	%	percent	$\triangle$	triangle
$\leq$	is less than or equal to	(arrow marks) or (double arrow marks)	sides are parallel	$\angle$	angle
$\approx$	is approximately equal to	(tick marks) or (double tick marks)	sides are congruent	∟	right angle
$\lvert -4 \rvert$	absolute value of negative 4	-7	negative 7	$^\circ$	degree

Formulas

Formula	Description
$C = 2 \times \pi \times r$	circumference of a circle
$C = \pi \times d$	circumference of a circle
$A = s^2$	area of a square
$A = \ell \times w$	area of a rectangle
$A = b \times h$	area of a parallelogram
$A = \frac{1}{2} \times b \times h$	area of a triangle
$A = \pi \times r \times r$	area of a circle
$A = \pi \times r^2$	area of a circle
$P = 2 \times \ell + 2 \times w$	perimeter of a rectangle
$SA = 2\ell w + 2\ell h + wh$	surface area of a rectangular prism
$V = \ell \times w \times h$	volume of a rectangular prism
$d = r \times t$	distance

Metric System of Measurement

Prefixes

kilo (k) = thousand
hecto (h) = hundred
deka (da) = ten
deci (d) = tenth
centi (c) = hundredth
milli (m) = thousandth

Length

1 centimeter (cm) = 10 millimeters (mm)
1 meter (m) = 100 centimeters or 1,000 millimeters
1 kilometer (km) = 1,000 meters

Mass

1 gram (g) = 1,000 milligrams (mg)
1 kilogram (kg) = 1,000 grams
1 metric ton (T) = 1,000 kilograms

Capacity

1 liter (L) = 1,000 milliliters (mL)
1 kiloliter (kL) = 1,000 liters

Customary System of Measurement

Length

1 foot (ft) = 12 inches (in.)
1 yard (yd) = 3 feet or 36 inches
1 mile (mi) = 1,760 yards or 5,280 feet

Weight

1 pound (lb) = 16 ounces (oz)
1 ton = 2,000 pounds

Capacity

1 cup (c) = 8 fluid ounces (fl oz)
1 pint (pt) = 2 cups
1 quart (qt) = 2 pints
1 gallon (gal) = 4 quarts

Glossary

A

absolute value 13.1 The distance of a number from zero on a number line.

acute angle 10.2 An angle that has a measure between 0° and 90°.

acute triangle 10.5 A triangle with three acute angles.

Addition Property of Equality 12.3 If a number is added to each side of an equation the two sides remain equal.

$$x - 8 = 12$$
$$x - 8 + 8 = 12 + 8$$
$$x = 20$$

additive identity 12.1 Adding zero to a number results in that number.

adjacent angles 10.3 Angles that are next to one another.

algebraic expression 1.4 An expression that includes at least one variable.

angle 10.2 Two rays with a common endpoint.

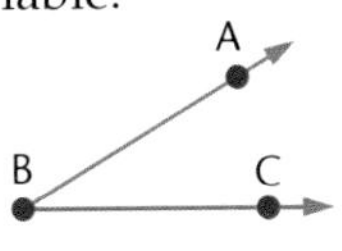

arc 10.4 A part of a circle that has two endpoints on the circle.

area 11.1 The number of square units needed to cover a region.

arithmetic sequence 14.5 A sequence in which each term is formed by adding the same number to or subtracting from the same number from the term before it. The sequence 2, 6, 10, 14 is an arithmetic sequence.

Associative Property 12.1 The property that states the way addends or factors are grouped does not change the sum or product.

B

bar graph 9.8 A graph that uses horizontal or vertical bars to display countable data.

base (of an exponent) 1.2 The number that is used as a factor. In 5^3, the base is 5.

base (of a three-dimensional figure) 11.8 A certain face of a three-dimensional figure.

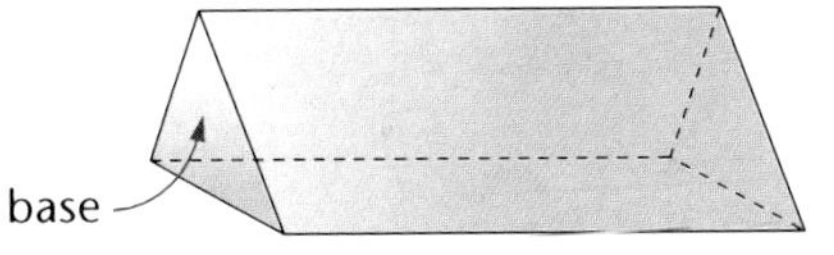

bisect 10.4 To separate into two congruent parts.

box-and-whisker plot 9.5 A graph that shows five main values from a set of data: the median, the highest and lowest values, and the values that separate the numbers into fourths.

C

center 11.5 The point in the plane from which all points in a circle are the same distance.

center of rotation 14.3 The point around which a figure is rotated.

certainty 8.1 An event whose probability is 1.

chord 11.5 A segment that has both endpoints on the circle.

circle 11.5 A curved figure in a plane. All points of a circle are the same distance from a given point in the plane called the center.

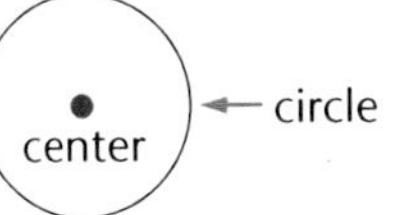

circle graph 9.8 A graph that compares parts of a whole.

circumference (*C*) 11.5 The distance around a circle.

collinear 10.1 Points on the same line.

common factor 4.5 A number that is a factor of two or more given whole numbers. Common factors of 24 and 30 are 1, 2, 3, and 6.

common multiple 4.9 A number that is a multiple of two or more given whole numbers. Some common multiples of 8 and 12 are 0, 24, 48, and 72.

common denominator 4.10 A common multiple of the denominators of two or more fractions.

Commutative Property 12.1 The property that states changing the order of the addends or factors does not change the sum or product.

compatible numbers 3.1 Numbers that are easy to compute mentally.

complement 8.3 All of the outcomes not included in a given event.

complementary angles 10.3 Two angles whose sum is 90°.

composite number 4.2 A whole number that is *not* prime. A composite number has more than two factors. Zero and 1 are neither prime nor composite.

compound event 8.5 An event consisting of two or more simple events.

congruent angles 10.3 Angles that are equal.

congruent figures 10.9 Figures that have the same size and shape.

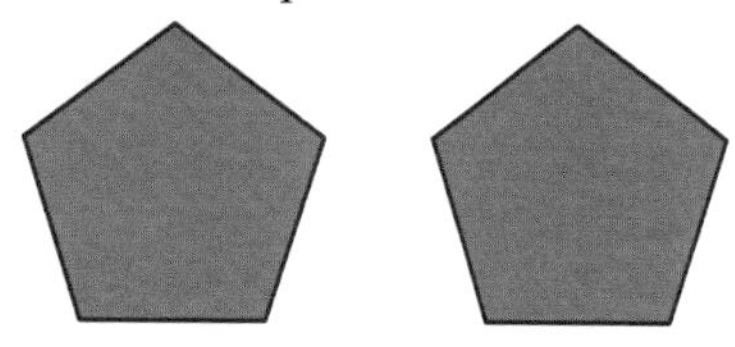

conjecture 2.1 A prediction about what may happen.

coordinate plane 14.1 The number plane formed by two perpendicular number lines.

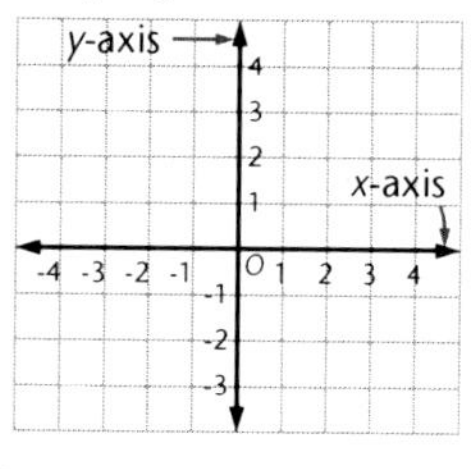

corresponding parts 10.9 The matching parts of two figures.

corresponding sides 7.4 The matching sides of two figures.

Counting Principle 8.2 If there are m ways of making one choice, and n ways of making a second choice, then there are $m \times n$ ways of making the first choice followed by the second choice.

cubic centimeter (cm^3) 11.10 A unit of volume. The space occupied by a cube with dimensions 1 cm by 1 cm by 1 cm.

D

decimal 2.2 Another way to write fractions and mixed numerals when the denominators are 10, 100, and so on; $1\frac{26}{100}$ written as a decimal is 1.26.

degree (°) 10.2 A unit for measuring angles.

diameter (d) 11.5 A line segment through the center of a circle with endpoints on the circle.

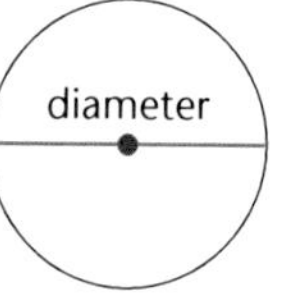

discount 7.12 A reduction of the regular price.

Distributive Property 1.5 It states that the product of a number and a sum is equal to the sum of the products.

divisible 4.1 A whole number is divisible by another if, upon division, the remainder is zero. When you divide 27 by 9, the remainder is zero. So 27 is divisible by 9.

Division Property of Equality 12.4 If you divide each side of an equation by the same nonzero number, the two sides remain equal.

$$4p = 12$$
$$\frac{4p}{4} = \frac{12}{4}$$
$$p = 3$$

double-bar graph 9.2 A graph that uses horizontal or vertical bars to display two sets of countable data.

E

edge of a three-dimensional figure 11.8 The line segment that is the intersection of two faces of the figure.

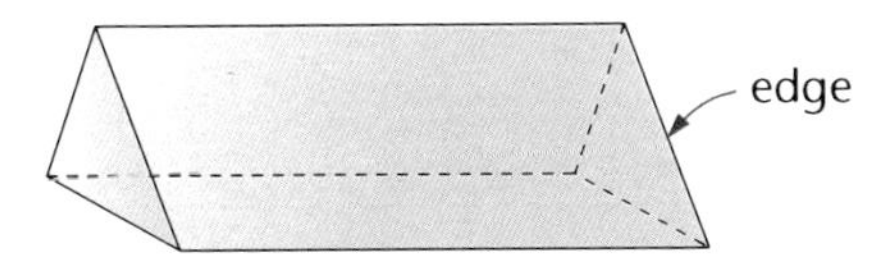

equation 1.6 A mathematical sentence with an equal sign.

$$h + 1 = 6$$

equilateral triangle 10.5 A triangle with all sides congruent.

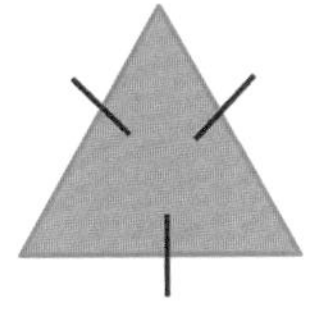

equivalent 2.4, 4.6 Two or more numerals are equivalent if they name the same number.

equivalent decimals: 0.6, 0.60, 0.600
equivalent fractions: $\frac{1}{5}, \frac{2}{10}, \frac{3}{15}$

even number 4.1 Any whole number that is divisible by 2. The even numbers are 0, 2, 4, 6, 8, 10, 12, and so on.

experimental probability 8.4 The probability based on experimental data or observations.

exponent 1.2 A number used to tell how many times the base is used as a factor. In 5^3, the exponent is 3.

F

face of a three-dimensional figure 11.8 One of the flat surfaces that form the three-dimensional figure.

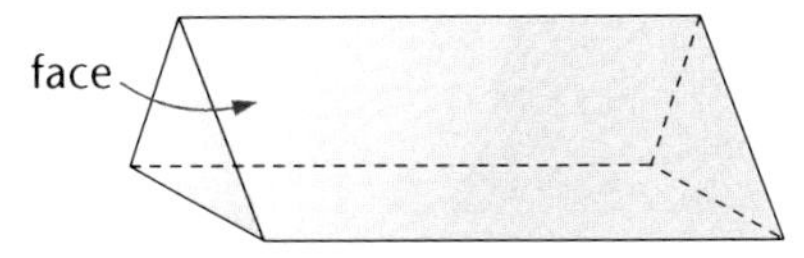

factor 4.1 Any one of the two or more numbers whose product is a given number. Since $3 \times 5 = 15$, the factors of 15 are 3 and 5. When you divide a whole number by one of its factors, the remainder is zero.

formula 7.6 An equation that shows how certain quantities are related.

frequency table 9.1 A table used for organizing numbers or items in a set of data. The frequency column shows the number of times each number or item occurs.

front-end estimation 2.7 A way of estimating a sum by adding the front-end digits.

function 14.6 A set of ordered pairs (x, y) in which each value of x is paired with exactly one value of y.

G

greatest common factor (GCF) 4.5 The greatest whole number that is a factor of two or more given whole numbers. The GCF of 12 and 18 is 6.

geometric sequence 14.5 A sequence formed by multiplying the term before it by the same number or dividing the term before it by the same number. The sequence 1, 3, 9 is a geometric sequence.

H

histogram 9.8 A bar graph that shows the frequency of a range of data.

I

image 14.2 The figure produced after a transformation.

improper fraction 4.8 A fraction in which the numerator is greater than or equal to the denominator. $\frac{7}{4}$, $\frac{6}{6}$, and $\frac{10}{5}$ are improper fractions.

independent event 8.5 One event does not affect the outcome of the other event.

inequality 13.7 A sentence that contains $<$, $>$, $\leq$, or $\neq$.

input 14.6 The number you substitute into a function.

integers 13.1 Whole numbers and their opposites.

. . . -3, -2, -1, 0, 1, 2, 3, . . .

intersecting lines 10.1 Lines that cross.

inverse operation 12.2 An operation that undoes another operation. Addition and subtraction are inverses, as are multiplication and division.

isosceles triangle 10.5 A triangle that has two congruent sides.

L

least common denominator (LCD) 4.10 The least common multiple of the denominators of two or more fractions.

least common multiple (LCM) 4.9 The least multiple, other than 0, common to sets of multiples. The LCM of 3 and 4 is 12.

line 10.1 A never-ending path that extends in both directions, such as $\overleftrightarrow{AB}$.

line graph 9.7 Uses a group of connected line segments to show changes in data.

line plot 9.8 A graph that shows the frequency of data by stacking Xs above each data value on a number line.

line of reflection 14.3 The line a figure is flipped over.

line of symmetry 10.10 The line drawn through a figure such that the two halves match exactly.

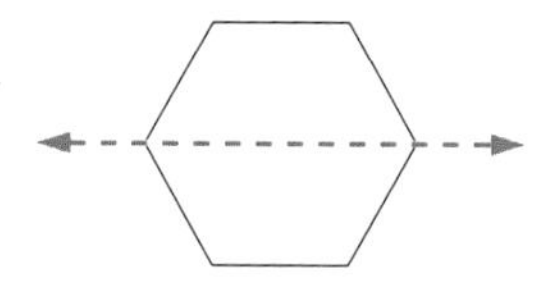

line segment 10.1 A part of a line that has two endpoints, such as $\overline{DE}$.

line symmetry 10.10 A figure has line symmetry when a line can be drawn through the figure such that the two halves match exactly.

logical reasoning 2.1 Allows you to make conclusions based on given information.

M

mean 9.4 The mean is found by adding all the numbers and dividing by the number of addends. It is often referred to as the average.

median 9.4 The middle number in a set of data arranged from smallest to largest.

19, 19, 21, 27, 39

mixed number 4.8 A number that consists of a whole number part and a fractional part, such as $1\frac{3}{10}$.

mode 9.4 The number that appears the most often in a set of data.

19, **19**, 21, 27, 39

multiple 4.9 One number from a list of products of a certain number.

Multiplication Property of Equality 12.4 If you multiply each side of an equation by the same number, the two sides remain equal.

$$\frac{x}{6} = 5$$
$$\frac{x}{6} - 6 = 5 - 6$$
$$x = 30$$

multiplicative identity 12.1 Multiplying a number by 1 results in that number.

N

net 11.9 A two-dimensional figure that can be folded into a three-dimensional figure.

noncollinear 10.1 Points that cannot be connected by one line.

O

obtuse angle 10.2 An angle that has a measure between 90° and 180°.

obtuse triangle 10.5 A triangle with an obtuse angle.

odd number 4.1 Any whole number that is not divisible by 2. The odd numbers are 1, 3, 5, 7, 9, 11, and so on.

opposites 13.1 Two numbers that are the same distance from zero in opposite directions.

ordered pair 14.1 A pair of numbers in a specific order. Ordered pairs indicate the location of a point on a coordinate plane.

order of operations 1.3 The order in which you should complete a problem.

origin 14.1 The point in the coordinate plane where the axes intersect. The ordered pair for the origin is (0, 0).

outcome 8.2 A possible result.

output 14.6 The value found using a function.

parallel lines 10.1 Lines in the same plane that never cross.

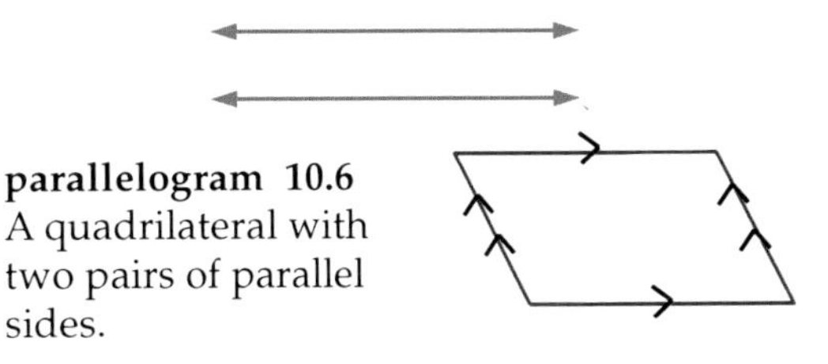

parallelogram 10.6 A quadrilateral with two pairs of parallel sides.

percent 7.7 A ratio that compares a number to 100. The symbol for percent is %.

percent proportion 7.11 A number that is compared to another number called the base (B). The rate (r) is a percent.

$$\frac{P}{B} = \frac{r}{100}$$

perfect number 4.2 A number in which the sum of all factors of the number, except the number itself, equals the number. The number 6 is a perfect number as the factors of 6 other than 6 are 1, 2, and 3 and $1 + 2 + 3 = 6$.

perimeter 11.1 The distance around a figure.

perpendicular lines 10.1 Two intersecting lines that form right angles.

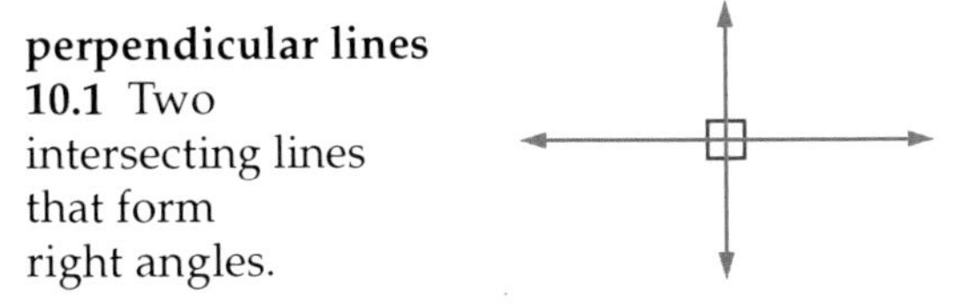

pi (π) 11.5 The ratio of the circumference of a circle to the diameter. Approximations for π are 3.14 and $\frac{22}{7}$.

plane 10.1 A never-ending flat surface.

point 10.1 An exact location.

polygon 10.7 A closed figure in a plane made up of line segments that meet, but do not cross.

power 1.2 A number that is written using an exponent.

prime factorization 4.3 A way to express a composite number as a product of prime numbers. The prime factorization of 60 is $2 \times 2 \times 3 \times 5$.

prime number 4.2 A whole number that has exactly two factors—itself and 1. Some prime numbers are 2, 3, 5, 7, and 11.

probability 8.1 The ratio of the number of ways an event can occur to the total number of possible outcomes. It describes the likeliness an event will occur.

proportion 7.3 An equation that states that two ratios are equivalent.

$$\frac{4}{7} = \frac{x}{21}$$

protractor 10.2 An instrument used for measuring angles.

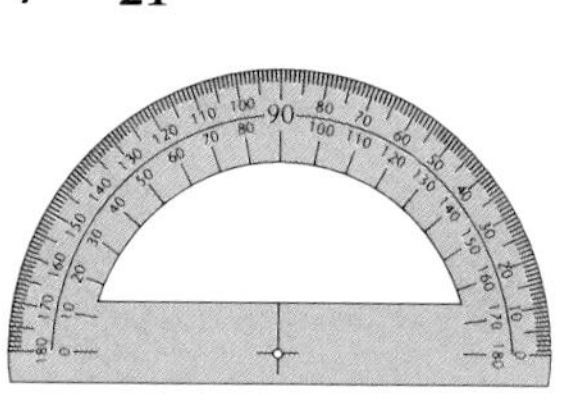

Q

quartiles 9.5 The values that separate the data into four equal parts.

R

radius (r) 11.5 A line segment from the center of a circle to any point on the circle.

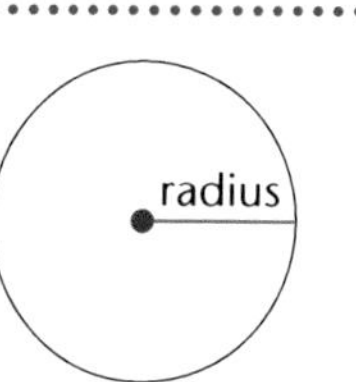

range 9.3 The difference between the least and greatest numbers in a set of data.

rate 7.2 A ratio that compares two quantities measured in different units.

ratio 7.1 A comparison of two numbers using division. The ratio of 3 to 5 may be stated as 3:5, 3 out of 5, or $\frac{3}{5}$.

ray 10.1 A never-ending straight path in one direction.

reciprocals 6.5 Two numbers whose product is 1. Since $\frac{2}{3} \times \frac{3}{2} = 1$, $\frac{2}{3}$ and $\frac{3}{2}$ are reciprocals.

rectangle 10.6 A parallelogram with all angles congruent.

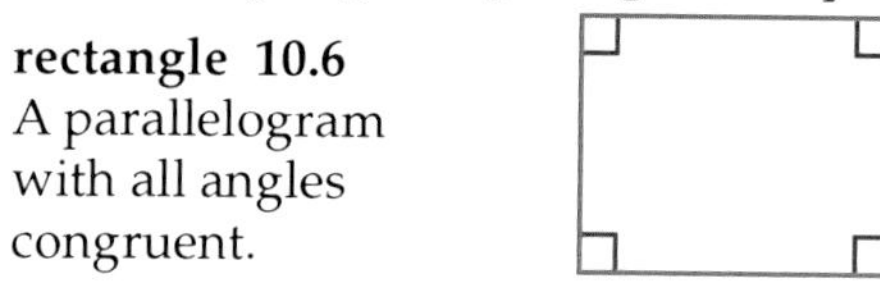

reflection 14.3 A transformation that flips a figure over a line.

regular polygon 10.7 A polygon that has all sides congruent and all angles congruent.

repeating decimal 3.4 A fraction for which the division of the numerator by the denominator of the fraction yields a quotient in which the digits repeat without end; $\frac{1}{3}$ can be expressed as the repeating decimal

$$0.333\ldots = 0.\overline{3}.$$

reversal primes 4.2 Prime numbers that make another prime number when reversed. An example is 17.

rhombus 10.6 A parallelogram with all sides congruent.

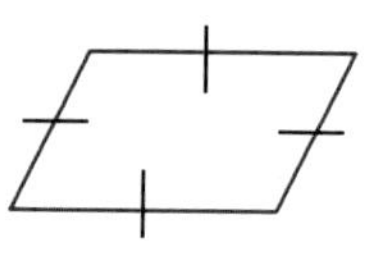

right angle 10.2 An angle that has a measure of 90°.

right triangle 10.5 A triangle with one right angle.

rotation 14.3 A transfromation that turns a figure about a fixed point.

S

scale drawing 7.5 An enlarged or reduced drawing of an object that is similar to the actual object.

sales tax 7.12 A percentage of any purchase added by the local government.

scalene triangle 10.5 A triangle that has no congruent sides.

scientific notation 3.7 A way of writing any number as the product of a factor and a power of ten. The factor must be 1 or greater but less than 10.

sequence 14.5 A set of numbers called terms that are in a specific order.

sides (of a polygon) 10.7 The line segments that form a polygon.

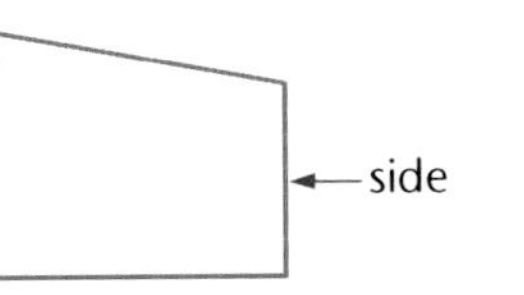

similar figures 7.4, 10.9 Figures that have the same shape but not the same size.

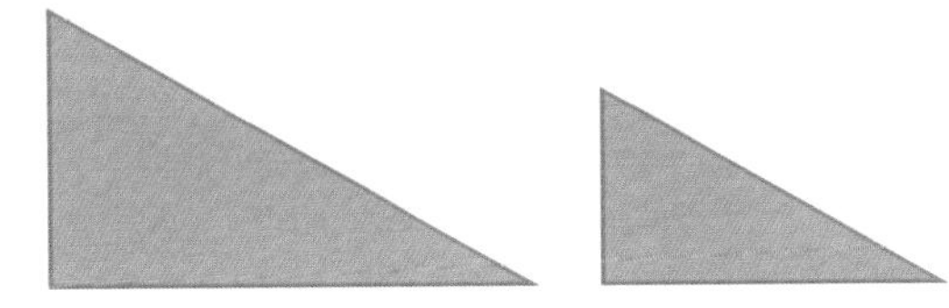

simplest form (of a fraction) 4.7 A fraction written so that the only common factor of the numerator and denominator is 1.

skew lines 10.1 Lines that do not intersect and are not parallel.

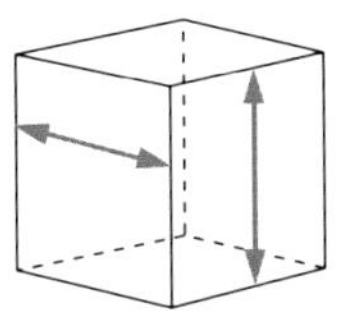

solution 1.6 A value that, when substituted for a variable in an equation, makes the equation true.

solution of an inequality 13.7 Any number that makes the inequality true.

square 10.6 A parallelogram with all sides and all angles congruent.

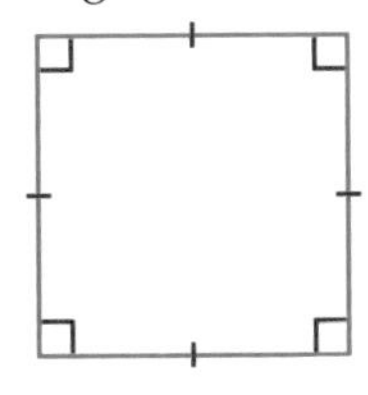

stem-and-leaf plot 9.3 A way to organize data that displays all numbers in the data set by place value. Numbers in the data set are ordered when a stem-and-leaf plot is completed.

2	1 1 38
3	0 2 4 555
4	3 5 7 7
5	1 2|1 = 21

supplementary angles 10.3 Two angles whose sum is 180°.

surface area 11.9 The sum of the area of all of the faces.

subtraction property of equality 12.2 If a number is subtracted from each side of an equation, the two sides remain equal.

$$k + 3 = 12$$
$$k + 3 - 3 = 12 - 3$$
$$k = 9$$

T

terminating decimal 3.4 A fraction for which the division of the numerator by the denominator of the fraction has a last remainder of 0. $\frac{1}{5}$ can be expressed by the terminating decimal 0.2.

theoretical probability 8.3 The number of favorable outcomes divided by the total number of possible outcomes.

three-dimensional figure 11.8 A figure that has three dimensions: length, width, and height.

tip 7.12 The percentage of the bill paid to the server in addition to the bill.

translation 14.2 A transformation that moves every point on a figure the same distance in the same direction.

trapezoid **10.6** A quadrilateral with only one pair of parallel sides.

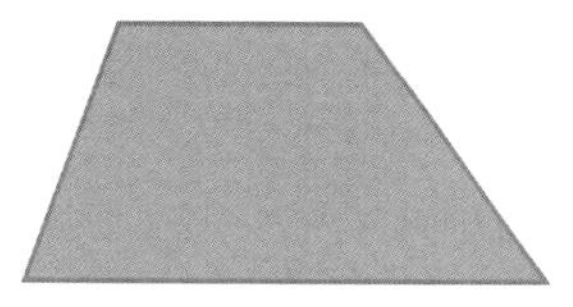

tree diagram **8.2** A diagram that uses branches to show all possible outcomes.

Penny	**Dime**	**Outcome**
H	H →	HH
	T →	HT
T	H →	TH
	T →	TT

twin primes **4.2** Two prime numbers that are consecutive odd intergers. Some twin primes are 3 and 5 and 5 and 7.

U

unit rate **7.2** The rate for one unit of a given quantity. It has a denominator of 1.

V

variable **1.4** A letter or other symbol used to represent a quantity.

Venn diagram **4.4** Overlapping circles enclosed in a rectangle that show how objects are classified.

vertex **10.2** The point where the sides meet to form an angle or a polygon.

vertical angles **10.3** Angles formed by two intersecting lines. Vertical angles have equal measures.

volume **11.10** The amount of space that a solid figure contains. Volume is equal to length times width times height.

W

work backward **6.7** To start at the end of the problem and work to the beginning.

Index

A

B

C

D

E

F

G

H

I

K

L

M

N

O

P

Q

R

S

T

U

V

W

Y

Z